PARTICLES AND NUCLEI

Volume 2, Part 3

PARTICLES AND NUCLEI

PARTICLES AND NUCLEI

N. N. Bogolyubov

Editor-in-Chief

Director, Laboratory for Theoretical Physics
Joint Institute for Nuclear Research
Dubna, USSR

A Translation of Problemy Fiziki Élementarnykh Chastits i Atomnogo Yadra
(Problems in the Physics of Elementary Particles and the Atomic Nucleus)

Volume 2, Part 3

CONSULTANTS BUREAU • NEW YORK-LONDON • 1973

The original Russian text, published by Atomizdat in Moscow in 1971 for the Joint Institute for Nuclear Research in Dubna, has been revised and corrected for the present edition. This translation is published under an agreement with Mezhdunarodnaya Kniga, the Soviet book export agency.

PROBLEMS IN THE PHYSICS OF ELEMENTARY PARTICLES AND THE ATOMIC NUCLEUS
PROBLEMY FIZIKI ÉLEMENTARNYKH CHASTITS I ATOMNOGO YADRA
Проблемы физики элементарных частиц и атомного

Library of Congress Catalog Card Number 72-83510
ISBN 978-1-4684-7552-4   ISBN 978-1-4684-7550-0 (eBook)
DOI 10.1007/978-1-4684-7550-0

Softcover reprint of the hardcover 1st edition 1973

A Division of Plenum Publishing Corporation
227 West 17th Street, New York, N. Y. 10011

United Kingdom edition published by Consultants Bureau, London
A Division of Plenum Publishing Company, Ltd.
Davis House (4th Floor), 8 Scrubs Lane, Harlesden,
London NW10 6SE, England

CONTENTS

Volume 2, Part 3

BEHAVIOR OF THE π-MESON FORM FACTOR AND A LIMIT ON ITS RADIUS

Dao Vong Duc and Nguyen Van Hieu

In the present review of theoretical work an investigation is made of the consequences of the general analytic properties of the form factor that can verified experimentally. The following problems are considered: 1) restrictions on the decrease of the form factor in the physical regions of the scattering and annihilation channels; 2) relationship between the behavior of the form factor in the physical region of the scattering channel and the behavior of its modulus on the cut; 3) exact sum rules for the form factor; 4) bounds on the radius of elementary particles.

INTRODUCTION

The lowest nontrivial order of perturbation theory can be used to study electromagnetic interactions of particles because the electromagnetic coupling constant is small compared with unity ($\hbar = c = 1$):

$$\frac{e^2}{4\pi} \approx \frac{1}{137}.$$

However, the matrix elements of electromagnetic interactions of hadrons calculated by the usual Feynman rules in quantum electrodynamics depend on strong interactions and therefore contain unknown scalar functions that depend on the particle momenta. These unknown scalar functions describe the contribution of strong interactions to the electromagnetic interaction of hadrons. In particular, in the lowest approximation in e, the matrix element of elastic scattering of an electron by a π-meson

$$e^- + \pi^+ \to e^- + \pi^+ \tag{I}$$

contains a function, $F_1(t)$, of the momentum transfer,

$$t = -(p-p')^2 = -(k-k')^2, \tag{1.1}$$

where k and p are the four-momenta of the electron and the π-meson in the initial state and k' and p' are, respectively, the same quantities for the particles in the final state (Fig. 1); namely,

$$M_1 = e^2 \bar{u}(k')\gamma_\mu u(k) \frac{1}{\sqrt{4p_0 p_0'}} (p+p')_\mu \frac{1}{(p-p')^2} F_1[-(p-p')^2]. \tag{1.2}$$

Similarly, in the lowest order in e the matrix element of the annihilation process

$$e^- + e^+ \to \pi^+ + \pi^- \tag{II}$$

contains a function, $F_2(s)$, of the square of the total reaction energy in the center of mass system,

$$s = -(p+\tilde{p})^2 = -(k+\tilde{k})^2, \tag{1.3}$$

Institute of Physics, Democratic Republic of Vietnam. Translated from Problemy Fiziki Élementarnykh Chastits i Atomnogo Yadra, Vol. 2, No. 3, pp. 533-582, 1972.

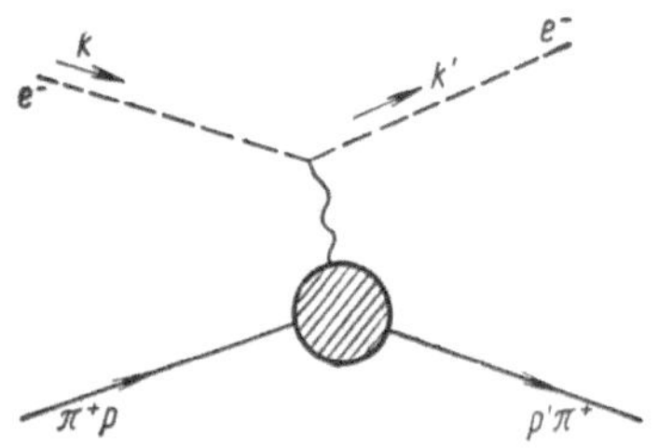

Fig. 1. Feynman diagram for the scattering process e + π → e + π in the lowest order of perturbation theory.

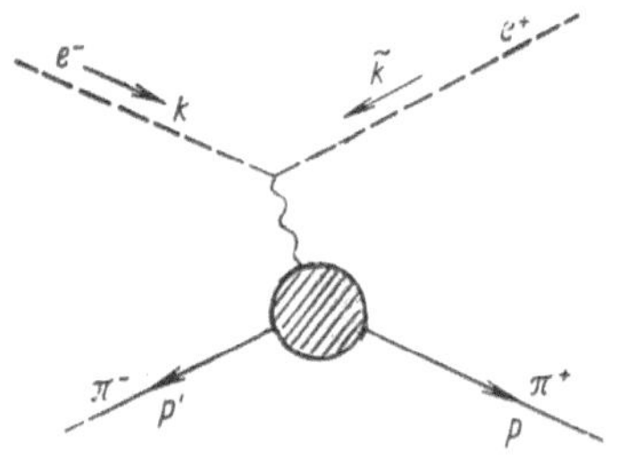

Fig. 2. Feynman diagram for the annihilation process $e^- + e^+ \to \pi^- + \pi^+$ in the lowest order of perturbation theory.

namely

$$M_2 = e^2\, v(-\tilde{k})\, \gamma_\mu\, u(k)\, \frac{1}{\sqrt{4p_0\tilde{p}_0}}\, (p-\tilde{p})_\mu\, \frac{1}{(p+\tilde{p})^2}\, F_2[-(p+\tilde{p})^2]. \qquad (1.4)$$

Here k and $\tilde{k}$ are the four-momenta of the electron and positron, p and $\tilde{p}$ are the four-momenta of the π^+ and π^- mesons (Fig. 2).

In the absence of a complete dynamic theory of strong interactions, the scalar functions in the matrix elements of processes in which hadrons participate cannot be determined. However, the general principles of field theory do entail certain properties for these functions, and these, in their turn, lead to experimentally verifiable consequences. For example, the dispersion relation for the amplitude of elastic scattering of a π-meson by a nucleon proved rigorously by Bogolyubov [1, 2] relates the imaginary and the real parts of the amplitude. For forward scattering these quantities can be completely determined in an experiment, and the dispersion relation is an experimentally verifiable consequence of the general principles of the theory – above all, the principle of microcausality. Rigorously proved analytic properties of the amplitude as a function of the energy and the momentum transfer also yield various asymptotic theorems [3-5], restrictions on the partial amplitudes, and on the growth and decrease of elastic and inelastic cross sections [6-10]. They are all amenable to experimental verification and constitute an extremely effective instrument for verifying the validity of the general principles of the theory.

A similar situation obtains for the functions F_1 and F_2 in the matrix elements (1.2) and (1.4). The general analytic properties of these functions have theoretical consequences that could be verified experimentally in the future. Here, we review the results of various papers that have explored the experimentally verifiable consequences of the analytic properties of F_1 and F_2.

It should be noted that the arguments t and s of the functions $F_1(t)$ and $F_2(s)$ in the matrix elements (1.2) and (1.4) vary in different intervals: the momentum transfer t, expressed in terms of the scattering angle θ and the cms momentum of the k-particles,

$$t = -2k^2(1-\cos\theta),$$

is always negative:

$$t \leqslant 0,$$

whereas the square of the cms total energy

$$s = (2E)^2$$

is always positive:

$$s \geqslant 4\mu^2,$$

where μ is the mass of the π-meson.

However, these functions are intimately related to each other. Such a relationship between matrix elements of pairs of crossed processes is well known in quantum electrodynamics as the Low substitution rule. In accordance with this rule, the matrix element of a process with the emission (or absorption) of a particle *a* with energy-momentum k can be obtained from the expression for the matrix element of a corresponding process with absorption (or emission) of the antiparticle $\bar{a}$ with energy-momentum $\tilde{k}$ by making the substitution $k \leftrightarrow -\tilde{k}$ and for Fermions $u(k) \leftrightarrow v(-\tilde{k})$. Let us apply this rule to the processes (I) and (II); then to obtain M_2 it is sufficient to take M_1 and then make the substitution $\bar{u}(k') \to \bar{v}(-\tilde{k})$, $k' \to -\tilde{k}$, $p' \to p$ and $p \to -\tilde{p}$ (see Figs. 1 and 2). In other words, $F_2(s)$ is obtained from $F_1(t)$ by replacing t by s.

In the case of quantum electrodynamics, when one considers only the interaction of leptons with the electromagnetic field, one can apply perturbation theory and the matrix elements can be calculated explicitly in each order in e. Then the Low substitution rule can be carried out simply. But if strong interactions are present, the matrix elements of the processes contain unknown functions and Low's rule remains meaningless until a method is specified that makes it possible to determine the values of these functions after the substitution. For example, to apply the Low rule for the processes (I) and (II) it is necessary to know what happens to the function $F_1(t)$ under the substitution $t \to s$. Putting it differently, one must be able in some manner to continue the function $F_1(t)$ from its domain of definition $t \leq 0$ into the region $t \geq 4\mu^2$. Now this can be done: there exists a function F(w), analytic in the complex w plane with the cut $w \geq 4\mu^2$, that is equal to $F_1(t)$ when $w = t \leq 0$ and tends to $F_2(s)$ as $w \to s$ from the upper half-plane: $w = s + i0$, $s \geq 4\mu^2$ [11]. In other words, each of the functions $F_1(t)$ and $F_2(s)$ is the analytic continuation of the other. These analytic properties give the Low substitution rule a transparent meaning.

Thus, the scalar functions $F_1(t)$ and $F_2(s)$ in the matrix elements of the processes (I) and (II) are the values of one and the same analytic function F(w) in different ranges of variation of w. The function F(w) is called the π-meson form factor. It is analytic in the complex w plane with the cut $w \geq 4\mu^2$, and for all real values of w outside the cut it is real:

$$\operatorname{Im} F(w) = 0, \quad w < 4\mu^2. \tag{1.5}$$

It follows that F(w) takes complex-conjugate values at complex-conjugate points (by the Riemann–Schwarz symmetry principle):

$$F(w^*) = F(w)^*. \tag{1.6}$$

In particular

$$F(s - i0) = F(s + i0)^*, \tag{1.7}$$

i.e., the discontinuity of F(w) on the transition from the upper to the lower edge of the cut is proportional to its imaginary part:

$$F(s + i0) - F(s - i0) = 2i \operatorname{Im} F(s + i0). \tag{1.8}$$

For the growth of F(w) there is also a bound. By virtue of the microcausality principle, F(w) as $w \to \infty$ cannot increase faster than any exponential function of $|w|^{1/2}$:

$$|F(w)| \leqslant \text{const}\, e^{\varepsilon |w|^{1/2}}, \quad w \to \infty \tag{1.9}$$

for any arbitrarily small $\varepsilon > 0$.

Now the specification of the values of an analytic function on an infinite set of points having a limit point within the analyticity domain of the function completely defines the function everywhere within the analyticity domain. In particular, the values of $F_1(t)$ and $F_2(s)$, which belong to the same analytic function in different intervals of the real axis, cannot be completely independent of one another. This suggests that the analytic properties of F(w) entail various relationships between the values of the functions $F_1(t)$ and $F_2(s)$ that determine the cross sections of the processes (I) and (II). These relations may be used to test the analyticity of F(w).

Analytic functions have another property: a restriction on their growth entails certain restrictions on their decrease. Since F(w) must satisfy the condition (1.9), it cannot decrease arbitrarily fast as $w \to \infty$. In other words, $F_1(t)$ and $F_2(s)$ cannot decrease too rapidly as $t \to -\infty$ and $s \to +\infty$, respectively. These restrictions on the decrease of the form factor are also amenable to experimental verification.

It should be noted that it is not possible to determine $F_2(s)$ itself in an experiment but only its modulus since the cross section of the process (II) is expressed in terms of this modulus. Therefore, in the study of the experimental consequences of analyticity of the form factor only assertions that apply directly to the modulus of $F_2(s)$, and not the function itself, have practical significance. The situation is different for $F_1(t)$; by virtue of the condition of reality (1.5) and the normalization condition

$$F_1(0) = F(0) = 1 \tag{1.10}$$

experimental data on the cross section of the process (I) can yield information about the values of this function itself.

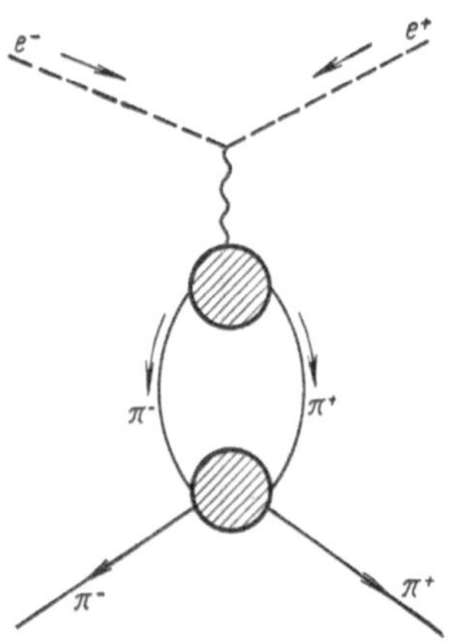

Fig. 3. Graphical representation of the two-particle unitarity condition for the annihilation process $e^- + e^+ \rightarrow \pi^- + \pi^+$.

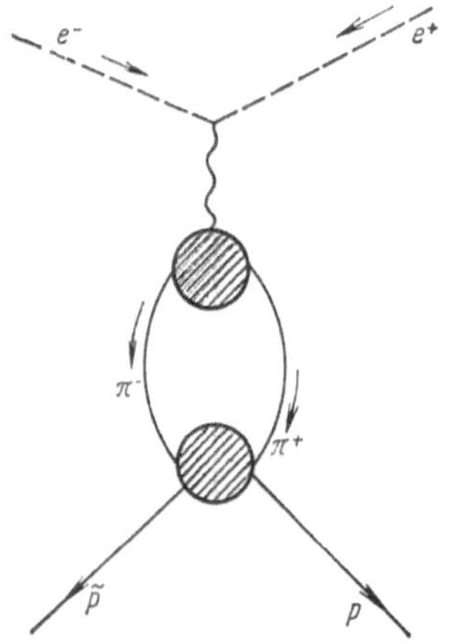

Fig. 4. Graphical representation of the two-particle unitarity condition for the annihilation process $e^- + e^+ \rightarrow p + \tilde{p}$.

This is the appropriate place for a remark. The cut $w \geq 4\mu^2$ arises because of two-pion intermediate states in the process (II). They give a contribution to the imaginary part of $F_2(s)$ determined by the condition of unitarity of the S matrix (Fig. 3). This cut coincides with the physical region $s \geq 4\mu^2$ of the process (II). For all the other processes of annihilation of an electron–positron pair into a pair of hadrons, for example,

$$e^- + e^+ \rightarrow K^+ + K^-; \quad e^- + e^+ \rightarrow p + \tilde{p}$$

etc., the corresponding form factors also have the cut $w \geq 4\mu^2$ since two-pion intermediate states also exist for these processes (Fig. 4). On the other hand, their physical regions are determined by the condition that the total energy $s^{1/2}$ cannot be less than $2m_K$ and $2m_N$, respectively, and therefore occupy only part of the cut $w \geq 4\mu^2$. In other words, for the latter processes the modulus of the form factor can be determined experimentally on only part and not all of the cut. Therefore, assertions about the modulus of the form factor on the whole of the cut cannot be verified experimentally. Moreover, for the form factors that occur in the matrix elements of the latter processes analytic properties similar to those of the π-meson form factor have not been proved.

In the following exposition we shall, in addition to the general analytic properties of the π-meson form factor F(w) formulated above, make certain additional assumptions. These are entirely concerned with observable quantities: $F_1(t)$ for $t \leq 0$ and $|F_2(s)|$ for $s \geq 4\mu^2$, and can therefore be verified experimentally. No approximations or assumptions of a dynamical nature will be made. This greatly restricts the possibility of obtaining definite predictions. Readers interested in the study of form factors in dynamical theories of strong interactions are referred to the reviews [12-14].

Our basic mathematical apparatus is the theory of analytic functions of a complex variable. Since the theorems employed are scattered throughout different books and monographs, and no single monograph has been published in which all the theorems are collected in one place, we have considered it desirable to quote them all, without proof, as an Appendix at the end of this review.

Before proceeding to the exposition of the results, let us formulate once more the general analytic properties of the π-meson form factor F(w) which will be used automatically in all that follows:

1) F(w) is analytic in the complex plane with the cut $w \geq 4\mu^2$;

2) F(w) satisfies the condition of reality

$$F(w^*) = F(w)^*;$$

3) F(w) increases slower than any exponential function of $|w|^{1/2}$:

$$|F(w)| \leqslant \text{const}\, e^{\varepsilon |w|^{1/2}}$$

for any $\varepsilon > 0$.

In addition, we assume in this review that on the cut $w \geq 4\mu^2$ the modulus of the form factor is a continuous function of w. This assumption can be experimentally verified.

2. ASYMPTOTIC BEHAVIOR

The cross section of the process (II) can be measured experimentally, and this makes it possible to determine the modulus of F(s) for $s \geq 4\mu^2$. The existing experimental data show that F(s) decreases as $s \rightarrow +\infty$. In what follows, we shall assume that F(s) in reality increases slower than some polynomial

$$|F(s)| \leqslant \text{const}\, s^n, \quad s \to +\infty. \tag{2.1}$$

Such an assumption can be verified experimentally. We shall now show that this assumption entails polynomial boundedness of F(w) in the whole of the w plane:

$$|F(w)| \leqslant \text{const}\, |w|^n, \quad w \to \infty. \tag{2.2}$$

To this end we apply the famous Phragmén–Lindelöf theorem in the theory of analytic functions (see Theorem 1 in the Appendix). This theorem states that if a function $f(z)$ which is analytic in the upper half-plane Im $z > 0$ and is continuous in the closed half-plane Im $z \geq 0$ is bounded on the real axis by some constant M

$$|f(x)| \leqslant M, \tag{2.3}$$

then everywhere in the upper half-plane

$$|f(z)| \leqslant M, \tag{2.4}$$

or there exists a sequence of points $z_n \to \infty$, such that

$$|f(z_n)| > \text{const}\, e^{a|z_n|}, \quad a > 0. \tag{2.5}$$

Using the change of variables $w = z^2$, we transform the w plane with cut into the upper half-plane Im $z > 0$ and set

$$f(z) = \frac{F(z^2)}{(z+i)^{2n}}. \tag{2.6}$$

The new function $f(z)$ is analytic in the upper half-plane and is bounded on the real axis because of the condition (2.1). In addition, since F(w) satisfies the condition (1.9),

$$|f(z)| \leqslant \text{const}\, e^{\varepsilon|z|}, \quad z \to \infty, \tag{2.7}$$

for any arbitrarily small $\varepsilon > 0$. Thus, the inequality (2.5) cannot hold for $f(z)$. It follows from the Phragmén–Lindelöf theorem that $f(z)$ is bounded in the whole of the upper half-plane of z, i.e., $F(z^2)$ increases not faster than a polynomial of degree 2n of z in the complex plane.

In particular, if $|F(w)|$ is bounded on the cut by a constant

$$|F(s)| \leqslant M, \quad s \geqslant 4\mu^2, \tag{2.8}$$

then it is bounded by the same constant on the whole of the complex plane:

$$|F(w)| \leqslant M. \tag{2.9}$$

Thus, we obtain a theorem.

<u>THEOREM 1.</u> If the form factor F(w) is bounded by some constant or is polynomially bounded on the cut

$$|F(s)| \leqslant M, \quad s \geqslant 4\mu^2$$

or

$$|F(s)| \leqslant \text{const}\, s^n, \quad s \to +\infty,$$

then it is bounded by the same constant or the same polynomial in the complex plane:

$$|F(w)| \leqslant M$$

or

$$|F(w)| \leqslant \text{const}\, |w|^n, \quad w \to \infty.$$

Although this theorem does not have practical significance, it is very helpful. In what follows, we shall frequently employ it to prove other assertions amenable to experimental verification.

As we have noted, a restriction on the growth of the form factor entails a restriction on its decrease. Thus, by virtue of the condition (1.9), F(w) cannot decrease as an exponential function of $-w^{1/2}$ as $w \to +\infty$. To prove this assertion, we shall use Carleson's theorem (see Theorem 2 in the Appendix). This theorem states that if a function $f(z)$ which is analytic in the upper half-plane Im z > 0 and continuous in the closed half-plane Im z ≥ 0 increases not faster than some linear exponential function

$$|f(z)| \leqslant \text{const}\, e^{k|z|}, \quad z \to \infty \tag{2.10}$$

and decreases exponentially on the real axis

$$|f(z)| \leqslant \text{const}\, e^{-a|z|}, \quad a > 0, \quad z \to \pm\infty, \tag{2.11}$$

it vanishes identically: $f(z) \equiv 0$. But if it does not vanish identically and satisfies the condition (2.10), it cannot satisfy the condition (2.11). Applying Carleson's theorem to the function $f(z) = F(z^2)$ and using the condition (1.9), we obtain the next theorem [15].

THEOREM 2. The form factor F(s) cannot decrease as $\exp(-a s^{1/2})$ as $s \to +\infty$ for any nonvanishing constant $a > 0$, i.e., for any given $a > 0$ there exists at least one sequence $s_n \to +\infty$ such that

$$|F(s_n)| > \text{const}\, e^{-a\sqrt{s_n}}.$$

Another restriction on the decrease, which is almost equivalent to Theorem 2, follows from Boas's theorem (Theorem 3 in the appendix): Suppose a function $f(z)$ which is analytic in the upper half-plane Im z > 0 and continuous in the closed half-plane Im z ≥ 0 satisfies the conditions*

$$\lim_{r \to \infty} \frac{1}{r} \cdot \max_{|z|=r} |f(z)| < \infty; \tag{2.12}$$

$$\int_{-\infty}^{\infty} \frac{\ln^+ |f(x)|}{1+x^2}\, dx < \infty, \tag{2.13}$$

and its zeros do not have finite limit points.

Then

$$\int_{-\infty}^{\infty} \frac{\ln^- |f(x)|}{1+x^2}\, dx < \infty. \tag{2.14}$$

We apply this theorem to the function

$$f(z) = F[4\mu^2(1 - z^2)].$$

By virtue of the restriction (1.9), the condition (2.12) is satisfied for $f(z)$. Suppose that F(s) is polynomially bounded on the cut:

$$|F(s)| \leqslant \text{const}\, s^n, \quad s \to +\infty. \tag{2.15}$$

Then the condition (2.13) is also satisfied. The inequality (2.14) shows that the integral

$$\int_{4\mu^2}^{\infty} \frac{\ln^- |F(s)|}{s\sqrt{s - 4\mu^2}}\, ds$$

converges. We thus arrive at a further theorem.

---

* $$\ln^+ x = \begin{cases} \ln x & \text{for } x \geqslant 1; \\ 0 & \text{for } x \leqslant 1; \end{cases}$$

$$\ln^- x = \begin{cases} -\ln x & \text{for } x \leqslant 1; \\ 0 & \text{for } x \geqslant 1. \end{cases}$$

THEOREM 3. Suppose the form factor F(s) is polynomially bounded on the cut

$$|F(s)| \leqslant \text{const}\, s^n, \quad s \to +\infty,$$

and vanishes nowhere on the cut. Then the integral

$$\int_{4\mu^2}^{\infty} \frac{\ln^- |F(s)|}{s\sqrt{s-4\mu^2}}\, ds < \infty$$

must converge and there is a lower bound on the decrease of F(s).

Thus, restrictions have been established on the decrease of F(w) for $w \to +\infty$, i.e., in the physical region of the annihilation channel. For the form factor in the physical region of the scattering channel a similar assertion holds. To obtain it, we apply the following theorem (Theorem 4 in the Appendix).

Suppose a function $f(z)$ which is analytic in the upper half-plane Im z > 0 and continuous in the closed half-plane Im z ≥ 0 satisfies the conditions (2.12) and (2.13). If

$$\frac{1}{y} \ln |f(iy)| \to -\infty, \tag{2.16}$$

as $y \to +\infty$, then $f(z)$ vanishes identically:

$$f(z) \equiv 0.$$

This means that in the limit $y \to \infty$ the function $f(iy)$ cannot decrease faster than any linear exponential function of y. If we apply this theorem to the function $f(z) = F(z^2)$, we obtain the next theorem [16].

THEOREM 4. If the form factor F(w) is polynomially bounded in the physical region of the annihilation channel:

$$|F(s)| \leqslant \text{const}\, s^n, \quad s \to +\infty,$$

then in the physical region of the scattering channel it can decrease, only not faster than some exponential function of $|t|^{1/2}$. In other words, there exists a constant $a > 0$ such that

$$|F(t_n)| \geqslant \text{const}\, e^{-a|t_n|^{1/2}},$$

for at least some sequence $t_n \to -\infty$.

There is a more intimate relationship between the behaviors of F(w) in the physical regions of the annihilation channel $w \geq 4\mu^2$ and the scattering channel $w \leq 0$. To establish the relationship we apply the Phragmén–Lindelöf theorem (see Theorem 1 in the Appendix). This theorem states that if a function $f(z)$ is analytic and bounded in the upper half-plane Im z > 0 the sets of its limit values as $z \to +\infty$ and $z \to -\infty$ (let them be E_+ and E_-, respectively) must possess the following property: either they have a common point or one of them surrounds the other; but if $f(z)$ tends to definite limits a_+ and a_- as $z \to +\infty$ and $z \to -\infty$, respectively, these limits are equal to each other.

Now suppose that as $w \to \pm\infty$ the form factor has the following behavior:

$$|F(s)| \approx a s^{\alpha} \quad \text{as} \quad s \to +\infty; \tag{2.17}$$

$$F(t) \approx b|t|^{\beta} \quad \text{as} \quad t \to -\infty, \tag{2.18}$$

where $a > 0$ and b is a real number. We introduce the new function

$$f(z) = \frac{F(z)}{(z+i)^{\beta}}. \tag{2.19}$$

It follows from (2.17) that

$$f(z) \to b e^{-i\pi\beta} \quad \text{as} \quad z \to -\infty \tag{2.20}$$

and

$$|f(z)| \approx a|z|^{\alpha-\beta} \quad \text{as} \quad z \to +\infty. \tag{2.21}$$

Using the Phragmén–Lindelöf theorem, we prove that β cannot exceed α. Suppose the opposite: $\beta > \alpha$. Then by (2.21)

$$|f(z)| \to 0 \quad \text{as} \quad z \to +\infty, \tag{2.22}$$

and the limit sets E_+ and E_- are two different points: $a_+ = 0$ and $a_- = b\,e^{-i\pi\beta} \neq 0$, which contradicts the theorem. Thus, $\alpha \geq \beta$.

Consider the special case $\alpha = \beta$. Then

$$|f(z)| \to a \quad \text{as} \quad z \to +\infty, \tag{2.23}$$

and the limit set E_+ is a subset of the circle with center at the origin and radius a. Since the other limit set is the point $be^{-i\pi\beta}$, we have only two possibilities: either $a = |b|$ and the limit sets E_+ and E_- have a common point or $a > |b|$ and E_+ surrounds E_-.

We thus have the conclusion: it is necessary that $\alpha \geq \beta$, and if $\alpha = \beta$, then $a \geq |b|$.

Similarly, one can consider the case when $F(t)$ decreases as an exponential function of $-|t|^{\gamma}$, $\gamma \leq 1/2$, as $t \to -\infty$. Suppose that

$$|F(s)| \approx Ae^{-as^{\alpha}}, \quad \alpha < \frac{1}{2} \quad \text{as} \quad s \to +\infty; \tag{2.24}$$

$$|F(t)| \approx Be^{-b|t|^{\beta}}, \quad \beta \leqslant \frac{1}{2} \quad \text{as} \quad t \to -\infty, \tag{2.25}$$

where $a \geq 0$, $b \geq 0$, $A > 0$, and B is a real number. Repeating the arguments we have just used, we can show that $\alpha \leq \beta$, and if $\alpha = \beta$ then $a \leq b \cos \pi\alpha$. In the special case when $\beta = 1/2$, we have $a = 0$. Thus, we have obtained the following theorem [15].

THEOREM 5. Suppose the form factor F has the following behavior in the physical regions of the annihilation and scattering channels:

$$|F(s)| \approx as^{\alpha}, \quad a > 0, \quad s \to +\infty$$

and

$$F(t) \approx b|t|^{\beta}, \quad t \to -\infty,$$

or

$$|F(s)| \approx Ae^{-as^{\alpha}}, \quad \alpha \leqslant \frac{1}{2}, \quad A > 0, \quad a > 0 \quad s \to +\infty,$$

and

$$F(t) \approx Be^{-b|t|^{\beta}}, \quad \beta \leqslant \frac{1}{2}, \quad b \geqslant 0, \quad t \to -\infty.$$

Then in the first case $\alpha \geq \beta$, and if $\alpha = \beta$, then $a \geq |b|$; in the second $\alpha \leq \beta$, and if $\alpha = \beta$, then $a \leq b \cos \pi\alpha$.

Finally, the result can be readily generalized to the case in which the form factor is asymptotically a product of an exponential and a power function.

3. LOWER BOUNDS IN THE PHYSICAL REGION OF THE SCATTERING CHANNEL

We now turn to the study of lower bounds for the form factor $F(t)$ in the physical region, $t \leq 0$, of the scattering channel $|F(s)|$ when some information is given about $|F(s)|$ in the physical region, $s \geq 4\mu^2$, of the annihilation channel. We shall use the normalization condition

$$F(0) = 1. \tag{3.1}$$

Suppose that the maximal value of $|F(s)|$ attained at a point of the cut is M:

$$\max_{s \geqslant 4\mu^2} |F(s)| = M. \qquad (3.2)$$

On the transition from the point t = 0 to points on the cut $s \geq 4\mu^2$, the form factor varies from 1 to M. The larger is M, the faster F(w) can vary and, therefore, the faster it can decrease between t = 0 and a point t < 0. This suggests that the values of the form factor at every point $t \leq 0$ of the physical region of the scattering channel have certain lower bounds that depend on M. Our aim is to establish these bounds.

To get a better understanding of the essence of the problem, consider a function $f(z)$ that is analytic in an annular region with center at the origin and radii R_1 and R_2, $R_1 < R_2$. Let M_1 and M_2 be the maxima of $|f(z)|$ on the circles $|z| = R_1$ and $|z| = R_2$, respectively. Then by Hadamard's three-circles theorem (Theorem 5 in the Appendix) $|f(z_0)|$, at a certain point z_0 within the analyticity domain of $f(z)$

$$\ln|f(z_0)| \leqslant \ln M_1 \frac{\ln|z_0|/R_2}{\ln R_1/R_2} + \ln M_2 \frac{\ln|z_0|/R_1}{\ln R_2/R_1}. \qquad (3.3)$$

If $|f(z_0)|$ and M_2 are specified, this formula makes it possible to establish a lower bound for M_1:

$$\ln M_1 \geqslant \left\{\ln|f(z_0)| - \ln M_2 \frac{\ln|z_0|/R_1}{\ln R_2/R_1}\right\} \frac{\ln R_1/R_2}{\ln|z_0|/R_2}. \qquad (3.4)$$

In the case of a form factor that is analytic in the w plane with the cut $w \geq 4\mu^2$, the role of the inner circle is played by some interval in the region $t \leq 0$ and that of the outer circle by the cut itself; t = 0 plays the role of the point z_0. Since the value of F(0) is known, a lower bound exists for F(t) for $t \leq 0$ if the maximum of $|F(s)|$ is given on the cut.

To apply Hadamard's theorem to the form-factor function, it is necessary to transform the w plane with cut $w \geq 4\mu^2$ and with some artificial cut lying within the interval $t \leq 0$ into a ring. Such a transformation is rather complicated. For simplicity we shall use only transformations that carry part of the w plane with the two cuts into a ring.

Using the change of variables

$$\xi = \left[\frac{w}{4\mu^2} + \alpha\right]^{1/2}, \qquad (3.5)$$

where α is a fairly large positive number, we first transform the w plane with cut $w \geq 4\mu^2$ into the upper half plane of ξ and set

$$F(w) = \Phi(\xi).$$

The function $\Phi(\xi)$ is analytic in the upper half-plane and takes real values on the interval $-\sqrt{1+\alpha} \leqslant \xi \leqslant \sqrt{1+\alpha}$. By the Riemann–Schwarz symmetry principle, it can be analytically continued into the lower half-plane of ξ. Thus, $\Phi(\xi)$ is analytic in the ξ plane with the two cuts $\xi \geqslant \sqrt{1+\alpha}$ and $\xi \leqslant -\sqrt{1+\alpha}$. Making the conformal mapping

$$\eta = \frac{\sqrt{1+\alpha}}{\xi}\left[\sqrt{1+\alpha} - \sqrt{1+\alpha-\xi^2}\right], \qquad (3.6)$$

we then transform the ξ plane into a disk C on the η plane with center at the origin and radius $a = \sqrt{1+\alpha}$. The point $\xi = \sqrt{\alpha}$, i.e., w = 0, is carried to the point $\eta = \eta_0$

$$\eta_0 = \frac{\sqrt{1+\alpha}}{\sqrt{\alpha}}\left(\sqrt{1+\alpha} - 1\right). \qquad (3.7)$$

Consider the symmetric points $\xi = \pm\sqrt{\alpha-\gamma}$, where γ is some positive number, $\gamma < \alpha$. Under the mapping (3.6), they are carried to the points $\eta = \pm c$:

$$c = \frac{\sqrt{1+\alpha}}{\sqrt{\alpha-\gamma}}\left[\sqrt{1+\alpha} - \sqrt{1+\gamma}\right]. \qquad (3.8)$$

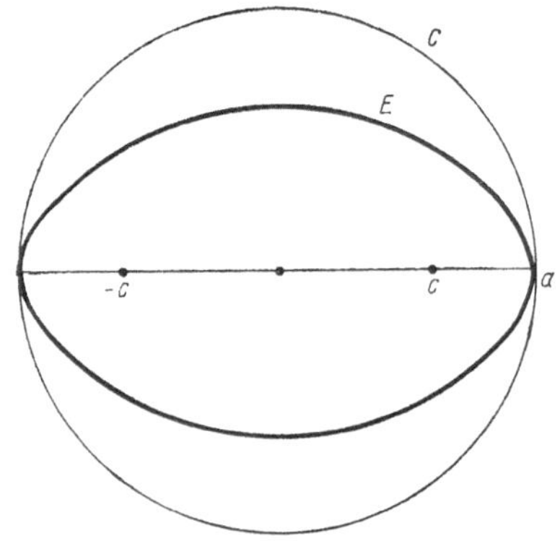

Fig. 5. The disk C and ellipse E with focuses at the points $\eta = \pm c$ and with semimajor axis $a = (1+\alpha)^{1/2}$.

The disk C contains an ellipse E with focuses at the points $\eta = \pm c$ and with semimajor axis a, where

$$a = \sqrt{1+\alpha} \tag{3.9}$$

(Fig. 5). Using the conformal mapping

$$z = \frac{1}{c}\left[\eta + \sqrt{\eta^2 - c^2}\right] \tag{3.10}$$

we transform the ellipse E with cut $-c \le \eta \le c$ into a ring with internal radius $R_1 = 1$ and external radius $R_2 = R$:

$$R = \frac{1}{c}\left[a + \sqrt{a^2 - c^2}\right] \tag{3.11}$$

and with center at the origin. The point $\eta = \eta_0$ is carried to the point $z_0 = r$:

$$r = \frac{1}{c}\left[\eta_0 + \sqrt{\eta_0^2 - c^2}\right]. \tag{3.12}$$

We set

$$\Phi(\xi) = f(z).$$

By hypothesis

$$\max_{|z|=R} |f(z)| \leqslant M. \tag{3.13}$$

We set

$$\max_{|z|=1} |f(z)| = m. \tag{3.14}$$

By Hadamard's three-circles theorem,

$$\ln|f(r)| \leqslant \left(1 - \frac{\ln r}{\ln R}\right)\ln m + \frac{\ln r}{\ln R}\ln M. \tag{3.15}$$

Since $f(r) = \Phi(\sqrt{\alpha}) = F(0) = 1$,

$$m \geqslant \left(\frac{1}{M}\right)^{\frac{\ln r/\ln R}{1-\ln r/\ln R}}. \tag{3.16}$$

Letting $\alpha \to \infty$ and using the expressions (3.11) and (3.12) for R and r and the expressions (3.7)-(3.9) for η_0, c, and a, we finally obtain

$$\max_{w \leqslant t < 0} |F(w)| \geqslant \left(\frac{1}{M}\right)^{\Psi\left(\frac{|t|}{4\mu^2}\right)}, \tag{3.17}$$

where

$$\Psi(\gamma) = \frac{[1-(1+\gamma)^{-1/2}]^{1/2}}{1-[1-(1+\gamma)^{-1/2}]^{1/2}}. \tag{3.18}$$

If F(t) decreases monotonically when the absolute value of t increases in the physical region of the scattering channel, (3.17) is replaced by

$$|F(t)| \geqslant \left(\frac{1}{M}\right)^{\Psi\left(\frac{|t|}{4\mu^2}\right)}. \tag{3.19}$$

Thus, we obtain a further theorem [15].

THEOREM 6. Suppose the form factor F(w) on the cut is bounded by a constant:

$$|F(s)| \leqslant M, \quad s \geqslant 4\mu^2.$$

Then there is a lower bound for its values in the physical region, $t \leq 0$, of the scattering channel:

$$\max_{w \leqslant t < 0} |F(w)| \geqslant \left(\frac{1}{M}\right)^{\Psi\left(\frac{|t|}{4\mu^2}\right)},$$

where

$$\Psi(\gamma) = \frac{[1-(1+\gamma)^{-1/2}]^{1/2}}{1-[1-(1+\gamma)^{-1/2}]^{1/2}}.$$

If F(t) decreases monotonically as $|t|$ increases in the region $t \leq 0$, then

$$|F(t)| \geqslant \left(\frac{1}{M}\right)^{\Psi\left(\frac{|t|}{4\mu^2}\right)}.$$

Let t_e be a positive value such that in the interval $-t_e \leq t \leq 0$ the form factor decreases monotonically by a factor of e. It then follows from the above results that

$$t_e \geqslant \frac{1}{(1+\ln M)^2-1}. \tag{3.20}$$

It is interesting to consider the function $\Psi(\gamma)$ defined by (3.18) for large γ, $\gamma \gg 1$. We have

$$\Psi(\gamma) \approx 2\sqrt{\gamma}, \quad \gamma \to \infty. \tag{3.21}$$

Thus, for large t the form factor must satisfy the condition

$$\max_{w \leqslant t < 0} |F(w)| \geqslant \text{const}\, e^{-2\sqrt{\frac{|t|}{4\mu^2}}\ln M}, \quad t \to -\infty \tag{3.22}$$

or, if F(t) decreases monotonically

$$|F(t)| \geqslant \text{const}\, e^{-2\sqrt{\frac{|t|}{4\mu^2}}\ln M}, \quad t \to -\infty. \tag{3.23}$$

This means that as $t \to -\infty$ the form factor cannot decrease faster than some exponential function of $-(|t|)^{1/2}$ and that the coefficient of $(|t|)^{1/2}$ in the exponential function is completely determined by specification of the maximum M. This result is a generalization of Theorem 4.

In proving the inequality (3.17) we have not exploited all the analytic properties that F(w) possesses. Indeed, after the conformal mappings (3.5) and (3.6) the form factor becomes a function of η that is analytic in the disk C. However, we have used only the analyticity of this function in the ellipse E contained within C; and, using the conformal mappings (3.10), we have transformed this ellipse (with the cut) into a ring and we have then applied Hadamard's theorem to the function $f(z)$, which is analytic in this ring. We shall now attempt to use the analyticity of F(w) in the whole of the complex w plane with the cut $w \geq 4\mu^2$. To this end we apply a theorem that is more general than Hadamard's theorem: the two-constants theorem. In formulating this theorem we require the concept of a harmonic measure.

Let D be some connected domain in the complex z plane whose boundary ∂D consists of a finite number of analytic arcs. Suppose further that α is some set of arcs of the boundary ∂D and β is the remaining part of the boundary: $\partial D = \alpha \cup \beta$. Let $\omega(z, \alpha, D)$ be a function that is harmonic in D, takes the value 1 at each internal point of the arcs of the set α, and vanishes at every internal point of the arcs belonging to β. We call this function, $\omega(z, \alpha, D)$, the harmonic measure at the point z of the set α with respect to D. It follows from the maximum and minimum principle for harmonic functions that $0 \leq \omega(z, \alpha, D) \leq 1$. (The Appendix contains more information about harmonic functions and the harmonic measure.) The two-constants theorem states the following (Theorem 8 in the Appendix). Suppose there is given a connected domain D with boundary ∂D consisting of a finite number of analytic arcs; that α is an open set consisting of several arcs on ∂D; and β is the set of remaining arcs on ∂D. Let β_0 be the interior part of β. Suppose also that a function $f(z)$ which is analytic in D and is continuous in the closed domain $\bar{D}$ with the exception of a finite number of points on ∂D satisfies the condition

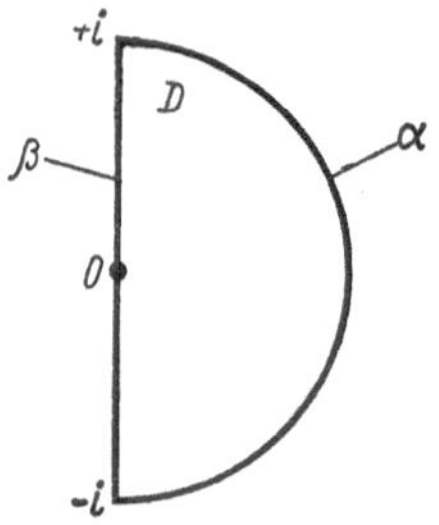

Fig. 6. The domain D consisting of the right-hand half-disk with radius equal to unity and centers at the origin; α is the right-hand semicircle; β is the interval [−i, i].

$$|f(z)| \leqslant M \quad \text{for} \quad z \in \alpha; \tag{3.24}$$

$$|f(z)| \leqslant m \quad \text{for} \quad z \in \beta_0. \tag{3.25}$$

Then for any point z in D

$$\ln |f(z)| \leqslant \omega(z, \alpha, D) \ln M + [1 - \omega(z, \alpha, D)] \ln m. \tag{3.26}$$

In the special case when D is a ring with radii R_1 and R_2, $R_1 < R_2$, and α is the outer circle with radius R_2 and β is the inner circle with radius R_1 the harmonic measure is

$$\omega(z, \alpha, D) = \frac{\ln |z|/R_1}{\ln R_2/R_1}. \tag{3.27}$$

Hadamard's three-circles theorem follows from the two-constants theorem.

Suppose that D is the right-hand half-disk with radius equal to unity and center at the origin; α is the right-hand semicircle on the boundary ∂D; and β is the interval [−i, i] on the imaginary axis (Fig. 6). Then

$$\omega(z, \alpha, D) = \frac{2}{\pi} \operatorname{arctg} \frac{2 \operatorname{Re} z}{1 - |z|^2}. \tag{3.28}$$

We now apply (3.26) with the harmonic measure (3.28) to the study of the form factor F(w). To this end we must transform the w plane with the cut conformally into the right-hand half-disk in Fig. 6. For some $\gamma > 0$ we first set

$$\xi = \frac{\sqrt{1+\gamma} - \sqrt{1 - w/4\mu^2}}{\sqrt{1+\gamma} + \sqrt{1 - w/4\mu^2}}. \tag{3.29}$$

Under the mapping $w \to \xi$ the w plane with the cut $w \geq 4\mu^2$ is transformed into the unit disk; the cut $w \geq 4\mu^2$ goes over into the unit circle; the point $w = -4\mu^2\gamma$ is carried to the center of the disk; and the interval $w \leq -4\mu^2\gamma$ on the real axis goes over into the interval [−1, 0]. The substitution

$$\xi = z^2; \quad z = \sqrt{\xi} \tag{3.30}$$

then transforms the unit disk into the right-hand unit half-disk in Fig. 6. It is obvious that the image of the cut $w \leq 4\mu^2$ under these last two mappings is the right-hand unit semicircle α and that the image of $w \leq -4\mu^2\gamma$ is the interval $\beta = [-i, i]$ on the imaginary axis. At the same time, the point w = 0 is carried to the point $z = z_0$:

$$z_0 = \sqrt{\xi_0} = \left[\frac{\sqrt{1+\gamma} - 1}{\sqrt{1+\gamma} + 1} \right]^{1/2}. \tag{3.31}$$

We set

$$F(w) = f(z).$$

By hypothesis, F(w) is bounded on the cut by the constant M. This means that

$$|f(z)| \leqslant M \quad \text{for} \quad z \in \alpha. \tag{3.32}$$

We write

$$m = \max_{z \in \beta} |f(z)| = \max_{t \leqslant -4\mu^2\gamma} |F(t)|. \tag{3.33}$$

It follows from the two-constants theorem that

$$\ln |f(z_0)| \leqslant \omega(z_0, \alpha, D) \ln M + [1 - \omega(z_0, \alpha, D) \ln m]. \tag{3.34}$$

On the other hand, by the normalization condition (3.1)

$$f(z_0) = F(0) = 1.$$

Thus,

$$m \geqslant \left(\frac{1}{M} \right)^{\frac{\omega(z_0, \alpha, D)}{1 - \omega(z_0, \alpha, D)}}. \tag{3.35}$$

Using the expression (3.28) for the harmonic measure $\omega(z, \alpha, D)$ and the expression (3.31) for z_0, we readily obtain

$$\omega(z_0, \alpha, D) = \frac{2}{\pi} \operatorname{arctg} \sqrt{\gamma}. \tag{3.36}$$

The inequality (3.35) can now be written in the form

$$\max_{w \leqslant t < 0} |F(w)| \geqslant \left(\frac{1}{M}\right)^{\varphi\left(\frac{|t|}{4\mu^2}\right)}, \tag{3.37}$$

where

$$\varphi(\gamma) = \frac{\frac{2}{\pi} \operatorname{arctg} \sqrt{\gamma}}{1 - \frac{2}{\pi} \operatorname{arctg} \sqrt{\gamma}}. \tag{3.38}$$

If F(t) decreases monotonically with increasing absolute values of t in the physical region, $t \leq 0$, of the scattering channel, then (3.37) is replaced by

$$|F(t)| \geqslant \left(\frac{1}{M}\right)^{\varphi\left(\frac{|t|}{4\mu^2}\right)}, \quad t < 0. \tag{3.39}$$

We now formulate this result in the form of a theorem [17].

THEOREM 7. Suppose that the form factor F(w) is bounded on the cut by some constant M:

$$|F(s)| \leqslant M, \quad s \geqslant 4\mu^2.$$

Then its values in the physical region, $t \leq 0$, of the scattering channel have a lower bound:

$$\max_{w \leqslant t < 0} |F(w)| \geqslant \left(\frac{1}{M}\right)^{\varphi\left(\frac{|t|}{4\mu^2}\right)},$$

where

$$\varphi(\gamma) = \frac{\frac{2}{\pi} \operatorname{arctg} \sqrt{\gamma}}{1 - \frac{2}{\pi} \operatorname{arctg} \sqrt{\gamma}}.$$

If F(t) decreases monotonically as $|t|$ increases in the region $t \leq 0$, then

$$|F(t)| \geqslant \left(\frac{1}{M}\right)^{\varphi\left(\frac{|t|}{4\mu^2}\right)}.$$

The function $\varphi(\gamma)$ for large γ has the asymptotic expression

$$\varphi(\gamma) \approx \frac{\pi}{2} \sqrt{\gamma}. \tag{3.40}$$

Recall that we have the factor $\pi/2$ instead of the factor 2 in (3.21). Therefore, instead of the inequalities (3.22) and (3.33) found by means of Hadamard's theorem, we now obtain

$$\max_{w \leqslant t < 0} |F(w)| \geqslant \text{const}\, e^{-\frac{\pi}{2} \sqrt{\frac{|t|}{4\mu^2}} \ln M}, \quad t \to -\infty, \tag{3.41}$$

or, if F(t) decreases monotonically,

$$|F(t)| \geqslant \text{const}\, e^{-\frac{\pi}{2}\sqrt{\frac{|t|}{4\mu^2}}\ln M}, \quad t \to -\infty. \tag{3.42}$$

Thus, the use of the more general theorem makes it possible to improve the results.

A question naturally arises: Can one essentially improve the results expressed by the inequality (3.41) or (3.42)? Since the form factor

$$F(w) = e^{[1-\sqrt{1-w/4\mu^2}]\,a} \tag{3.43}$$

for any $a > 0$ has the properties

$$\ln M = a;$$

$$|F(t)| \approx e\, e^{-\sqrt{\frac{|t|}{4\mu^2}}\ln M},$$

in the most favorable case one can only obtain bounds of the form (3.41) or (3.42) in which $\pi/2$ is replaced by unity if one does not consider the constant in front of the exponential function.

Thus, we have established lower bounds for the form-factor function in the physical region, $t \leq 0$, of the scattering channel in the case when the form factor is bounded on the cut. Our treatment can be readily generalized to the case when F(w) increases polynomially on the cut. These generalizations are given in [17]. A result analogous to Theorem 7 has also been obtained independently in [18].

4. SUM RULES

Dispersion sum rules have been widely used in recent years in the theory of strong interactions. They are a very general instrument for studying the relationship between the asymptotic behavior of an amplitude and its behavior in the resonance region [19]. In the present section of the review we derive a number of dispersion sum rules for the π-meson form factor which are consequences of its general analytic properties. These sum rules contain only the modulus of the form factor in the physical regions of the scattering channel and annihilation channel, i.e., only observable quantities, and they can therefore be used to verify analyticity experimentally.

To illustrate the main idea, consider first the simple case when F(w) does not have zeros in the plane. Then not only F(w) but also ln F(w) is an analytic function in the complex w plane and the real part of the latter

$$\operatorname{Re}\ln F(w) = \ln |F(w)|$$

is a harmonic function. The sum rules are a consequence of | F(w) | being harmonic and the normalization condition

$$F(0) = 1. \tag{4.1}$$

Using the conformal transformation

$$z = \frac{1-\sqrt{1-w/4\mu^2}}{1+\sqrt{1-w/4\mu^2}} \tag{4.2}$$

we transform the w plane with the cut $w \geq 4\mu^2$ into the unit disk: the point $w = 0$ is carried to the center $z = 0$ and the cut $w \geq 4\mu^2$ goes over into the unit circle. We set

$$f(z) = F(w).$$

By hypothesis, the function $f(z)$, which is analytic in the unit disk, has no zeros. Therefore, the function

$$u(z) = \ln |f(z)| \equiv \ln |F(w)|$$

is harmonic inside this disk. The function u(z) is continuous in the unit disk. Then

$$u(0) = \frac{1}{2\pi}\int_0^{2\pi} u(e^{i\varphi})\, d\varphi \tag{4.3}$$

or

$$\ln|f(0)| = \frac{1}{2\pi}\int_0^{2\pi} \ln|f(e^{i\varphi})|\,d\varphi. \tag{4.4}$$

Because of the normalization condition (4.1), the left-hand side of (4.4) vanishes. Returning to the old variable w, we obtain

$$\int_{4\mu^2}^{\infty} \frac{\ln|F(s)|}{s\sqrt{s-4\mu^2}}\,ds = 0. \tag{4.5}$$

Note that if F(s) is polynomially bounded on the cut, then the integral on the left-hand side of (4.5) must converge by Theorem 3.

In the general case, F(w) can have zeros – even infinitely many. They are logarithmic poles of the function

$$u(z) = \ln|f(z)|,$$

and this function is not harmonic in the unit disk. It also need not be continuous in this closed disk. However, if $f(z)$ is an analytic function, $\ln|f(z)|$ is subharmonic. On the other hand, for any function u(z) that is subharmonic in the unit disk

$$u(0) \leqslant \frac{1}{2\pi}\int_0^{2\pi} u(e^{i\varphi})\,d\varphi \tag{4.6}$$

(see the Appendix for more details on harmonic and subharmonic functions). Therefore, in the general case, Eq. (4.4) is replaced by

$$\int_0^{2\pi} \ln|f(e^{i\varphi})|\,d\varphi \geqslant 0. \tag{4.7}$$

Reverting to the variable w, we obtain

$$\int_{4\mu^2}^{\infty} \frac{\ln|F(s)|}{s\sqrt{s-4\mu^2}}\,ds \geqslant 0. \tag{4.8}$$

This result was obtained in [20] and then independently in [21]. The proof, which is based on the property of subharmonicity, is very simple.

The result can be readily generalized so as to obtain upper bounds for the form factor at all points outside the cut and, in particular, in the physical region, $t \leq 0$, of the scattering channel if the values of the form-factor modulus are given on the cut. To see this, make the change of variables

$$w_1 = w - t \tag{4.9}$$

for some $t < 0$. The point $w = t$ is then carried into the point $w_1 = 0$. Set

$$F_1(w_1) = F(w) = F(w_1 + t)$$

and apply the above arguments to this function. Then

$$\ln|F_1(0)| \leqslant \frac{\sqrt{4\mu^2 - t}}{\pi}\int_{4\mu^2 - t}^{\infty} \frac{\ln|F_1(w_1)|}{w_1\sqrt{w_1 - (4\mu^2 - t)}}\,dw_1 \tag{4.10}$$

or

$$\ln|F(t)| \leqslant \frac{\sqrt{4\mu^2 - t}}{\pi}\int_{4\mu^2}^{\infty} \frac{\ln|F(s)|}{(s-t)\sqrt{s-4\mu^2}}\,ds. \tag{4.11}$$

This result is obtained in [22] by different methods.

The values of the modulus of the form factor in the physical region of the annihilation channel occur on the left-hand side of (4.8). The expression (4.2) is a sum rule. There are also sum rules that contain simultaneously the values of the modulus of the form factor in the physical regions of the scattering and the annihilation channel. To obtain one of these sum rules, we use the conformal mapping

$$z=\frac{\sqrt{1+w/t_0}-\sqrt{1-w/4\mu^2}}{\sqrt{1+w/t_0}+\sqrt{1-w/4\mu^2}} \tag{4.12}$$

instead of (4.2). Under this mapping, the w plane is transformed into the unit disk and the cut $w \geq 4\mu^2$ together with the semiinfinite interval $w \leq -t_0$ goes over into the unit circle.

We set $f(z) = F(w)$ and, since $\ln|f(z)|$ is subharmonic,

$$\int_{-\infty}^{-t_0}\frac{\ln|F(t)|}{|t|\sqrt{(t-4\mu^2)(t+t_0)}}\,dt+\int_{4\mu^2}^{\infty}\frac{\ln|F(s)|}{s\sqrt{(s-4\mu^2)(s+t_0)}}\,ds\geqslant 0. \tag{4.13}$$

This result is obtained in [23] together with other sum rules of the same nature.

We now formulate these sum rules in the form of a theorem.

THEOREM 8. For the values of the modulus of the form factor in the physical regions of the annihilation and the scattering channel the following sum rules hold:

$$\int_{4\mu^2}^{\infty}\frac{\ln|F(s)|}{s\sqrt{s-4\mu^2}}\,ds\geqslant 0;$$

$$\int_{-\infty}^{-t_0}\frac{\ln|F(t)|}{|t|\sqrt{(t-4\mu^2)(t+t_0)}}\,dt+\int_{4\mu^2}^{\infty}\frac{\ln|F(s)|}{s\sqrt{(s-4\mu^2)(s+t_0)}}\,dt\geqslant 0;$$

$$\ln|F(t)|\leqslant\frac{\sqrt{4\mu^2-t}}{\pi}\int_{4\mu^2}^{\infty}\frac{\ln|F(s)|}{(s-t)\sqrt{s-4\mu^2}}\,ds.$$

5. BOUND ON THE RADIUS FOR A GIVEN VALUE OF THE MAXIMUM OF THE FORM-FACTOR MODULUS ON THE CUT

In studying the lower bounds for the form factor in the physical region of the scattering channel for a given value of the maximum of its modulus on the cut, we have noted that the form factor can change more rapidly, the larger is M. The quantity that characterizes the rapidity of this variation is the derivative of the form factor. This suggests that for a given maximum M the derivative of the form factor at the origin cannot exceed some quantity that depends on M. On the other hand, the derivative of the form factor at the origin determines the mean square radius of the corresponding particle:

$$\langle r^2\rangle=6F'(0). \tag{5.1}$$

We may therefore say that, because of the analyticity of the form factor, a given value of the maximum M of the form-factor modulus on the cut entails a certain upper bound, which depends on M, for the radius of the corresponding particle. We shall show that this is indeed the case and obtain concrete expressions for the upper limit of the radius.

To obtain a clearer understanding of the results, let us first consider the dispersion relation for F(w). Suppose F(w) is bounded. Then one can write down a dispersion relation with one subtraction:

$$F(w)=F(0)+\frac{w}{\pi}\int_{4\mu^2}^{\infty}\frac{\operatorname{Im}F(w')}{w'(w'-w)}\,dw'. \tag{5.2}$$

It follows that

$$F'(0) = \frac{1}{\pi} \int_{4\mu^2}^{\infty} \frac{\operatorname{Im} F(w')}{(w')^2} dw'. \qquad (5.3)$$

Then, replacing Im F(w') by its upper bound M on the right-hand side of (5.3), we obtain

$$|F'(0)| \leqslant \frac{M}{\pi} \int_{4\mu^2}^{\infty} \frac{dw'}{(w')^2} = \frac{M}{4\pi\mu^2}. \qquad (5.4)$$

This result shows that an upper limit for F'(0) does indeed exist. However, the upper limit (5.4) is far from being a least upper bound and is of no practical interest. We shall now improve this estimate.

In the present section we assume that the form factor is bounded on the cut by a constant:

$$|F(s)| \leqslant M, \quad s \geqslant 4\mu^2. \qquad (5.5)$$

However, our arguments can be readily generalized to the case when F(s) increases polynomially as $s \to +\infty$. We shall also make frequent use of the normalization condition

$$F(0) = 1. \qquad (5.6)$$

We shall also assume that in the physical region, $t \leq 0$, of the scattering channel

$$|F(t)| \leqslant 1, \quad t \leqslant 0. \qquad (5.7)$$

This assumption can be completely verified in an experiment. The existing experimental data indicate that the condition (5.7) is well satisfied.

To illustrate the method, let us first consider the case when F(w) has no zeros. Then ln F(w) is also an analytic function. Using the analyticity of this function, we obtain for

$$F'(0) = \left(\frac{d}{dw}[\ln F(w)]\right)_{w=0} \qquad (5.8)$$

an upper limit proportional to ln M and not to M, and we therefore improve the result (5.4).

To prove this, use the conformal mapping (4.2) to transform the w plane with cut into the unit disk and set

$$f(z) = F(w). \qquad (5.9)$$

It is readily verified that

$$F'(0) = \frac{1}{16\mu^2} f'(0). \qquad (5.10)$$

Since $f(z)$ has no zeros, $\ln f(z)$ is also analytic in the unit disk. Moreover, by virtue of the normalization condition (5.6), we have $f(0) = 1$ and $\ln f(0) = 0$, and therefore $\ln f(z)/z$ is also analytic. Since

$$f'(0) = \lim_{z \to 0} \frac{f(z)}{z},$$

it follows from the principle of the modulus maximum (see the Appendix) that

$$|f'(0)| \leqslant \max_{|z|=r\leqslant 1} \left|\frac{\ln f(z)}{z}\right| = \frac{1}{r} \max_{|z|=r\leqslant 1} |\ln f(z)|. \qquad (5.11)$$

Note that $|\ln f(z)|$ is the modulus of the analytic function $\ln f(z)$ and $\ln|f(z)|$ is its real part. By Carathéodory's theorem (Theorem 6 in the Appendix),

$$\max_{|z|=r\leqslant 1} |g(z)| \leqslant \frac{2r}{1-r} \max_{|z|=1} \operatorname{Re} g(z) \qquad (5.12)$$

for any function g(z) that is analytic in the unit disk, and therefore,

$$\max_{|z|=r\leqslant 1} |\ln f(z)| \leqslant \frac{2r}{1-r} \max_{|z|=1} \ln|f(z)|. \tag{5.13}$$

Substituting (5.13) into the right-hand side of (5.11), we obtain

$$|f'(0)| \leqslant \frac{2}{1-r} \max_{|z|=1} \ln|f(z)|. \tag{5.14}$$

By the assumption (5.5),

$$\max_{|z|=1} |f(z)| = \max_{s\geqslant 4\mu^2} |F(s)| \leqslant M.$$

On the other hand, we can always make the passage to the limit r → 0 on the right-hand side of (5.14). Then

$$|f'(0)| \leqslant 2\ln M, \tag{5.15}$$

and therefore

$$|F'(0)| \leqslant \frac{1}{8\mu^2} \ln M. \tag{5.16}$$

In the general case, when F(w) has zeros, $\ln f(z)$ is not an analytic function in the unit disk. To apply the method, it is first necessary to find a disk of smaller radius in the z plane in which there are no zeros of $f(z)$ or a disk in the w plane in which there are no zeros of F(w). All the arguments can then be applied to this disk. Note that F(w) is equal to unity for w = 0. For given maximum M, the rapidity of variation of F(w) is bounded, and therefore F(w) cannot vanish at points near w = 0. This suggests that there exists a disk with center at w = 0 in which F(w) vanishes nowhere. We now find such a disk with maximal radius.

Set

$$w' = 1 + \frac{w}{\alpha}, \quad \alpha \gg 4\mu^2 \tag{5.17}$$

and consider in the w' plane an ellipse E with focuses at the points w = ±1 and semimajor axis

$$a = 1 + \frac{4\mu^2}{\alpha}. \tag{5.18}$$

In E, the function G(w') = F(w) is analytic since this ellipse has only one common point with the cut $w \geq 4\mu^2$. By means of the conformal mapping

$$z = w' + \sqrt{(w')^2 - 1} \tag{5.19}$$

we transform the ellipse E with the cut $-1 \leq w' \leq 1$ into a ring with inner radius equal to unity and outer radius R,

$$R = a + \sqrt{a^2 - 1} \tag{5.20}$$

and we set $f(z) = F(w)$.

By the assumptions (5.5) and (5.7)

$$\max_{|z|=R} |f(z)| \leqslant M; \tag{5.21}$$

$$\max_{|z|=1} |f(z)| \leqslant 1. \tag{5.22}$$

It follows from Hadamard's three-circles theorem that

$$|F(w)| = |f(z)| \leqslant M^{\frac{\ln|z|}{\ln R}}. \tag{5.23}$$

Take α sufficiently large; then

$$\frac{\ln|z|}{\ln R} \approx \sqrt{\frac{|w|}{4\mu^2}}, \tag{5.24}$$

and

$$|F(w)| \leqslant M^{\sqrt{\frac{|w|}{4\mu^2}}}. \tag{5.25}$$

Consider the disk $|w| \le r \le 4\mu^2$ in the w plane. The function F(w) is analytic in this disk and

$$\max_{|w|\leqslant r} |F(w)| \leqslant M^{\sqrt{\frac{r}{4\mu^2}}}. \tag{5.26}$$

There is a well-known theorem in the theory of analytic function according to which a function $f(z)$ which is analytic in the disk $|z| \le r$, is equal to unity at the origin, and has maximum modulus M, cannot vanish within the disk

$$|z| < \frac{r}{M} \tag{5.27}$$

(see Theorem 7 in the Appendix). Applying this theorem to F(w), regarded as an analytic function in the disk $|w| \le r$, and noting (5.26), we conclude that F(w) does not have zeros in a disk with radius

$$w_0 = r e^{-\sqrt{\frac{r}{4\mu^2}} \ln M}, \quad r \leqslant 4\mu^2. \tag{5.28}$$

We choose r such that w_0 attains its greatest value. It is readily shown that this greatest value of w_0 is

$$w_0 = \begin{cases} \dfrac{4\mu^2}{M}, & \text{if} \quad \ln M \leqslant 2; \\ \dfrac{4\mu^2}{e^2}\left(\dfrac{2}{\ln M}\right)^2, & \text{if} \quad \ln M \geqslant 2. \end{cases} \tag{5.29}$$

Thus, F(w) does not have zeros in the disk $|w| \le w_0$, where w_0 is determined by (5.29). In this disk, ln F(w) is an analytic function; moreover, because of the normalization condition (5.6), the function ln F(w)/w is also analytic. Applying the maximum principle to this function and then Carathéodory's theorem to ln F(w) (as above), we obtain

$$|F'(0)| \leqslant \max_{|w|\leqslant r \leqslant w_0} \left|\frac{\ln F(w)}{w}\right| = \frac{1}{r} \max_{|w|=r\leqslant w_0} |\ln F(w)| \leqslant \frac{2}{w_0 - r} \max_{|w|=w_0} \ln|F(w)|. \tag{5.30}$$

Then, letting r tend to zero and using (5.26), we find

$$|F'(0)| \leqslant \frac{1}{\mu\sqrt{w_0}} \ln M. \tag{5.31}$$

Substituting the value (5.29) for w_0 into this relation, we finally obtain [24]

$$|F'(0)| \leqslant \begin{cases} \dfrac{\sqrt{M}\ln M}{2\mu^2}, & \text{if} \quad \ln M \leqslant 2; \\ \dfrac{e(\ln M)^2}{4\mu^2}, & \text{if} \quad \ln M \geqslant 2. \end{cases} \tag{5.32}$$

THEOREM 9. Suppose that the form factor F(w) is bounded on the cut, $|F(w)| \le M$, $s \ge 4\mu^2$, and in the physical region of the scattering channel $|F(t)| \le 1$, $t \le 0$. Then

$$|F'(0)| \leqslant \begin{cases} \dfrac{\sqrt{M}\ln M}{2\mu^2}, & \text{if} \quad \ln M \leqslant 2; \\ \dfrac{e(\ln M)^2}{4\mu^2}, & \text{if} \quad \ln M \geqslant 2. \end{cases}$$

Note that the inequality (5.26) is important for finding the maximal radius w_0. However, this inequality was derived solely from the analyticity of G(w') in E and this function is in reality analytic in the whole of the w plane with the cut $w \geq 4\mu^2$. Let us now attempt to use all the analytic properties at our disposal.

To this end we transform the w plane into the right-hand unit half-disk by means of the conformal transformations

$$\xi = \frac{1-\sqrt{1-w/4\mu^2}}{1+\sqrt{1-w/4\mu^2}} \tag{5.33}$$

and

$$z = \sqrt{\xi}. \tag{5.34}$$

Under these mappings, the cut $w \geq 4\mu^2$ is carried into the right-hand unit semicircle and the physical region, $w \leq 0$, of the scattering channel goes over into the interval $[-i, i]$ on the imaginary axis. We set $f(z) = F(w)$. In accordance with the assumptions (5.5) and (5.7),

$$|f(z)| \leqslant M, \quad z \in \alpha, \tag{5.35}$$

where α is the right-hand unit semicircle, and

$$|f(z)| \leqslant 1, \quad z \in [-i, i]. \tag{5.36}$$

Let $\omega(z, \alpha, D)$ be the harmonic measure of the right-hand unit half-disk D at the point z with respect to α. Then

$$\omega(z, \alpha, D) = \frac{2}{\pi} \operatorname{arctg} \frac{2 \operatorname{Re} z}{1-|z|^2}. \tag{5.37}$$

Applying the two-constants theorem to $f(z)$ [see (3.34) and the Appendix], we obtain

$$|f(z)| \leqslant M^{\omega(z, \alpha, D)}, \tag{5.38}$$

i.e.,

$$|f(z)| \leqslant e^{\frac{2}{\pi} \operatorname{arctg} \frac{2 \operatorname{Re} z}{1-|z|^2} \ln M} \leqslant e^{\frac{2}{\pi} \operatorname{arctg} \frac{2|z|}{1-|z|^2} \ln M}. \tag{5.39}$$

To shorten the expressions, we set

$$\rho = \frac{2}{\pi} \ln M. \tag{5.40}$$

Using the inequality (5.39) to find the maximal radius of the half-disk in which $f(z)$ does not vanish and repeating all the foregoing arguments, we obtain [17]

$$|F'(0)| \leqslant \begin{cases} \dfrac{M \ln M}{4\pi\mu^2} \operatorname{arctg} \dfrac{2\sqrt{M}}{M-1}, & \text{if} \quad \ln M \leqslant \pi; \\ \dfrac{\ln M}{4\pi\mu^2} \dfrac{1}{\Phi(\rho)} \operatorname{arctg} \dfrac{2\sqrt{\Phi(\rho)}}{1-\Phi(\rho)}, & \text{if} \quad \ln M \geqslant \pi, \end{cases} \tag{5.41}$$

where

$$\Phi(\rho) = \frac{\rho^2-2-\sqrt{(\rho^2-2)^2-4}}{2} e^{-\rho \operatorname{arctg} \frac{2}{\sqrt{\rho^2-4}}}. \tag{5.42}$$

THEOREM 10. Under the conditions of Theorem 9

$$|F'(0)| \leqslant \begin{cases} \dfrac{M \ln M}{4\pi\mu^2} \operatorname{arctg} \dfrac{2\sqrt{M}}{M-1}, & \text{if} \quad \ln M \leqslant \pi; \\ \dfrac{\ln M}{4\pi\mu^2} \cdot \dfrac{1}{\Phi(\rho)} \operatorname{arctg} \dfrac{2\sqrt{\Phi(\rho)}}{1-\Phi(\rho)}, & \text{if} \quad \ln M \geqslant \pi, \end{cases}$$

where

$$\Phi(\rho) = \frac{\rho^2 - 2 - \sqrt{(\rho^2-2)^2-4}}{2}\, e^{-\rho \operatorname{arctg} \frac{2}{\sqrt{\rho^2-4}}};$$

$$\rho = \frac{2}{\pi} \ln M.$$

Theorems 9 and 10 can be readily generalized to the case when F(s) increases polynomially as $s \to \infty$. Such generalizations can be found in [17].

Using the harmonic measure principle, we can now also prove an inequality that determined an upper limit for F'(0) under the same assumptions as in Theorems 9 and 10. Using the conformal mapping (5.33), we transform the w plane into the unit disk and set

$$g(\xi) = \frac{1}{M} F(w). \tag{5.43}$$

The function $g(\xi)$ satisfies

$$|g(\xi)| \leqslant 1. \tag{5.44}$$

Thus, the image of the unit disk C_ξ on the ξ plane under the mapping

$$\xi \to \eta = g(\xi) \tag{5.45}$$

is a subset of the unit disk C_η on the η plane. By the condition (5.7)

$$|g(\xi)| \leqslant \frac{1}{M} \quad \text{for } \xi \in [-1, 0], \tag{5.46}$$

since the interval $w \leq 0$ goes over under the transformation (5.33) into the interval [−1, 0] on the ξ plane. This means that the image of the interval

$$\beta = [-1, 0] \tag{5.47}$$

is contained in the interval [−1/M, 1/M] on the η plane and certainly in the interval [−1, 1/M]. It is assumed that M > 1.

We introduce the new function

$$h(\xi) = \frac{g(\xi) - \frac{1}{M}}{1 - \frac{1}{M} g(\xi)}. \tag{5.48}$$

Under the mapping

$$\eta = g(\xi) \to \zeta = h(\xi) \tag{5.49}$$

the unit disk C_η is transformed into the unit disk C_ζ and the interval [−1, 1/M] on the η plane goes over into the interval β = [−1, 0] on the ζ plane. Thus, the mapping

$$\xi \to \zeta = h(\xi) \tag{5.50}$$

has the following properties: it transforms the unit disk C_ξ into a subset of the unit disk C_ζ and the interval β = [−1, 0] on the ξ plane into a subset of the interval β = [−1, 0] on the ζ plane.

Let $\omega(\xi, \beta, C_\xi)$ and $\omega(\zeta, \beta, C_\zeta)$ be the harmonic measures of the disks C_ξ and C_ζ at the points ξ and ζ with respect to the intervals β on the ξ and ζ planes, respectively, the intervals β also being regarded as parts of the boundaries of the disks C_ξ and C_ζ (disks with a cut along one of the radii). In accordance with the harmonic measure principle (see the Appendix)

$$\omega(\xi, \beta, C_\xi) \leqslant \omega(\zeta, \beta, C_\zeta). \tag{5.51}$$

For the harmonic measures in the last inequality, we have the expressions

$$\left.\begin{aligned}\omega(\xi,\beta,C_\xi)&=1-\frac{2}{\pi}\operatorname{arctg}\frac{2\operatorname{Re}\sqrt{\xi}}{1-|\xi|};\\ \omega(\zeta,\beta,C_\zeta)&=1-\frac{2}{\pi}\operatorname{arctg}\frac{2\operatorname{Re}\sqrt{\zeta}}{1-|\zeta|}.\end{aligned}\right\} \tag{5.52}$$

Substituting them into (5.51), we obtain

$$\operatorname{arctg}\frac{2\operatorname{Re}\sqrt{\xi}}{1-|\xi|}\geqslant\operatorname{arctg}\frac{2\operatorname{Re}\sqrt{\zeta}}{1-|\zeta|}. \tag{5.53}$$

For infinitesimally small ξ, this inequality gives

$$|\xi|\geqslant|\zeta|. \tag{5.54}$$

On the other hand, by the definition (5.48) and (5.49)

$$\zeta\approx\frac{1}{1-\frac{1}{M^2}}g'(0)\,\xi,\ \xi\to 0. \tag{5.55}$$

It follows from (5.54) and (5.55) that

$$|g'(0)|\leqslant 1-\frac{1}{M^2}. \tag{5.56}$$

Using (5.33) and (5.43), we find

$$g'(0)=16\,\mu^2\frac{1}{M}F'(0). \tag{5.57}$$

Substituting (5.57) into (5.56), we finally obtain

$$|F'(0)|\leqslant\frac{1}{16\mu^2}\,\frac{M^2-1}{M}. \tag{5.58}$$

This result was found in [25] in the case when F(w) has a finite number of zeros. Following [26], we have proved it without any assumptions concerning the zeros of F(w), but with the additional assumption (5.7), which can be experimentally verified.

A different, very elegant proof of the inequality (5.58) has been given in [27] without the assumption (5.7). It is noted in [27] that the function

$$\varphi(\xi)=\frac{h(\xi)}{\xi}=\frac{1}{\xi}\cdot\frac{g(\xi)-\frac{1}{M}}{1-\frac{1}{M}g(\xi)}, \tag{5.59}$$

is analytic in the unit disk $|\xi|<1$ and

$$|\varphi(\xi)|\leqslant 1,\ |\xi|\leqslant 1. \tag{5.60}$$

Its value at $\xi=0$ is

$$\varphi(0)=\frac{g'(0)}{1-\frac{1}{M^2}}. \tag{5.61}$$

Therefore, the condition (5.60) gives

$$|g'(0)|\leqslant 1-\frac{1}{M^2},$$

i.e., we again obtain (5.56) and consequently (5.58) without the assumption (5.7).

THEOREM 11. If the form factor is bounded on the cut by a constant:

$$|F(s)| \leqslant M, \quad s \geqslant 4\mu^2,$$

then

$$|F'(0)| \leqslant \frac{1}{16\mu^2} \cdot \frac{M^2-1}{M}.$$

6. BOUNDS ON THE RADIUS FOR GIVEN VALUES OF THE FORM-FACTOR MODULUS ON THE CUT

We have shown that if the maximum of the form-factor modulus on the cut is given its derivative at the origin cannot have modulus exceeding a certain limit. However, it is in principle experimentally possible to determine not only the maximal value of the form-factor modulus in the physical region of the annihilation channel but also the values of the modulus itself for different values of the argument. It follows that the use of the results obtained in the foregoing section to estimate a bound for the radius enables us to exploit only part of the information which we have at our disposal from annihilation experiments. It is desirable to generalize our results in such a way that the expression for the upper bound of the radius contains the values of the form factor on the whole of the cut.

As in the foregoing section, we shall find it convenient to work with functions that are analytic in a disk. Therefore, as before, we shall use the conformal mapping (4.2) and transform the plane with the cut $w \geq 4\mu^2$ into the unit disk.

We recall that under this mapping the cut $w \geq 4\mu^2$ is carried into the unit circle. We set

$$F(w) = f(z). \tag{6.1}$$

Since the values of $|F(s)|$ for $s \geq 4\mu^2$ can be determined experimentally, we shall assume that these values are specified and bounded. Suppose that the values of $|f(z)|$ on the unit circle are given:

$$|f(e^{i\varphi})| = \rho(\varphi), \tag{6.2}$$

where $\rho(\varphi)$ is a given and bounded function of φ. Our problem reduces to finding the maximum of $|f'(0)|$ for all functions that are analytic in the unit disk and satisfy (6.2).

We recall that everywhere in this review we assume that $|F(s)|$ for $s \geq 4\mu^2$ is continuous; this can be verified experimentally. This means that $\rho(\varphi)$ is a continuous function for all φ.

We now introduce an auxiliary function g(z) as follows:

$$g(z) = \exp\left\{\frac{1}{2\pi}\int_{-\pi}^{\pi} \frac{e^{i\varphi}+z}{e^{i\varphi}-z} \ln \rho(\varphi)\, d\varphi\right\}. \tag{6.3}$$

Obviously it is analytic in the unit disk. Note that the real part of the function in the argument of the exponential function,

$$u(z) = \frac{1}{2\pi}\int_{-\pi}^{\pi} \frac{1-|z|^2}{1+|z|^2-2|z|\cos(\theta-\varphi)} \ln \rho(\varphi)\, d\varphi; \tag{6.4}$$

$$z = |z|\, e^{i\theta},$$

is harmonic in the unit disk and its limit values as $|z| \to 1$ are equal to $\ln|f(e^{i\varphi})|$:

$$\lim_{r\to 1} u(re^{i\varphi}) \approx \ln \rho(\varphi) \equiv \ln |f(e^{i\varphi})|. \tag{6.5}$$

On the other hand,

$$u(z) = \ln |g(z)|. \tag{6.6}$$

Therefore on the unit circle

$$\lim_{r\to 1} |g(re^{i\varphi})| = \rho(\varphi) = |f(e^{i\varphi})|. \tag{6.7}$$

In addition, the function g(z) has no zeros in the unit disk, as can be seen from the definition (6.3). Hence

$$h(z) = g(0)\frac{f(z)}{g(z)} \tag{6.8}$$

is analytic in the unit disc. The coefficient g(0) is introduced for convenience.

In accordance with the normalization condition,

$$F(0) = 1, \tag{6.9}$$

the function h(z) is equal to unity at the origin:

$$h(0) = 1. \tag{6.10}$$

It follows from (6.7) that on the unit circle

$$\lim_{r\to 1} |h(re^{i\varphi})| = |g(0)|. \tag{6.11}$$

For all other values of z the modulus of h(z) also does not exceed $|g(0)|$:

$$|h(z)| \leqslant g(0), \quad |z| \leqslant 1 \tag{6.12}$$

(Theorem 9 in the Appendix).

We now apply to h(z) all the arguments given in the proof of Theorem 11. Then

$$|h'(0)| \leqslant \frac{M^2-1}{M}, \tag{6.13}$$

where

$$M = |g(0)|. \tag{6.14}$$

Noting (6.10) and using the definition (6.8), we can write h'(0) in the form

$$h'(0) = \left[\frac{d}{dz}\ln h(z)\right]_{z=0} = \left[\frac{d}{dz}\ln f(z)\right]_{z=0} - \left[\frac{d}{dz}\ln g(z)\right]_{z=0} = \frac{f'(0)}{f(0)} - \frac{g'(0)}{g(0)} = f'(0) - \frac{g'(0)}{g(0)}. \tag{6.15}$$

On the other hand, it follows from (6.8) that

$$g(0) = \exp\left\{\frac{1}{2\pi}\int_{-\pi}^{\pi} \ln\rho(\varphi)\,d\varphi\right\} = \exp\left\{\frac{1}{2\pi}\int_{-\pi}^{\pi} \ln|f(e^{i\varphi})|\,d\varphi\right\}; \tag{6.16}$$

$$\frac{g'(0)}{g(0)} = \frac{1}{\pi}\int_{-\pi}^{\pi} \cos\varphi \ln\rho(\varphi)\,d\varphi = \frac{1}{\pi}\int_{-\pi}^{\pi} \cos\varphi \ln|f(e^{i\varphi})|\,d\varphi. \tag{6.17}$$

We set

$$N = \frac{g'(0)}{g(0)}. \tag{6.18}$$

Taking into account (6.15), we obtain from (6.13)

$$|f'(0) - N| \leqslant \frac{M^2-1}{M},$$

i.e.,

$$N - \frac{M^2-1}{M} \leqslant f'(0) \leqslant N + \frac{M^2-1}{M}. \tag{6.19}$$

Thus, if the function $\rho(\varphi)$ is given, M and N are completely determined and $f'(0)$ has the upper bound $N + [(M^2 - 1)/M]$ and the lower bound $N - [(M^2 - 1)/M]$. Reverting to the variable w and the function F(w), we obtain the following theorem.

THEOREM 12. If the modulus of the form factor has given bounded values on the cut, its derivative at the origin satisfies

$$\frac{1}{16\mu^2}\left[N-\frac{M^2-1}{M}\right] \leqslant F'(0) \leqslant \frac{1}{16\mu^2}\left[N+\frac{M^2-1}{M}\right],$$

where

$$M = \exp\frac{1}{\pi}\int_1^\infty \frac{1}{\nu\sqrt{\nu-1}} \ln|F(4\mu^2\nu)|\,d\nu;$$

$$N = \frac{2}{\pi}\int_1^\infty \frac{2-\nu}{\nu^2\sqrt{\nu-1}} \ln|F(4\mu^2\nu)|\,d\nu.$$

Upper and lower limits for F'(0) are obtained in [28] under certain assumptions on the continuity of the integrals of $\ln|f(z)|$. Here we have followed [29] and established the limits without such additional assumptions. Note that by virtue of the sum rule (4.8) the constant M defined above is not less than unity.

The upper and lower limits established in Theorem 12 are the best estimates for the given values of $|F(s)|$ for $s \geq 4\mu^2$. This follows from the existence of explicit examples of a form factor F(w) that satisfies the condition (6.2) for which F'(0) takes the maximal value

$$\frac{1}{16\mu^2}\left[N+\frac{M^2-1}{M}\right]$$

or the minimal value

$$\frac{1}{16\mu^2}\left[N-\frac{M^2-1}{M}\right].$$

In other words,

$$\sup F'(0) = \frac{1}{16\mu^2}\left[N+\frac{M^2-1}{M}\right]; \tag{6.20}$$

$$\inf F'(0) = \frac{1}{16\mu^2}\left[N-\frac{M^2-1}{M}\right]. \tag{6.21}$$

These expressions for F'(0) correspond to the solutions of the following extremal problem: to find the greatest and the least of all possible values of F'(0) for given values of $|F(s)|$, $s \geq 4\mu^2$.

To prove these assertions, it is sufficient to construct a form factor F(w) explicitly having F'(0) equal to (6.20) or (6.21).

Let z_0 be some real number whose modulus is less than unity. Consider the function

$$\tilde{g}(z) = \frac{z_0+z}{1+z_0 z} g(z), \tag{6.22}$$

where g(z) is the function defined in (6.3). Since the modulus of the factor in front of g(z) on the right-hand side of (6.22) is equal to unity on the unit circle,

$$|\tilde{g}(e^{i\varphi})| = |g(e^{i\varphi})|. \tag{6.23}$$

With allowance for (6.7)

$$|\tilde{g}(e^{i\varphi})| = |f(e^{i\varphi})| = \rho(\varphi). \tag{6.24}$$

On the other hand

$$\tilde{g}(0) = z_0 g(0). \tag{6.25}$$

If we take

$$z_0 = \frac{1}{g(0)} = \frac{1}{M}, \tag{6.26}$$

then the function $\tilde{g}(z)$ satisfies the normalization condition

$$\tilde{g}(0) = 1, \tag{6.27}$$

like the function $f(z)$. We recall once more that $M \geq 1$; the case when $M = 1$, $z_0 = 1$, is trivial.

Thus, the function $\tilde{g}(z)$ defined in accordance with (6.22) with allowance for (6.3) and (6.26) satisfies all the conditions imposed on $f(z)$: it is analytic in the unit disk; is equal to unity at the origin; and its modulus takes the given values $|f(e^{i\varphi})|$ on the unit circle.

If we take $f(z)$ to be this function $\tilde{g}(z)$, it follows from (6.22) that

$$f'(0) = \frac{g'(0)}{g(0)} + \frac{1}{z_0} - z_0. \tag{6.28}$$

Substituting the value (6.26) for z_0 into the expression, we find the value of $f'(0)$ corresponding to the upper bound (6.20).

Thus, the extremal form factor for which $F'(0)$ has the greatest value corresponds to the function

$$f(z) = \frac{z_0 + z}{1 + z_0 z} g(z). \tag{6.29}$$

Similarly, for the function

$$f(z) = \frac{z_0 - z}{1 - z_0 z} g(z) \tag{6.30}$$

we have

$$f'(0) = \frac{g'(0)}{g(0)} + z_0 - \frac{1}{z_0}, \tag{6.31}$$

i.e., its derivative at the origin takes the value corresponding to the lower bound (6.21). Thus, we have proved our assertions – the values (6.20) and (6.21) are extremal.

It is difficult to use Theorem 12 to estimate bounds for the radius of a particle because the values of $|F(s)|$ are as yet known only in a small range of energies. Moreover, although it is in principle possible to determine $|F(s)|$ for all s in the physical region $s \geq 4\mu^2$, of the annihilation channel, it is in practice never possible to measure $|F(s)|$ beyond some maximal energy. However, judging from the experimental data, we can draw certain conclusions concerning the behavior of the form factor at energies greater than this maximal energy. Thus, if the form factor decreases with increasing s over a large range of s values lying below the maximal value s_{max}, we can assume that $|F(s)|$ for $s \geq s_{max}$ does not exceed its value at $s = s_{max}$. The result found in this manner is more reliable than the result obtained by making some assumption about the explicit form of the form-factor modulus for $s \geq s_{max}$.

Thus, let us solve the following problem. Given the values of $|F(s)|$ in the interval $4\mu^2 \leq s \leq s_{max}$ on the cut. Suppose also that for $s \geq s_{max}$ the values of $|F(s)|$ do not exceed $|F(s_{max})|$. To find the upper and lower limits of $F'(0)$, we perform the following.

Let φ_{max} be the values of the angle φ into which the point s_{max}, taken on the upper edge of the cut, is carried under the mapping (6.1). In terms of the function $f(z) = F(w)$, which is analytic in the unit disk, our assertions mean that

$$\left.\begin{aligned}|f(e^{i\varphi})|&=f(\varphi); \quad |\varphi|\leqslant\varphi_{max};\\ |f(e^{i\varphi})&\leqslant f(\varphi_{max}); \quad |\varphi|\geqslant\varphi_{max},\end{aligned}\right\} \tag{6.32}$$

where $f(\varphi)$ is a given function. We set

$$f_1(\varphi)=\begin{cases} f(\varphi); & |\varphi|\leqslant\varphi_{max};\\ f(\varphi_{max}), & |\varphi|\geqslant\varphi_{max}\end{cases} \tag{6.33}$$

and, proceeding from this function rather than $f(\varphi)$, we construct a function $g_1(z)$ analogous to $g(z)$ in (6.3)

$$g_1(z)=\exp\left\{\frac{1}{2\pi}\int_{-\pi}^{\pi}\frac{e^{i\varphi}+z}{e^{i\varphi}-z}\ln f_1(\varphi)\,d\varphi\right\}. \tag{6.34}$$

Instead of the function h(z) defined by (6.8), we now consider the corresponding function

$$h_1(z)=g_1(0)\frac{f(z)}{g_1(z)}. \tag{6.35}$$

It is also analytic in the unit disk and its modulus does not exceed

$$M_1=g_1(0)=\exp\left\{\frac{1}{2\pi}\int_{-\pi}^{\pi}\ln f_1(\varphi)\,d\varphi\right\}. \tag{6.36}$$

Applying to $h_1(z)$ all the arguments invoked above in our study of h(z), we obtain

$$|h_1'(0)|=\left|f'(0)-\frac{g'(0)}{g(0)}\right|\leqslant\frac{M_1^2-1}{M_1} \tag{6.37}$$

and therefore

$$N_1-\frac{M_1^2-1}{M_1^2}\leqslant f'(0)\leqslant N_1+\frac{M_1^2-1}{M_1}, \tag{6.38}$$

where M_1 is defined in accordance with (6.36) and

$$N_1=\frac{g_1'(0)}{g_1(0)}=\frac{1}{\pi}\int_{-\pi}^{\pi}\cos\varphi\ln f_1(\varphi)\,d\varphi. \tag{6.39}$$

Reverting to the original variable w, we obtain another theorem.

THEOREM 13. Suppose we are given the values of the modulus of the form factor F(s) in some interval $4\mu^2 \leq s \leq s_{max}$ of the physical region of the annihilation channel and that for $s \geq s_{max}$ the values of $|F(s)|$ do not exceed $|F(s_{max})|$. Then the possible values of the derivative of the form factor at the origin lie between the following limits:

$$\frac{1}{16\mu^2}\left[N_1-\frac{M_1^2-1}{M_1^{\prime}}\right]\leqslant F'(0)\leqslant\frac{1}{16\mu^2}\left[N_1+\frac{M_1^2-1}{M_1^2}\right],$$

where

$$M_1=\exp\left\{\frac{1}{\pi}\int_1^{\nu_{max}}\frac{1}{\nu\sqrt{\nu-1}}\ln|F(4\mu^2\nu)|\,d\nu+\ln|F(4\mu^2\nu_{max})|\frac{1}{\pi}\int_{\nu_{max}}^{\infty}\frac{1}{\nu\sqrt{\nu-1}}\,d\nu\right\};$$

$$N_1=\frac{2}{\pi}\int_1^{\nu_{max}}\frac{2-\nu}{\nu^2\sqrt{\nu-1}}\ln|F(4\mu^2\nu)|\,d\nu+\ln|F(4\mu^2\nu_{max})|\frac{2}{\pi}\int_{\nu_{max}}^{\infty}\frac{2-\nu}{\nu^2\sqrt{\nu-1}}\,d\nu,$$

and

$$\nu_{max} = \frac{s_{max}}{4\mu^2}\,.$$

Finally note that under these conditions the upper and lower limits given in Theorem 19 are the extremal values.

CONCLUSIONS

Let us now summarize the paper. We have investigated the experimental consequences of the general analytic properties of the π-meson form factor. In an experiment on the elastic scattering of an electron by a π-meson, $e + \pi \rightarrow e + \pi$, it is possible to measure the values of the form factor $F(t)$ for $t \leq 0$ and, in particular, to determine $F'(0)$; an experiment on the annihilation of an electron–positron pair into a pair of charged π-mesons, $e^+ + e \rightarrow \pi^+ + \pi^-$, makes it possible to determine the values of $|F(s)|$ for $s \geq 4\mu^2$. All our assertions are related directly to these observable quantities alone.

In Sec. 2 we have investigated the asymptotic behavior of $F(w)$ for large values of w in the physical regions of the scattering and annihilation processes and we have proved the following propositions:

1. The form factor $F(s)$ as $s \rightarrow +\infty$ can decrease only more slowly than $\exp(-\varepsilon s^{1/2})$ for any arbitrarily small $\varepsilon > 0$.

2. If the form factor is polynomially bounded on the cut and does not vanish at any point, the integral

$$\int_{4\mu^2}^{\infty} \frac{\ln^-|F(s)|}{s\sqrt{s-4\mu^2}}\,ds$$

must converge. This imposes a restriction on the decrease of the form factor as $s \rightarrow +\infty$ that is almost equivalent to condition 1.

3. If the form factor $F(s)$ is polynomially bounded as $s \rightarrow +\infty$, then as $t \rightarrow -\infty$ its value $F(t)$ can decrease only not faster than some exponential function $\exp[-a(|t|)^{1/2}]$.

4. The behavior of the form factor $F(w)$ as $w \rightarrow +\infty$ is related to its behavior as $w \rightarrow -\infty$ (in the physical regions of the annihilation and the scattering channel) in the following manner: if

$$|F(s)| \approx as^{\alpha};\ a > 0 \quad \text{as} \quad s \rightarrow +\infty$$

and

$$F(t) \approx b|t|^{\beta} \quad \text{as } t \rightarrow -\infty$$

or

$$|F(s)| \approx Ae^{-as^{\alpha}};\ \alpha \leqslant \frac{1}{2};\ A > 0,\ a \geqslant 0 \quad \text{as} \quad s \rightarrow +\infty$$

and

$$F(t) \approx Be^{-b|t|^{\beta}},\ \beta \leqslant \frac{1}{2} \quad \text{as } t \rightarrow -\infty,$$

then it is necessary that $\alpha \geq \beta$ in the first case, and if $\alpha = \beta$ then $a \geq |b|$; in the second case, $\alpha \leq \beta$, and, if $\alpha = \beta$, then $a \leq b \cos \pi\alpha$.

Section 3 is devoted to the study of a lower bound on the form factor in the physical region, $t \leq 0$, of the scattering channel for a given value of the maximum of its modulus in the physical region of the annihilation channel. We have shown that if

$$|F(s)| \leqslant M \text{ for } s \geqslant 4\mu^2,$$

then

$$\max_{w \leqslant t \leqslant 0} |F(w)| \geqslant \left(\frac{1}{M}\right)^{\varphi\left(\frac{|t|}{4\mu^2}\right)},$$

where $\varphi(\gamma)$ can be taken to be one of the following functions:

$$\varphi(\gamma)=\frac{[1-(1+\gamma)^{-1/2}]^{1/2}}{1-[1-(1+\gamma)^{-1/2}]^{1/2}}$$

or

$$\varphi(\gamma)=\frac{\frac{2}{\pi}\operatorname{arctg}\sqrt{\gamma}}{1-\frac{2}{\pi}\operatorname{arctg}\sqrt{\gamma}}.$$

In particular, for large γ

$$\varphi(\gamma)\approx\frac{\pi}{2}\sqrt{\gamma}.$$

But if the form factor decreases monotonically with increasing absolute value of the momentum transfer in the physical region, $t \le 0$, of the scattering channel, its values at each point of this region satisfy

$$|F(t)|\geqslant\left(\frac{1}{M}\right)^{\varphi\left(\frac{|t|}{4\mu^2}\right)}.$$

In Sec. 4 we have obtained various exact sum rules that explicitly contain the modulus of the form factor in observable regions:

$$\int_{4\mu^2}^{\infty}\frac{\ln|F(s)|}{s\sqrt{s-4\mu^2}}\,ds\geqslant 0;$$

$$\int_{-\infty}^{-t_0}\frac{\ln|F(t)|}{|t|\sqrt{(t-4\mu^2)(t+t_0)}}\,dt+\int_{4\mu^2}^{\infty}\frac{\ln|F(s)|}{s\sqrt{(s-4\mu^2)(s+t_0)}}\,ds\geqslant 0;$$

$$\ln|F(t)|\leqslant\frac{\sqrt{4\mu^2-t}}{\pi}\int_{4\mu^2}^{\infty}\frac{\ln|F(s)|}{(s-t)\sqrt{s-4\mu^2}}\,ds;\ t\leqslant 4\mu^2.$$

For a given value of the maximum of the form-factor modulus on the cut (in the physical region of the annihilation channel for the case of the π-meson), the radius of the corresponding particle cannot exceed a certain limit that depends on this maximum. Various bounds forthis limit were obtained in Sec. 5. We have shown that if $|F(s)|$ for $s \ge 4\mu^2$ is not greater than a constant M: $|F(s)| \le M$, $s \ge 4\mu^2$, and its modulus for $t \le 0$ is not greater than unity, $|F(t)| \le 1$, then $F'(0)$ satisfies the inequality $|F'(0)| \le \xi(M)$, where the $\xi(M)$ can be taken as

$$\xi(M)=\begin{cases}\dfrac{\sqrt{M}\ln M}{2\mu^2}, & \text{if}\quad \ln M\leqslant 2;\\[2mm] \dfrac{e(\ln M)^2}{4\mu^2}, & \text{if}\quad \ln M\geqslant 2,\end{cases}$$

or

$$\xi(M)=\begin{cases}\dfrac{M\ln M}{4\pi\mu^2}\operatorname{arctg}\dfrac{2\sqrt{M}}{M-1}, & \text{if}\quad \ln M\leqslant\pi;\\[2mm] \dfrac{\ln M}{4\pi\mu^2}\cdot\dfrac{1}{\Phi(\rho)}\operatorname{arctg}\dfrac{2\sqrt{\Phi(\rho)}}{1-\Phi(\rho)} & \text{if}\quad \ln M\geqslant\pi;\end{cases}$$

where

$$\rho=\frac{2}{\pi}\ln M,$$

$$\Phi(\rho)=\frac{\rho^2-2-\sqrt{(\rho^2-2)^2-4}}{2}\,e^{-\rho\operatorname{arctg}\frac{2}{\sqrt{\rho^2-4}}}.$$

Without the assumption that the form-factor modulus in the physical region of the scattering channel does not exceed unity, we have obtained the following upper bound:

$$|F'(0)| \leqslant \frac{1}{16\mu^2} \cdot \frac{M^2-1}{M},$$

where M is the maximum of the modulus of F(s) for $s \geq 4\mu^2$.

Finally, in the last section, we have obtained upper and lower bounds for F'(0) for given values of the form-factor modulus in the physical region of the annihilation channel:

$$\frac{1}{16\mu^2}\left[N - \frac{M^2-1}{M}\right] \leqslant F'(0) \leqslant \frac{1}{16\mu^2}\left[N + \frac{M^2-1}{M}\right],$$

where

$$M = \exp\left\{\frac{1}{\pi}\int_1^\infty \frac{1}{\nu\sqrt{\nu-1}} \ln|F(4\mu^2\nu)|\,d\nu\right\};$$

$$N = \frac{2}{\pi}\int_1^\infty \frac{2-\nu}{\nu^2\sqrt{\nu-1}} \ln|F(4\mu^2\nu)|\,d\nu.$$

In practice, it is possible to determine $|F(s)|$ only in a finite region up to some s_{max}. Suppose that

$$|F(s)| \leqslant |F(s_{max})| \text{ for } s \geqslant s_{max}.$$

Then F'(0) has the following upper and lower limits, which do not depend on the actual behavior of $|F(s)|$ for $s \geq s_{max}$:

$$\frac{1}{16\mu^2}\left[N_1 - \frac{M_1^2-1}{M_1}\right] \leqslant F'(0) < \frac{1}{16\mu^2}\left[N_1 + \frac{M_1^2-1}{M_1}\right],$$

where

$$M_1 = \exp\left\{\frac{1}{\pi}\int_1^{\nu_{max}} \frac{1}{\nu\sqrt{\nu-1}} \ln|F(4\mu^2\nu)|\,d\nu + \ln|F(s_{max})|\frac{1}{\pi}\int_{\nu_{max}}^\infty \frac{1}{\nu\sqrt{\nu-1}}\,d\nu\right\};$$

$$N_1 = \frac{2}{\pi}\int_1^{\nu_{max}} \frac{2-\nu}{\nu^2\sqrt{\nu-1}} \ln|F(4\mu^2\nu)|\,d\nu + \ln|F(s_{max})|\frac{2}{\pi}\int_{\nu_{max}}^\infty \frac{2-\nu}{\nu^2\sqrt{\nu-1}}\,d\nu;$$

$$\nu_{max} = \frac{s_{max}}{4\mu^2}.$$

Experiments on the annihilation of an electron–positron pair into a charged π-meson pair are being carried out at present. In these it will be possible to determine $|F(s)|$ for $s \geq 4\mu^2$. The values of F(t) for $t \leq 0$ and, in particular, F'(0) can be determined either directly in an experiment on the elastic scattering of an electron by a meson or indirectly in various different experiments. This will make it possible to verify all our results.

8. APPENDIX

One of the most remarkable properties of analytic functions of a complex variable is the well-known maximum principle: if a function $f(z)$ is analytic in a connected domain D and continuous in the closed domain $\bar{D}$, its modulus can attain a maximum only on the boundary ∂D of this domain. The maximum principle admits a generalization to the case when the boundary ∂D has a set of points in the neighborhood of which $f(z)$ can be unbounded. Let us consider a special case when such a set consists of a single point and D is simply connected [30, 31].

Generalized Maximum Principle. Suppose $f(z)$ is analytic in a simply connected domain D and is continuous in the closed domain $\bar{D}$ with the exception of a point ξ on ∂D. Further, suppose that at all points of ∂D that differ from ξ the modulus of $f(z)$ is bounded by a constant:

$$|f(z)| \leqslant M, \quad z \in \partial D, \quad z \neq \xi.$$

Suppose that there exists a function $\omega(z)$ with the following properties:

1) $\omega(z)$ is analytic in D;

2) $|\omega(z)| < 1$ in D;

3) $\omega(z) \neq 0$ in D;

4) for any positive ε the function $[\omega(z)]^{\varepsilon} f(z)$ is bounded at ξ by any number that exceeds M. Then, in the whole domain, $|f(z)| \leq M$.

Using the generalized maximum principle, we can prove the following theorem.

THEOREM 1. (Phragmén–Lindelöf). Suppose $f(z)$ is analytic in the upper half-plane $\mathrm{Im}\, z > 0$; is continuous in the closed half-plane $\mathrm{Im}\, z \geq 0$; and its modulus is bounded by some constant M on the real axis:

$$|f(x)| \leqslant M, \quad -\infty < x < \infty.$$

Then there are two possibilities: either

$$|f(z)| \leqslant M$$

at all points z in the upper half-plane, or there exists a sequence of points $z_n \to \infty$, $\mathrm{Im}\, z_n > 0$, such that

$$|f(z_n)| \geqslant \mathrm{const}\; e^{a|z_n|}.$$

for some constant $a > 0$.

If the modulus of $f(z)$ is smaller than any exponential function $\exp(|\varepsilon z|)$ for arbitrarily small $\varepsilon > 0$ as $z \to \infty$, $\mathrm{Im}\, z > 0$, the first possibility must be realized, i.e., $f(z)$ is bounded in the whole of the half-plane.

Suppose $f(z)$ is analytic and bounded in the upper half-plane. Let E_+ and E_- be the sets of limit values of $f(z)$ as $z \to +\infty$ and $z \to -\infty$ on the real axis, respectively. Then there are two possibilities: either the sets E_+ and E_- have at least one common point, or one of them surrounds the other, separating it from the circle with radius M and with center at the origin. In particular, if there exist finite limits a_+ and a_- as $z \to +\infty$ and $z \to -\infty$, respectively, these limits must be equal to each other. The Phragmén–Lindelöf theorem has the following important corollary [32].

THEOREM 2 (Carleson). Suppose $f(z)$ is analytic in the upper half-plane $\mathrm{Im}\, z > 0$; is continuous in the closed half-plane $\mathrm{Im}\, z \geq 0$; and increases not faster than $\exp(k|z|)$ as $z \to \infty$. If $f(z)$ decreases exponentially on the real axis:

$$|f(z)| \leqslant e^{-\alpha|z|}, \quad \alpha > 0, \quad z \to \pm\infty,$$

it vanishes identically.

Carleson's theorem imposes a restriction on the decrease of an analytic function with given maximal growth. Other restrictions follow from the next theorems [33].

THEOREM 3. Suppose a function $f(z)$ is analytic in the upper half-plane $\mathrm{Im}\, z > 0$; is continuous in the closed half-plane $\mathrm{Im}\, z \geq 0$; and is of exponential type, i.e.,

$$\alpha = \lim_{r \to \infty} \frac{1}{r} \max_{|z|=r} \ln|f(z)| < \infty,$$

and the zeros of $f(z)$ do not have finite limit points. Further, suppose that

$$\int_{-\infty}^{\infty} \frac{\ln^+|f(x)|}{1+x^2}\, dx < \infty.$$

Then

$$\int_{-\infty}^{\infty} \frac{\ln^- |f(x)|}{1+x^2}\, dx < \infty.$$

THEOREM 4. Suppose $f(z)$ is analytic in the upper half-plane $\operatorname{Im} z > 0$; is continuous in the closed half-plane $\operatorname{Im} z \geq 0$; and is of exponential type, i.e.,

$$\alpha = \lim_{r\to\infty} \frac{1}{r} \max_{|z|=r} \ln |f(z)| < \infty,$$

and the integral

$$\int_1^r \frac{1}{x^2} \ln |f(x) f(-x)|\, dx,$$

as a function of r is bounded. Then if

$$\frac{1}{y} \ln |f(iy)| \to 0 \quad \text{as} \quad y \to \infty,$$

we have

$$f(z) \equiv 0.$$

We now consider some corollaries of the maximum modulus principle.

THEOREM 5 (Hadamard). Suppose $f(z)$ is analytic in the annular region D: $R_1 < |z| < R_2$ and continuous in $\bar{D}$, its modulus on the circle $|z| = R_i$ not exceeding M_i, $i = 1, 2$. Then

$$\ln \frac{R_1}{R_2} \ln |f(z)| \leqslant \ln \frac{|z|}{R_2} \ln M_1 - \ln \frac{|z|}{R_1} \ln M_2.$$

THEOREM 6 (Carathéodory). Let $f(z)$ be a function that is analytic in the disk $|z| < R$ and continuous in the closed disk $|z| \leq R$ and $f(0) = 0$. Set

$$M(r) = \max_{|z|=r} |f(z)|;\ A(r) = \max_{|z|=r} \operatorname{Re} f(z).$$

Then

$$M(r) \leqslant \frac{2r}{R-r} A(R).$$

THEOREM 7. Suppose $f(z)$ is a function that is analytic in the disk $|z| < R$; is continuous in the closed disk $|z| \leq R$; and is equal to unity for $|z| = 0$. If its modulus has a maximum equal to M, it cannot have zeros within the disk $|z| < R/M$.

Theorems 5 and 6 are proved in [31, 32]. Theorem 7 is proved in [34].

The concept of an analytic function is intimately related to the concept of a harmonic function. We recall that a function $u(x, y)$ of two real variables x and y is said to be harmonic if it is twice differentiable with respect to these variables and satisfies the Laplace equation

$$\frac{\partial^2 u(x, y)}{\partial x^2} + \frac{\partial^2 u(x, y)}{\partial y^2} = 0.$$

The real and the imaginary parts of an analytic function are harmonic functions. Like analytic functions, the value of a harmonic function at every point within its harmonicity domain is completely determined by specifying the values of this function on any contour surrounding the point and lying in the analyticity domain of the function. In particular, for the harmonic function $u(x, y) = u(z)$ we have

$$u(z_0)=\frac{1}{2\pi}\int_{-\pi}^{\pi} u(z_0+re^{i\varphi})\,d\varphi.$$

The concept of a subharmonic function [35] generalizes the concept of a harmonic function. A function h(x, y) = h(z) is said to be subharmonic in a domain D if it satisfies the following conditions:

1) it is defined and continuous at all points of D with the exception of possibly a finite number of points or points of a sequence z_n that does not have limit points in D and for each point z_n

$$\lim_{z\to z_n} h(z)=-\infty,$$

and one can therefore set

$$h(z_n)=-\infty;$$

2) for every point z within D and for sufficiently small r

$$h(z)\leqslant\frac{1}{2\pi}\int_{-\pi}^{\pi} h(z+re^{i\varphi})\,d\varphi.$$

Examples of subharmonic functions are $|f(z)|$ and $\ln|f(z)|$, where $f(z)$ is an analytic function in D. Thus,

$$|f(z)|\leqslant\frac{1}{2\pi}\int_{0}^{2\pi}|f(z+re^{i\varphi})|\,d\varphi;$$

$$\ln|f(z)|\leqslant\frac{1}{2\pi}\int_{0}^{2\pi}\ln|f(z+re^{i\varphi})|\,d\varphi$$

for sufficiently small r and for all z within D.

Harmonic functions prove to be very useful in the study of analytic functions. The concept of a harmonic measure plays an especially important role [30, 31].

Let D be a connected domain with boundary ∂D consisting of a finite number of analytic arcs. Further, suppose α is some set of arcs of the boundary ∂D and β is the remaining part: $\alpha \cup \beta = \partial D$. Let $\omega(z, \alpha, D)$ be a function that is harmonic in D; is equal to unity at each internal point of the set α; and vanishes at each internal point of β. It is called the harmonic measure of the set α at the point z with respect to D. In accordance with the maximum and minimum principle for harmonic functions, $0 \le \omega(z, \alpha, D) \le 1$ everywhere in D.

For an annular domain with radii $R_1 < R_2$

$$\omega(z,\alpha,D)=\frac{\ln|z|/R_1}{\ln R_2/R_1},$$

if α is to be taken to be the larger circle:

$$\alpha=\{z:|z|=R_2\}.$$

If D is the right-hand unit half-disk and α is the right-hand unit semicircle:

$$D=\left\{z:|z|\leqslant 1,\ |\arg z|\leqslant\frac{\pi}{2}\right\};$$

$$\alpha=\left\{z:|z|=1,\ |\arg z|\leqslant\frac{\pi}{2}\right\},$$

then $\omega(z, \alpha, D) = (2/\pi)\tan^{-1}[2\operatorname{Re} z/(1-|z|^2)]$.

Using the concept of the harmonic measure, we now formulate the two-constants theorem which generalized Hadamard's three-circles theorem to the case of an arbitrary domain.

THEOREM 8. Let D be a connected domain with boundary ∂D consisting of a finite number of analytic arcs; α be an open set consisting of the points of a finite number of arcs of ∂D; β be the complement of α in ∂D; and β_0 be the interior of the set. Further, suppose that $f(z)$ is analytic in D; is continuous at all points of $\bar{D}$ with the exception of a finite number of points; and satisfies

$$|f(z)| \leqslant M, \quad z \in \alpha,$$
$$|f(z)| \leqslant m, \quad z \in \beta_0.$$

Then for any point in D

$$\ln |f(z)| \leqslant \omega(z, \alpha, D) \ln M + [1 - \omega(z, \alpha, D)] \ln m.$$

Finally, let us consider a further property of functions that are analytic and bounded in the unit disk $|z| < 1$. Let $f(z)$ be a function that is analytic for $|z| < 1$ and takes finite limiting values $f(e^{i\varphi})$ as $z \to e^{i\varphi}$ along nontangential directions at all points of the boundary of the disk with the exception of a finite number of points. We set

$$F(z) = \exp\left[\frac{1}{2\pi}\int_{-\pi}^{\pi} \frac{e^{i\varphi}+z}{e^{-i\varphi}-z} \ln |f(e^{i\varphi})|\, d\varphi\right].$$

The function F(z) is analytic in $|z| < 1$ and has no zeros in this disk and $|F(z)|$ takes the limiting values $|f(e^{i\varphi})|$ as $z \to e^{i\varphi}$ along nontangential directions. Therefore, the function

$$g(z) = \frac{f(z)}{F(z)}$$

is also analytic in the unit disk. As $z \to e^{i\varphi}$, it tends to limit values whose modulus is equal to unity. For all other z in the disk its modulus also does not exceed unity.

Thus, we have the following theorem.

THEOREM 9. All functions $f(z)$ that are analytic in the disk $|z| < 1$ and take finite limit values $f(e^{i\varphi})$ on the unit circle along nontangential directions at all points with the exception of a finite number of points can be represented in the form

$$f(z) = \Phi(z) \exp\left\{\frac{1}{2\pi}\int_{-\pi}^{\pi} \frac{e^{i\varphi}+z}{e^{i\varphi}-z} \ln |f(e^{i\varphi})|\, d\varphi\right\},$$

where $\Phi(z)$ are functions that are analytic in $|z| < 1$ and whose moduli do not exceed unity:

$$|\Phi(z)| \leqslant 1; \quad |z| \leqslant 1.$$

This theorem is proved in a more general form in the book [36].

LITERATURE CITED

1. N. N. Bogolyubov and D. V. Shirkov, Introduction to the Theory of Quantized Fields, Interscience (1959).
2. N. N. Bogolyubov, B. V. Medvedev, and M. K. Polivanov, Questions of the Theory of Dispersion Relations [in Russian], Fizmatgiz, Moscow (1958).
3. I. Ya. Pomeranchuk, Zh. Éksp. Teor. Fiz., 34, 725 (1958).
4. L. Van Hove, Phys. Lett., 5, 252 (1963).
5. A. A. Logunov et al., Phys. Lett., 7, 69 (1963).
6. L. Lukaszuk and A. Martin, Nuovo Cimento, 52A, 122 (1967).
7. M. Froissart, Phys. Rev., 123, 1053 (1961).
8. A. A. Logunov, M. A. Mestvirishvili, and Nguen (Nguyen) Van Hieu, Proceedings of the International Conference on Particles and Fields, Interscience, New York (1967).

9. A. A. Logunov and Nguen (Nguyen) Van Hieu, Proceedings of the Topical Conference on Hadron Collisions at High Energies, CERN, Geneva (1968).
10. Nguen Van Hieu, ZhÉTF Pis. Red., 7, 391 (1968).
11. S. S. Schweber, An Introduction to Relativistic Quantum Field Theory, Evanston, Ill. (1961).
12. S. D. Drell and F. Zachariasen, Electromagnetic Structure of Nucleons, London (1961).
13. G. Barton, Introduction to Dispersion Techniques in Field Theory, New York (1965).
14. P. S. Isaev, Problemy Fiziki Élementarnykh Chastits i Atomnogo Yadra (Particles and Nuclei), 2, No. 1 (1971).
15. Nguen Van Hieu, Dokl. Akad. Nauk SSSR, 182, 1303 (1968).
16. A. Martin, Nuovo Cimento, 37, 671 (1955).
17. V. Baluni, Nguen Van Hieu, and V. A. Suleimanov, Yad. Fiz., 2, 635 (1961).
18. G. G. Volkov, V. V. Ezhela, and M. A. Mestvirishvili, Yad. Fiz., 2, 857 (1969).
19. A. A. Logunov, L. D. Soloviev, and A. N. Tavkhelidze, Phys. Lett., 24B, 181 (1967).
20. B. V. Geshkenbein and B. L. Ioffe, Zh. Éksp. Teor. Fiz., 46, 902 (1964).
21. Nguen (Nguyen) Van Hieu, Preprint JINR, E2-3509 [in English] (1967).
22. Truong Nguen (Nguyen) Tran and R. Mau Vinh, Phys. Rev., 177, 2494 (1969).
23. Nguen Thi Hong, Nucl. Phys., B11, 127 (1969).
24. Nguen Van Hieu, Yad. Fiz., 7, 1111 (1968).
25. B. V. Geshkenbein, Yad. Fiz., 2, 1282 (1969).
26. V. Baluni and Nguen Van Hieu, Preprint JINR R2-4508 [in Russian], Dubna (1969).
27. Nguen Thi Hong, Yad. Fiz., 13, 409 (1971).
28. Dao Vong Duc and Nguen Van Hieu, Teor. Mat. Fiz., 3, 178 (1970).
29. Dao Vong Duc and Nguen Van Hieu, Report at International Conference on High-Energy Physics [in Russian], Kiev (1970).
30. R. H. Nevanlinna, Analytic Functions, Springer, Berlin (1970).
31. S. Stoilow, The Theory of Functions of a Complex Variable [Russian translation], IL (1962).
32. E. C. Titchmarsh, The Theory of Functions, Oxford, New York (1952).
33. R. P. Boas, Entire Functions, Academic Press, New York (1954).
34. J. D. Bessis, Nuovo Cimento, 45, 974 (1966).
35. A. I. Markusevich, Theory of Functions of a Complex Variable, Vol. I, Englewood Cliffs, New Jersey (1965).
36. K. M. Hoffman, Banach Spaces of Analytic Functions, Prentice-Hall, London (1962).

THE VECTOR DOMINANCE MODEL AND AN EXPERIMENTAL CHECK OF THIS MODEL ON THE BASIS OF VECTOR-MESON DECAY TO AN ELECTRON – POSITRON PAIR

M. N. Khachaturyan

An analysis is made of the methods and results of studying $V \rightarrow e^+e^-$ decay of ρ, ω, and φ mesons produced in hadron–hadron and electron–positron collisions.

INTRODUCTION

One of the basic problems of high-energy physics is to determine the properties of vector mesons. Light was first shed on the exceptional role of vector mesons in nature and their connection with conservation laws and with the universal nature of interactions in the basic study by Yang and Mills [1], which has greatly affected the development of the physics of elementary particles. Among the first studies in this field was that by Sakurai [2], who attempted to construct a theory for strong interactions on the basis of the concepts of conserved currents and universality. A direct generalization of the Yang–Mills concept to conservation of baryon charge and hypercharge led Sakurai to predict isosinglet vector mesons and an isotriplet of vector mesons. Construction of an appropriate Lagrange formalism led to a universal theory for strong interactions; here "universality" means that (in complete analogy with electrodynamics) a common or universal interaction constant can be introduced to describe the interaction of the vector field with all fields having the appropriate charge. As in electrodynamics, the charge has two functions: it is an integral of motion and a coupling constant. This combination of roles is obviously not trivial and clearly distinguishes this theory from those in which new interaction constants are associated with each new field.

Sakurai [2] treats the entire set of strong interactions as various manifestations of three fundamental interactions between vector mesons and baryon, hypercharge, or isospin currents. After analyzing a vast amount of experimental information on the basis of these fundamental concepts and determining several regularities, he was also successful in formulating several important predictions (e.g., the existence of ω and φ mesons) which have subsequently been confirmed experimentally.

The general theory of gauge fields was derived by Uttijama [3] before Sakurai's studies appeared; Uttijama showed that any conservation laws (or, more precisely, any symmetry groups of a field system) can be associated with gauge fields characterized by a universal interaction. In particular, Uttijama showed that the Lorentz transformation group also corresponds to a gauge field – the gravitational field.

The universality of weak interactions has also been discussed in the literature in connection with the conservation of lepton charge and the vector nature of weak interactions [4]. Accordingly, the concept of local gauge invariance affects all the fundamental interactions and is associated with the dynamics of the conservation of such quantities as isospin, hypercharge, baryon charge, and lepton charge; this concept clearly leads to a special class of fields – gauge fields, characterized by a universal interaction.

Before Sakurai's study appeared, Nambu [4] predicted vector mesons in an attempt to explain the electromagnetic structure of the nucleon. On the basis of the concept of the pion sheath of the nucleon

---

Joint Institute for Nuclear Research, Dubna. Translated from Problemy Fiziki Élementarnykh Chastits i Atomnogo Yadra, Vol. 2, No. 3, pp. 583-634, 1972.

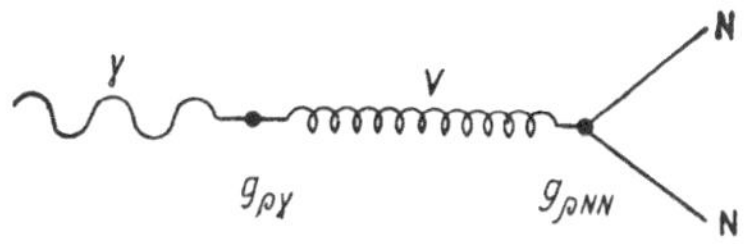

Fig. 1. Diagram for the interaction of a vector meson V with a photon and nucleon according to Nambu [4].

which was current at that time, it was concluded that the mean square radii of the proton and neutron were related by

$$\langle r^2\rangle_p = -\langle r^2\rangle_n.$$

Experiments clearly proved this relation to be wrong. It was necessary to introduce an electromagnetic interaction which would have the same sign for the neutron and proton in order to offset the "negative pion sheath." According to the Nambu theory, the vector meson must interact with a photon and nucleon according to the diagram in Fig. 1.

Frazer and Fulko [5] analyzed the isovector analog of this diagram in their dispersion theory for the nucleon form factor. The universality concept and the theory of electromagnetic form factors developed independently before the publication of Gell-Mann and Zachariasen's suggestion [6] that diagrams like that in Fig. 1 should predominate for all strong-interacting particles (the "vector-dominance model").

Two conclusions can be drawn: 1) The universality of the vector-meson interaction follows from the vector-dominance hypothesis and from the universality of electromagnetic interactions. 2) The effective constant $g_{\gamma\rho} = e(m_\rho^2 / f_\rho)$ for the photon-vector meson transition is related to the universal strong-interaction constant. It is this relationship which allows various experimental tests of the idea of the universality of strong interactions. The model combining universality and vector dominance can be formulated as the following postulate [7]:

$$j_\mu^\alpha(x) = \frac{m_\rho{}^2}{f_\rho}\rho_\mu^\alpha(x), \tag{1}$$

where $j_\mu^\alpha(x)$ is the density of the isovector component of the electromagnetic current, and $\rho_\mu^\alpha(x)$ is the component of the operator corresponding to the ρ-meson field.

Up to this point, the universality concept has been discussed only with respect to a single integral of motion — the isospin. Basis relation (1) can be generalized to mesons associated with hypercharge and baryon charge (ω and φ mesons) in an obvious manner: the isoscalar part of the electromagnetic current $j_\mu(x)$ is related to a combination of operators of the ω and φ fields [8]. Analytically, the vector-dominance model can be written as a relation between the electromagnetic hadron current and the phenomenological meson fields $\rho_\mu(x)$, $\omega_\mu(x)$, and $\varphi_\mu(x)$:

$$j_\mu(x) = -\left[\frac{m_\rho{}^2}{8f_\rho}\rho_\mu(x) + \frac{m_\omega{}^2}{8f_\omega}\omega_\mu(x) + \frac{m_\varphi{}^2}{8f_\varphi}\varphi_\mu(x)\right], \tag{2}$$

where μ are the space–time coordinates; j_μ is the operator corresponding to the electromagnetic-current density of the hadrons; $\rho_\mu(x)$, $\omega_\mu(x)$, and $\varphi_\mu(x)$ are the operators corresponding to the ρ-, ω-, and φ-meson fields; the f_V are the coupling constants between the vector mesons and the γ ray; and m_V is the mass of the vector meson.

In the vector-dominance model the coupling constants f_V are important for an understanding of many phenomena, e.g., the electromagnetic form factors of pseudoscalar mesons and nucleons, electromagnetic meson decays, etc. The only direct method for determining the coupling constant f_V between vector mesons and γ rays is through measurement of the relative probability for lepton decay of vector mesons.

The coupling constants f_V and the partial decay width $\Gamma(V \to e^+e^-)$ are related by

$$\Gamma(V \to e^+e^-) = \frac{\alpha^2}{3}\left(\frac{4\pi}{f_v{}^2}\right)m_v. \tag{3}$$

The coupling constant f_V can also be calculated on the basis of symmetry theory; e.g., in SU(6) symmetry we find the ratios

$$f_\rho^{-2} : f_\omega^{-2} : f_\varphi^{-2} = 9:1:2. \tag{4}$$

Analogous predictions can be found on the basis of the quark model.

By measuring the partial widths for $V \to e^+e^-$ decays one can also solve the problem of testing the hypothesis of $\omega-\varphi$ mixing [9]. This hypothesis was advanced because the Gell-Mann–Okubo mass-splitting equations could not explain the difference in the masses of the vector-meson octet within the framework of SU(3) symmetry.

The Gell-Mann–Okubo mass equation in first order in the interaction violating SU(3) symmetry is

$$M = a + bY + c\,[Y^2/4 - I(I+1)]. \tag{5}$$

Let us consider the application of Eq. (5) to the vector-meson octet. This octet contains 1) an isotopic triplet, I = 1, Y = 0; 2) two isotopic doublets, I = 1/2, Y = +1, and I = 1/2, Y = −1; and 3) an isotopic (but nonunitary) singlet, I = Y = 0. The corresponding particles are: 1) the triplet ρ^+, ρ^-, ρ^0; and 2) the doublets (K^{*+}, K^{*0}) and (K^{*-}, $\tilde{K}^{*0}$). Two particles can play the role of the isotopic singlet: the φ meson (M = 1019 MeV) and the ω meson (M = 783 MeV). From the Gell-Mann–Okubo equation we can find the mass m_{φ_8}, a member of the unitary octet:

$$\varphi_8 = \frac{4}{3}K^* - \frac{1}{3}\rho = 925\ \text{MeV}. \tag{6}$$

Accordingly, φ_8 cannot be identified with either the ω meson (m_ω = 783 MeV) or with the φ meson (m_φ = 1019 MeV). Sakurai [9] predicted that the physical particles ω and φ were a mixture of members of the unitary singlet φ_1 and of the unitary octet φ_8 which, if there is no violation of SU(3) symmetry, are degenerate (or nearly so) with respect to mass. To find the experimental values of m_φ = 1019 MeV and m_ω = 783 MeV, a $\omega-\varphi$ mixing angle of $\theta \approx 40°$ must be selected.

There is another way to introduce the mixing angles, which can be described in the following manner in the language of the quark model. We assume that ω consists only of the nonstrange quarks

$$\omega = \frac{1}{\sqrt{2}}(\tilde{p}p + \tilde{n}n) \tag{7}$$

and

$$\varphi = \tilde{\Lambda}\Lambda,$$

where ω and φ are the wave vectors of the ω and φ mesons in unitary space. This requirement unambiguously determines the mixing angles:

$$\operatorname{tg}\theta = \frac{1}{\sqrt{2}};\quad \theta \approx 35°. \tag{8}$$

However, the agreement in terms of mass is slightly poorer:

$$m_\omega = m_\rho = 762\ \text{MeV instead of 783 MeV};$$

$$m_\varphi = 1002\ \text{MeV instead of 1019 MeV}.$$

It is thus difficult to draw any definite conclusions about the mixing of ω and φ from the mass equations. Even a slight change in the mixing angles can have a qualitative effect; e.g., for an ideal mixing angle of $\tan\theta = 2^{-1/2}$, the reactions

$$\left.\begin{array}{l} \varphi \to \rho + \pi, \\ \varphi \to \pi^0 + \gamma,\ \text{etc.} \end{array}\right\} \tag{9}$$

are strictly forbidden. To unambiguously determine the mixing angle, we must measure the ratio

$$\frac{\Gamma(\varphi \to e^+e^-)}{\Gamma(\omega \to e^+e^-)} \approx \operatorname{ctg}\theta. \tag{10}$$

Accordingly, by measuring the partial widths for e^+e^- decay of vector mesons, we can determine the universal-interaction constant in its most pure form and test the applicability of the concept of local gauge invariance to strong interactions.

Measurement of the constants of the vector meson-photon transition is important for determining the range of applicability of Maxwell-Dirac quantum electrodynamics, since in the momentum-transfer range in which these transitions are important quantum electrodynamics does not hold.

1. METHODS AND RESULTS OF STUDIES OF ELECTRON – POSITRON DECAY OF VECTOR MESONS

Studies published up to 1966, carried out primarily by the chamber method, incorporated only estimates of the upper boundary for $V \to e^+e^-$ decay [10-14]. These studies are now of only historical interest and will not be discussed below.

A very fruitful period in the study of $V \to e^+e^-$ decay was that between 1966 and 1968, when most of the studies forming the basis for the solution of this interesting problem were carried out. Experimental detection of $V \to e^+e^-$ decay is an exceptionally complicated problem: According to theoretical predictions, the probability for lepton decay of a vector meson is on the order of 10^{-5}, or 10^{-7}, according to some estimates [15-18]. A further complication is that this process is similar to a vast number of background reactions having cross sections many orders of magnitude greater than that for the production of electron pairs. The $V \to e^+e^-$ ($V \to \rho, \omega, \varphi$) has been studied in three reactions:

$$\pi^- + p \to V^0 + n, \quad V^0 \to e^+e^-; \tag{11a}$$

$$e^+ + e^- \to V^0, \quad V^0 \to \text{hadrons}; \tag{11b}$$

$$\gamma + C \to V^0 + C, \quad V^0 \to e^+e^- \tag{11c}$$

Use has been made of both direct spectrometry of the resonance mass through detection of electron–positron pairs (by means of the Cerenkov mass spectrometer at Dubna and the DESY-MIT magnetic spectrometer) and indirect methods based on time-of-flight spectrometry of the recoil particles (CERN and Rutherford Laboratory).

The simplicity (in terms of detecting apparatus) of experiments with clashing electron–positron beams allows relatively simple systems consisting of spark chambers and scintillation counters to be used.

The first studies establishing the existence of $\rho \to e^+e^-$ and $\varphi \to e^+e^-$ decays, in which the partial widths of these decays were measured, were carried out on the 10-GeV proton synchrotron of the Joint Institute for Nuclear Research, Dubna [29-31]. These results were subsequently confirmed by experiments at Novosibirsk (USSR), Orsay (France), CERN (Switzerland), DESY-MIT (FRG-USA), and the Rutherford Laboratory (England).

Two types of apparatus have been used to study $V \to e^+e^-$ decays in $\pi^-p \to Vn$ reactions: 1) a Cerenkov mass spectrometer (Dubna), and 2) a neutron time-of-flight spectrometer (Rutherford Laboratory and CERN).

2. STUDY OF $V \to e^+e^-$ DECAYS BY MEANS OF A CERENKOV MASS SPECTROMETER. PROCEDURE AND RESULTS

Summary of the Procedure

New, efficient methods had to be developed to experimentally detect $V \to e^+e^-$ decays because of their extreme complexity.

A new instrument, a Cerenkov mass spectrometer, was conceived and developed in 1964 in the High-Energy Laboratory of the Joint Institute for Nuclear Research. This instrument was capable of measuring the effective mass for many types of decay (e^+e^-, $\gamma\gamma$, $\pi^0\gamma$, $\pi^0\pi^0$, $\pi^0e^+e^-$, etc.) and of identifying the secondary particles (electrons and γ rays) [19, 20]. The effective mass can be determined from

$$M^2 = (\sum E_i)^2 - (\sum \mathbf{p}_i)^2, \tag{12}$$

where E_i and $\mathbf{p}_i$ are the total energies and momenta of the decay products.

For $X \to e^+e^-$, $\gamma\gamma$ decays Eq. (12) takes the simpler form

$$M^2 = 2E_1E_2(1 - \cos\theta). \tag{13}$$

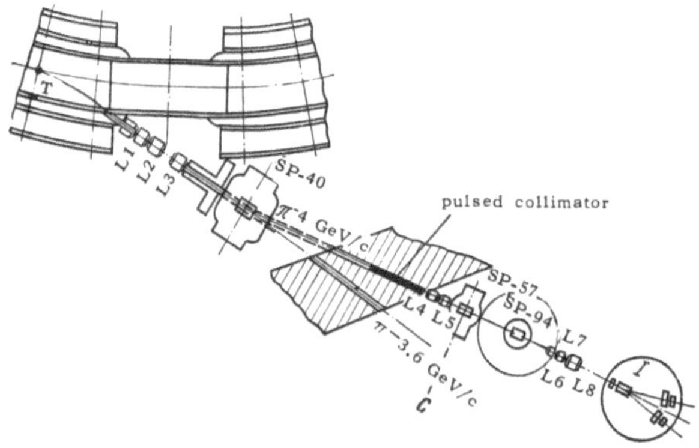

Fig. 2. Experimental setup (JINR, Dubna). L1-L8) Magnetic lenses; DM) deflecting magnets; T) target.

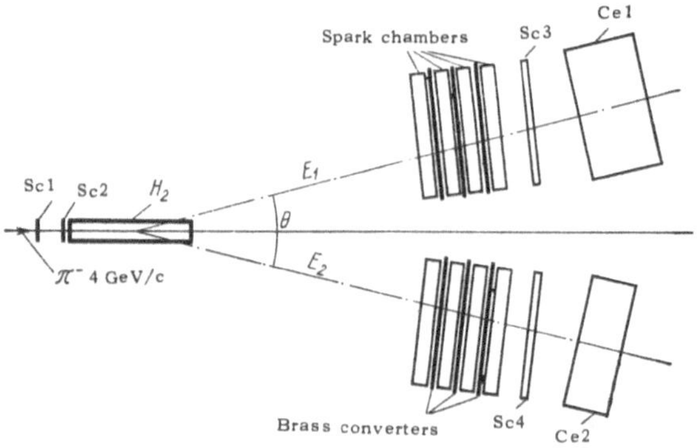

Fig. 3. Diagram of the Cerenkov mass spectrometer. Sc1-Sc4) Scintillation counters; Ce1, Ce2) Cerenkov total-absorption γ spectrometers; H_2) liquid-hydrogen target.

We see from Eq. (13) that a determination of the square of the effective mass requires measurements of three parameters: the energies E_1 and E_2 of the electrons or γ rays and the angle θ between them. These requirements are met by the Cerenkov mass spectrometer, which uses spark chambers to measure the angle between the electrons and total-absorption Cerenkov spectrometers to measure the electron energies.

Study of the $\pi^- p \to e^+e^- n$ reaction, whose effective cross section is on the order of 10^{-33} cm^2, places stiff requirements on the parameters of the apparatus. One of the basic characteristics of the apparatus is the accuracy with which the effective mass can be measured. The relative rms error in the effective mass is

$$\Delta M/M = \pm 0.5\,[(\Delta E_1/E_1)^2 + (\Delta E_2/E_2)^2 + (\Delta\theta/\mathrm{tg}^{\theta}/_2)^2]^{1/2}, \tag{14}$$

where E_i, ΔE_i, θ, and $\Delta\theta$ are the energies of the electrons and positrons, the errors in these energies, the emission angles, and the errors in the emission angles, respectively.

The Cerenkov γ spectrometer developed in the Joint Institute for Nuclear Research is capable of measuring electron energies within ±5% (between 1 and 4 GeV). If a spark in a spark chamber can be localized within ±0.5 μ ($\Delta\theta$ = 1.7 mrad with a 50-cm base line), the mass error for a particle having the mass of a ρ meson is ΔM = ±27 MeV.

Experimental Apparatus

Figures 2-4 show the layout of the apparatus, the experimental geometry, and a block diagram of the electronics. Scintillation counters Sc1 and Sc2, 7 × 7 × 1 cm in size, are used to monitor the π^- beam. Spark chambers Sp1 and Sp2 are two identical units of four modules having a working region of 50 × 50 cm. The first spark chambers encountered by the beam are connected in anticoincidence.

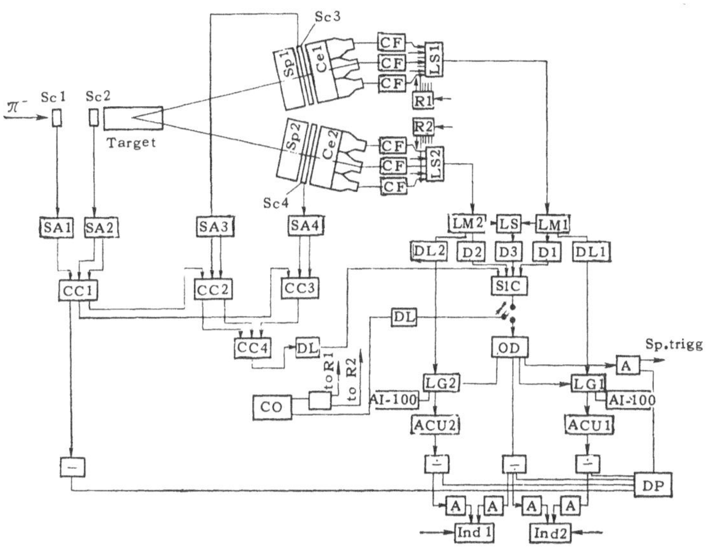

Fig. 4. Block diagram of the Cerenkov mass spectrometer. SA) Shaping amplifier; CC1-CC4) coincidence circuits; SlC) slow-coincidence circuit; DL) delay; CF) cathode follower; LS) linear summer; LM) linear multiplier; D) discriminator; LG) linear gate; OD) output discriminator; CO) control oscillator; AI-100) amplitude analyzer; $\div$) scalar; A) amplifier; ACU) amplitude-conversion unit; DP) digital printer; Ce1, Ce2) Cerenkov γ spectrometers; Sc) scintillation counter; Sp) spark chamber; R1, R2) multipliers; Ind1, Ind2) spark-chamber indicators.

Brass converters having a total thickness of 1.2 radiation lengths are placed in front of the second and subsequent chambers to convert the shower particles (electrons and γ rays). The distributed converter system (with three converters per channel) significantly reduces the angular error in the determination of the track direction due to multiple scattering (in the detection of γ rays).

The triggering efficiency is increased by placing scintillation counters Sc3 and Sc4, connected in coincidence, between the spark chambers and the Cerenkov γ spectrometers.

The Cerenkov total-absorption spectrometers Ce1 and Ce2 have lead-glass radiators 50 × 50 × 30 cm and 50 × 50 × 20 cm in size. Light is gathered in each spectrometer by nine photomultipliers having a photocathode diameter of 17 cm.

The spark chambers are triggered when there is a coincidence of pulses from counters Sc1-Sc4 and Ce1 and Ce2 provided that

1) $$E_1 = E_2 \geqslant E_t = 0.5\ \text{GeV}; \tag{15}$$

2) the sum of E_1 and E_2 satisfies

$$E_1 + E_2 \geqslant E_\Sigma = 3.5\ \text{GeV}. \tag{16}$$

The thresholds E_t and E_Σ are chosen on the basis of a kinematic analysis of the reaction taking account of the energy resolutions of the spectrometers. The optimum experimental conditions are chosen through a computer simulation of reaction (11a).

Analysis of the kinematics of (11a) shows that with the optimum geometrical setup the electron energy spectra reach maxima at $0.5\,E_{\pi^-}$ and lie between 0.5 and 3.5 GeV, where E_{π^-} is the energy of the incident π meson (Fig. 9). The sum of the electron energies $E_1 + E_2$ is essentially constant and equal, within an error corresponding to the momentum transferred to the nucleon, to the energy of the π meson (Fig. 8).

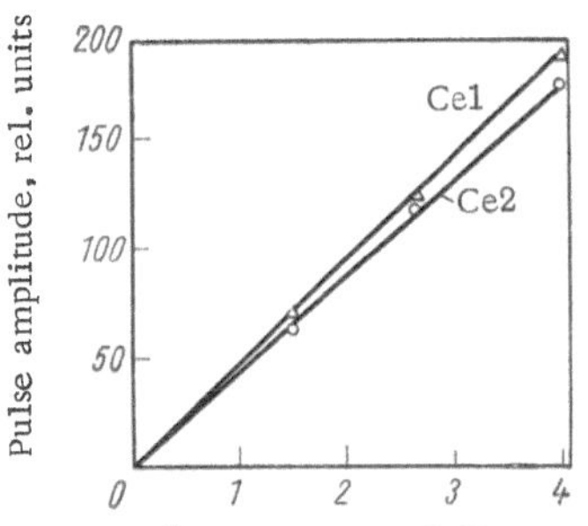

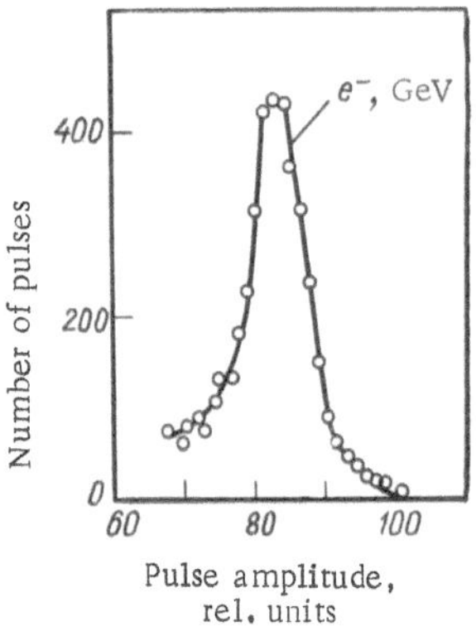

Fig. 5 Fig. 6

Fig. 5. Electron-energy dependence of the pulse amplitude of the Cerenkov γ spectrometer.

Fig. 6. Amplitude distribution of the pulses from the Cerenkov γ spectrometer produced by a 4-GeV electron beam.

These kinematic relations allow the spark chambers to be triggered by a logic circuit [see (15) and (16)] which significantly reduces the number of background operations of the apparatus.

The long times required for measurements of these very rare events (the frequency at which $\varphi \to e^+e^-$ decays are detected averages one per 7 days of continuous accelerator operation) and the nature of the apparatus require a constant monitoring of the basic parameters of the apparatus. The Cerenkov mass spectrometer is periodically energy calibrated in an electron beam. The apparatus is calibrated for effective mass by means of the $\pi^-p \to \eta^0 n$, $\eta^0 \to \gamma\gamma$ reaction. The calibration results are shown in Figs. 5-7.

Experimental Conditions

Two series of measurements were carried out. In the first series, vector mesons were produced by 4.0-GeV/c π mesons with $\Delta p/p = \pm 1.5\%$ in a liquid-hydrogen target 50 cm long through reaction (11a). The angle chosen between the axes of the two detectors in this series of measurements, 26°, provided maximum efficiency for the detection of $V \to e^+e^-$ decays ($\varepsilon \approx 10\%$) in the mass region centered near 750 MeV.

Under these experimental conditions, it is possible to simultaneously detect, in the mass range 300-1300 MeV, a large number of processes allowed by the logic:

$$\pi^- + p \to \begin{array}{l} \gamma + \gamma + n \\ \pi^0 + \gamma + n \\ \pi^0 + \pi^0 + n \\ \pi^0 + e^+ + e^- + n \text{ etc.} \end{array} \qquad (17)$$

The overwhelming majority of events are due to the first two reactions. About 20,000 photographs were taken during the measurements. Those photographs showing tracks of a charged particle in the first and second channels which intersected within the effective target volume in the Cerenkov spectrometers were selected for analysis. The e^+e^- pairs were analyzed on the basis of energy criteria through the use of computer-simulated energy spectra for electrons and positrons.

Figures 8 and 9 show these distributions for processes (11a); Fig. 8 shows that the distribution for the sum of the electron and positron energies is grouped above 3.5 GeV. Below 3.5 GeV there is a negligible contribution. Events having energies above 3.5 GeV were selected for analysis.

In the first series of measurements the background events are due primarily to conversion of γ rays in the hydrogen walls of the target and in the spark chambers. The experimental spectrum of $\gamma\gamma$ events (which are detected at the same time as e^+e^- events) is used to determine the magnitude of the conversion background. This spectrum consists of the sum of all possible processes producing conversion pairs.

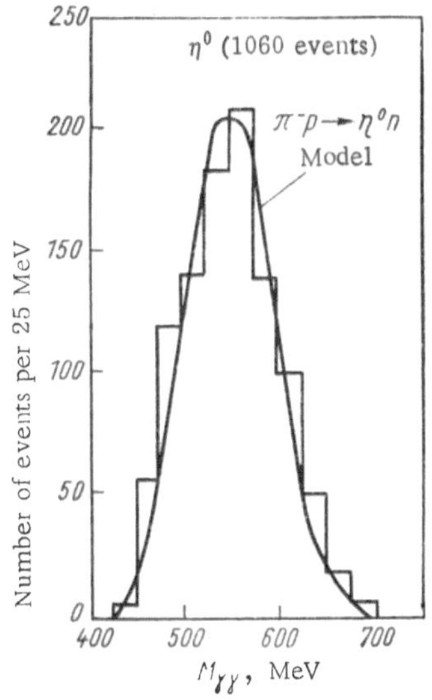

Fig. 7

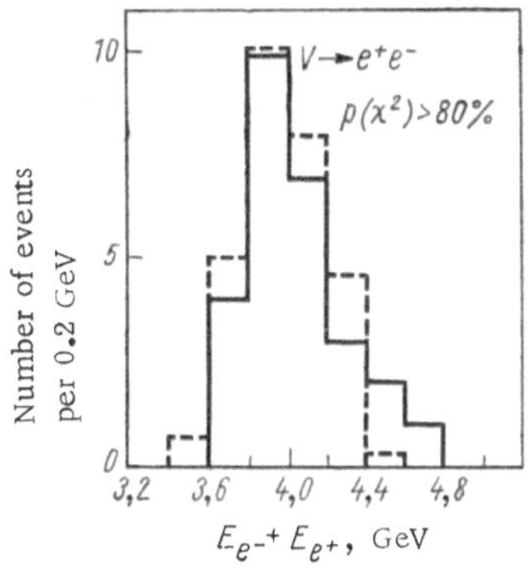

Fig. 8

Fig. 7. Effective-mass spectrum from $\eta^0 \rightarrow \gamma\gamma$ decays according to the Cerenkov mass spectrometer. The curve is the result of a computer simulation of the reaction $\pi^-p \rightarrow \eta^0 n$, $\eta^0 \rightarrow \gamma\gamma$.

Fig. 8. Histogram of the sum of the electron and positron energies for the identified $\rho^0 \rightarrow e^-e^+$ events. The broken lines show the results of a computer simulation of the $\pi^-p \rightarrow \rho^0 n$, $\rho^0 \rightarrow e^+e^-$ reaction at $p_\pi = 4$ GeV/c. The $p(\chi^2)$ test shows that the experimental histogram agrees with the computer results.

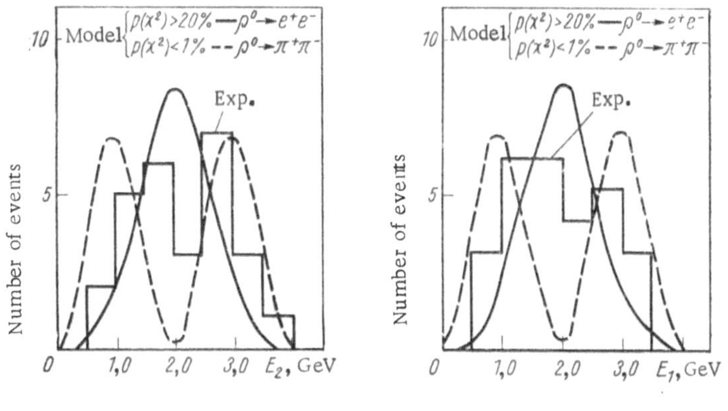

Fig. 9. Experimental electron and positron energy spectra for identified events. The curves were obtained by a computer simulation of $\rho^0 \rightarrow e^+e^-$ and $\rho^0 \rightarrow \pi^+\pi^-$ decays. The $p(\chi^2)$ test shows that the simulation results agree with experiment.

Analysis of these results shows that less than 25% of the background events fall in the range 650-850 MeV, i.e., in the range corresponding to the ρ^0 mass; above 850 MeV the relative number of background events is negligible.

Integration of the area under the histogram (after the background is subtracted) yields about 25 events in the ρ^0 mass range and three events with masses equal to that of the φ meson. The numbers of ρ and φ

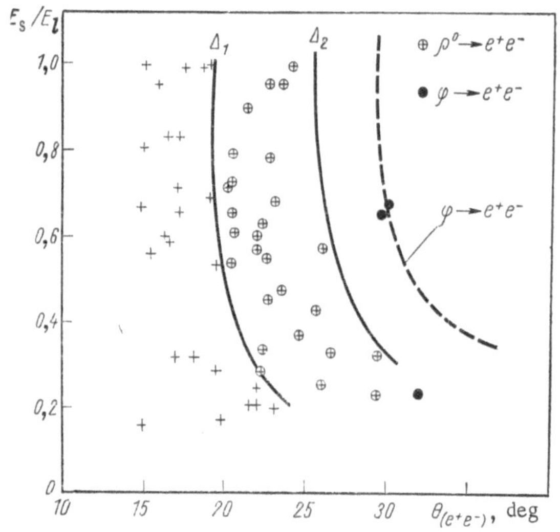

Fig. 10

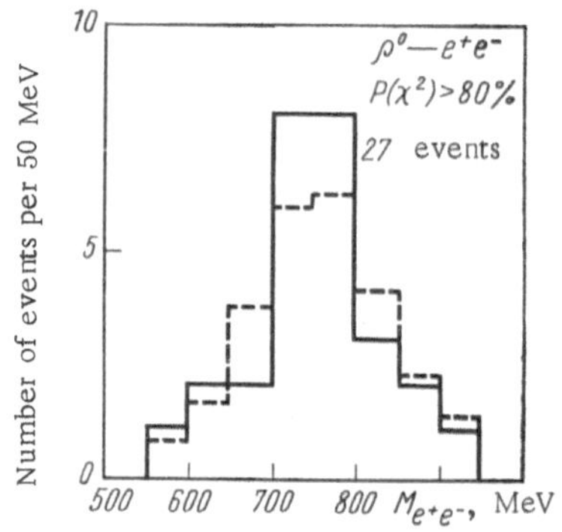

Fig. 11

Fig. 10. Analysis of e^+e^- pairs on the basis of the kinematics for the production and decay of V particles in the reaction $\pi^-p \to V^0 \to e^+e^-$. Curves Δ_1 and Δ_2 bound the region of allowed values of the function $E_s/E_l = f(\theta)$ for $\rho \to e^+e^-$ decays with an account of the errors in the angles and energies and of the width Γ_ρ. Here $E_s/E_l \leq 1$ and θ are, respectively, the ratio of the small product-particle energy to the large energy and the emission angle of the electron in the laboratory coordinate system. The broken curve was calculated from Eq. (13) for $\varphi \to e^+e^-$. The region of allowed values of the function $E_s/E_l = f(\theta)$ is not indicated for φ mesons.

Fig. 11. Effective-mass spectrum for $\rho^0 \to e^+e^-$ events. The broken lines show the histogram resulting from the computer simulation of the $\pi^-p \to \rho^0 n$, $\rho^0 \to e^+e^-$ reaction.

mesons were also found by a second method based on kinematic equation (13), which relates the electron and positron energies and emission angles to the resonance mass. The results of this analysis are shown in Fig. 10. Curves Δ_1 and Δ_2 in Fig. 10 show the allowed values of the function $E_1/E_2 = f(\theta)$ for $\rho^0 \to e^+e^-$ decays, calculated with an account of the errors in the measured quantities (angles and energies) and of the width of the ρ^0 meson. Twenty-seven events grouped within the kinematic corridor defined by Δ_1 and Δ_2 satisfy the kinematics for $\rho^0 \to e^+e^-$ decay, and three events satisfy the kinematics for $\varphi \to e^+e^-$ decays. Events on the left of Δ_1 are background events resulting from γ-ray conversion (primarily from the $\eta^0 \to \gamma\gamma$ reaction).

Comparing the numbers of $V \to e^+e^-$ decays found by the different methods of analysis, i.e., by subtracting the background, and from the kinematics of the decay, we find a difference no greater than 8%, within the experimental error.

Figure 11 compares histograms of the experimental and computer-simulated effective-mass distributions for the $\pi^-p \to \rho^0 n$, $\rho^0 \to e^+e^-$ reaction.

The χ^2 value shown in Fig. 11 is evidence that the computer results agree with experiment. One of the background processes considered was simulation of e^+e^- pairs by $\pi^+\pi^-$ pairs. This possibility was studied experimentally; the results are shown in Table 1, where W_{π^-} is the probability that a π^- meson will be detected at the energy threshold E_0.

With this threshold a π^- meson will be detected if an energy greater than E_0 is evolved in the spectrometer. Here ε_{e^-} is the probability that an electron having a momentum equal to that of a π meson will be detected with the same energy threshold, E_0, and R_e is the energy resolution of the spectrometer. The experimental data were analyzed with an energy threshold corresponding to detection of 80% of the elec-

TABLE 1

$P_{\pi^- - e^-} = 4.0$ GeV/c, $R_e = \pm 6.1\%$		$P_{\pi^- - e^-} = 4.0$ GeV/c, $R_e = \pm 12.5\%$		$P_{\pi^- - e^-} = 1.1$ GeV/c, $R_e = \pm 14\%$	
ε_{e^-}, %	W_{π^-}, 10^{-4}	ε_{e^-}, %	W_{π^-}, 10^{-4}	ε_{e^-}, %	W_{π^-}, 10^{-4}
50	1	50	7	50	9
80	2,0	80	16	80	31
85	2,5	85	21	85	47
90	4,0	90	27	90	83
95	8,0	95	39	95	132
~100	20	~100	75	~100	200

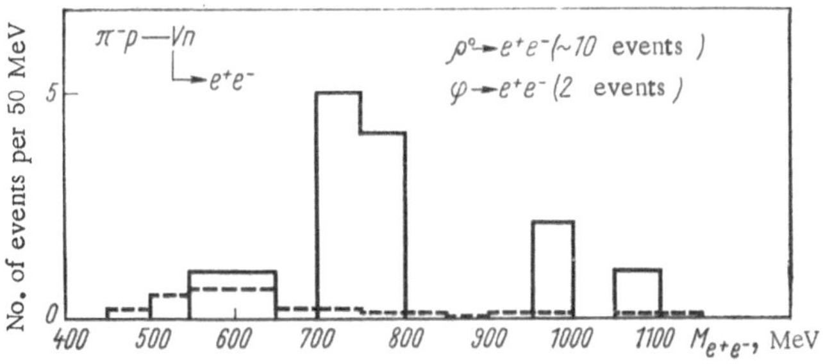

Fig. 12. Effective-mass spectra for $V \rightarrow e^+e^-$ decay (solid lines) and conversion background (broken lines) obtained in the second series of measurements with a target 25 cm long.

trons, so the probability that a π meson was detected as an electron was no greater than $2 \cdot 10^{-4}$ for a single channel and no greater than $(2 \cdot 10^{-4})^2 = 4 \cdot 10^{-8}$ for two channels.

Second Series of Experiments

The experimental geometry was modified for this series of experiments to allow an increase in the intensity of the π beam, an increase in the efficiency of the detection of the $V \rightarrow e^+e^-$ decays, and an extension of this detection to higher effective masses. The angle between the axes of the two detectors in the second series of measurements was selected to be 29°; with this angle the efficiency reaches a maximum at 900 MeV, and the electron decays of ρ and φ mesons can be simultaneously detected with an efficiency of about 10%.

Criteria analogous to those of the first series of measurements yielded 13 candidates for e^+e^- pairs. Figures 12 and 13 show the effective-mass distribution and the results of a kinematic analysis of these events. Comparison of the corresponding distributions of the first and second series of measurements shows these distributions to be markedly different. Use of a target with a high exit aperture and reduction of the wall thickness reduced the background of γ-ray conversion to a level not exceeding 6% of the basic effect (at m_ρ).

After the background was subtracted, 9.5 and two events remained in the mass ranges corresponding to ρ^0 and φ mesons. The analogous results obtained on the basis of the kinematics of the decay are essentially the same: ten $\rho^0 \rightarrow e^+e^-$ events and two $\varphi \rightarrow e^+e^-$ events (Fig. 13). The basic results of the first and second series of measurements are shown in Table 2.

Figure 14 shows the effective-mass distribution according to the results of both series of experiments after the background was subtracted. The broken lines in Fig. 14 result from a normalization of the basic histogram for detection efficiency.

Production and Decay of V Mesons

Experimentally, the differential cross section for process (11a) is measured over a certain angular range, so in analyzing the data one must take into account the angular distributions for the production and decay of vector particles.

TABLE 2

Parameter	First series	Second series
Length of liquid-hydrogen target, cm	50	25
Number of π^- mesons detected, 10^9	3.36	3.53
Number of identified $\rho^0 \to e^+e^-$ decays	27	10
Number of identified $\varphi \to e^+e^-$ decays	3	2
Detection efficiency for $\rho^0 \to e^+e^-$ decay	9.0%	9.5%
Detection efficiency for $\varphi \to e^+e^-$ decay	2.6%	4.1%
Angle between detector axes, deg	26	29
Range of 4-momentum transfer, $(\mathrm{GeV}/c)^2$	0-0.3	
Range of V emission angles in the lab. system, deg	0-8	
Range of V emission angles in c.m. system, deg	0-30	
Range of V energies in lab. system, GeV	3.83-4.0	
Effective-mass resolution, MeV	$\Delta M_\rho = \pm 27$ $\Delta M_\varphi = \pm 36$	

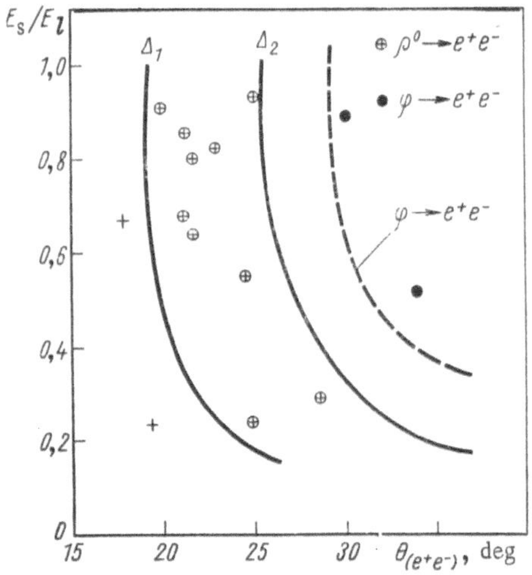

Fig. 13

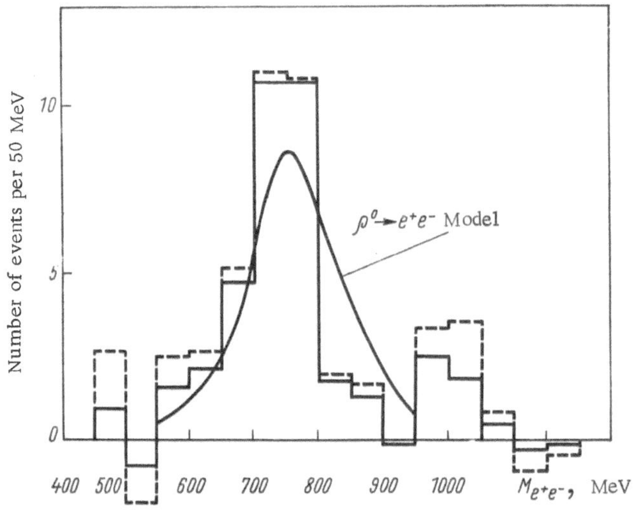

Fig. 14

Fig. 13. Analysis of e^+e^- pairs on the basis of the kinematics of the production and decay of V particles in the reaction $\pi^-p \to Vn$, $V \to e^+e^-$ for the second series of measurements. Curves Δ_1 and Δ_2 bound the region of allowed values of the function $E_s/E_l = f(\theta)$ for $\rho \to e^+e^-$ decays. For the kinematic curve for $\varphi \to e^+e^-$ decay (broken curve) the region of allowed $E_s/E_l = f(\theta)$ values is not indicated.

Fig. 14. Effective-mass spectrum of e^+e^- pairs according to the results of both the first and second series of measurements after subtraction of the background. The broken lines show the histogram obtained from the basic normalization for detection efficiency. The curve was found by computer simulation of $\rho^0 \to e^+e^-$ decay.

The differential cross section for the reaction $\pi^-p \to e^+e^-n$ can be written

$$d\sigma = d\sigma_V \, d\Omega_q \frac{\Gamma_\alpha}{\Gamma} \cdot \frac{3}{4\pi} \cdot \frac{1}{2} [1 - W_{00}(\theta^*, \varphi)] f(m^2) \, dm^2, \tag{18}$$

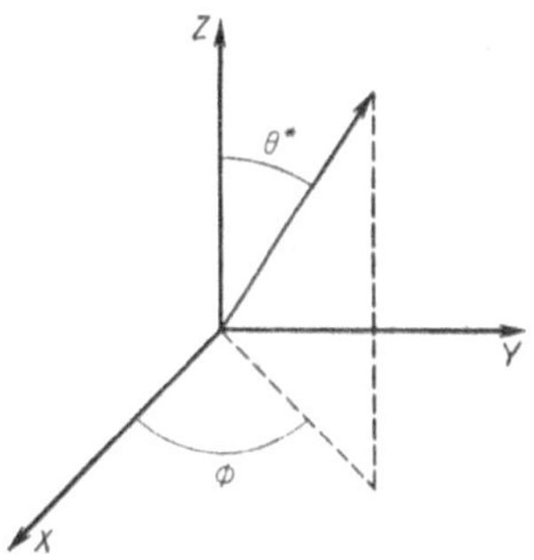

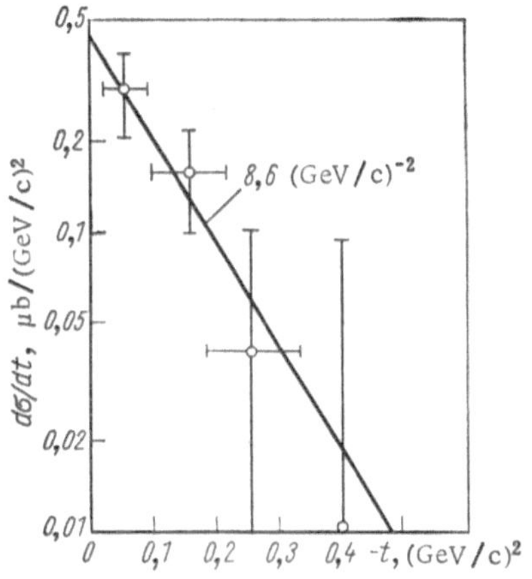

Fig. 15 Fig. 16

Fig. 15. Coordinate axes in the rest system of the V meson. The Z axis is parallel to the π beam, and the Y axis is along the normal to the reaction plane.

Fig. 16. Differential cross section for the reaction $\pi^- p \to \rho^0 n$, $\rho^0 \to e^+e^-$ as a function of $-t$.

where $d\sigma_V$ is the cross section for the production of vector particles, Γ_α is the quantity measured in the experiment, and

$$\beta = \frac{3}{4\pi} \cdot \frac{1}{2} [1 - W_{00}(\theta^*, \varphi)]\, d\Omega_q \tag{19}$$

is the angular distribution of e^+e^- decay products in the rest system of the vector particle. The Z axis is chosen parallel to the π beam, and the Y axis is chosen along the normal to the reaction plane (Fig. 15). Here also we have

$$W_{00} = \rho_{00}\cos^2\theta^* + \rho_{11}\sin^2\theta^* - \rho_{1-1}\sin^2\theta^*\cos 2\varphi - \sqrt{2}\,\mathrm{Re}\,\rho_{10}\sin 2\theta^*\cos\varphi;$$

ρ_{mn} are the elements of the spin density matrix of the vector meson; and $f(m^2)dm^2$ is the mass distribution of the V meson. Equation (18) is found from a kinematic analysis and the single assumption that the amplitude for the process can be written as the product of the amplitudes for the production and decay of the unstable particle.

Use of a model can significantly simplify the equation for the cross section. For example, if the V mesons are produced by high-energy π^- mesons, states of both even and odd G values may occur in single-meson exchange.

In the case of an odd G value, i.e., the single-meson mechanism for ρ^0 production, we have $\rho_{00} = 1$ and $\rho_{ij} = 0$, where i, j = 0. Then we can write

$$W_{\rho\to\pi^+\pi^-}(\theta^*, \varphi) = \frac{3}{4\pi}\cos^2\theta^*; \tag{20}$$

$$W_{\rho\to e^+e^-}(\theta^*, \varphi) = \frac{3}{8\pi}\sin^2\theta^*; \tag{21}$$

i.e., in this case the angular distributions do not depend on φ. In the case of even-G exchange, i.e., in the case of ω production (exchange of ρ^0 meson) we have $\rho_{11} = \rho_{-1-1}$, $\rho_{1-1} \neq 0$ and $\rho_{ij} = 0$. For the angular distribution for the $\omega \to e^+e^-$ reaction we have

$$W_{\omega\to e^+e^-}(\theta^*, \varphi) = \frac{1 - 0.5\sin^2\theta^*(1 - 2\rho_{1-1}\cos 2\varphi)}{8\pi/3}. \tag{22}$$

TABLE 3

Parameter	$\rho^0 \to e^+e^-$, 37 events		
	first series	second series	average of series I and II
$B(\rho^0 \to e^+e^-)\cdot 10^5$	5,9±1,5	4,5±1,3	5,1±1,0
$\Gamma(\rho^0 \to e^+e^-)$, keV	7,1±1,8	5,5±1,7	6,2±1,2
$f_\rho^2/4\pi$	2,16±0,56	3,16±0,96	2,4±0,48

TABLE 4*

Parameter	$\varphi \to e^+e^-$, 5 events
$B(\varphi \to e^+e^-)\cdot 10^5$	66^{+44}_{-28}
$\Gamma(\varphi \to e^+e^-)$, keV	$2,5^{+1,7}_{-1,1}$
$f_\varphi^2/4\pi$	$7,2^{+4,8}_{-3,2}$

* The errors indicated are statistical.

Averaging over the azimuthal angle φ, we find

$$W_{\omega\to e^+e^-}(\theta^*, \varphi) = \frac{3}{16\pi}(1+\cos^2\theta^*). \quad (23)$$

For $\varphi \to e^+e^-$, we have

$$W_{\varphi\to e^+e^-}(\theta^*, \varphi) = \frac{3}{16\pi}(1+\cos^2\theta^*). \quad (24)$$

The most detailed results available are those for the production cross section and density matrix of ρ mesons. For values of $-t$ up to $30m_\pi^2$, where m_π is the mass of the π meson, the differential cross section for ρ^0 production in the reaction $\pi^-p \to \rho^0 n$ at 4 GeV/c is described well by the experimental function [21]

$$d\sigma/dt = A\exp(8.53t). \quad (25)$$

The production cross section for ρ^0 mesons was measured in [21-24], where information was also reported on the production and decay mechanisms for the ρ^0 meson. The production cross sections for ω mesons were measured at momenta of 3.25, 3.65, and 5.1 GeV/c [25-28]. The results show that ω mesons are produced with larger momentum transfers than in the production of ρ^- mesons, whose distribution essentially vanishes at $|t| = 15m_\pi^2$. The differential cross section for ω production can be described by the function $d\sigma/dt \sim \exp(4t)$. The cross section for the process $\pi^-p \to \varphi n$ was evaluated in [27].

Results

The relative probabilities, coupling constants, and partial widths for the reactions $\rho^0 \to e^+e^-$ and $\varphi \to e^+e^-$ are shown in Tables 3 and 4 [29-31].

From these results we can calculate the differential cross section for the reaction $\pi^-p \to \rho^0 n \to ne^+e^-$ at 4 GeV/c (Fig. 16). The line drawn through the experimental points in Fig. 16 by the method of least squares can be approximated by the equation $d\sigma/dt = Be^{-At}$ with the parameters $A = 8.6 \pm 3.3$ $(\text{GeV/c})^{-2}$ and $B = (d\sigma/dt)_{t=0} = 0.46 \pm 0.15$ $\mu b/(\text{GeV/c})^2$ with $p(\chi)^2 > 0.50$.

The cross section for the reaction $\pi^-p \to \rho^0 n \to e^+e^-n$ is equal to $\sigma_t = 59 \pm 18$ μb; the mass and width of the ρ^0 meson found from the results of the first and second series of measurements are, respectively,

$$M_{\rho^0} = (742 \pm 12)\ \text{MeV},$$
$$\Gamma_{\rho^0} = (120 \pm 20)\ \text{MeV}.$$

Since five $\varphi \to e^+e^-$ events were identified, it is crucial to consider whether these events could be simulated by background events. Estimates show that the probability that five such decay events could be simulated as a result of fluctuations in the average background level is less than 0.1%. The probability for simulation by $\rho^0 \to e^+e^-$ decays due to the large width of the ρ meson is no greater than 0.1%, according to a calculation based on the use of the Breit–Wigner p wave equation for the mass distribution of the μ meson. Corrections were made to the experimental data for: 1) the μ-meson impurity in the beam (about 10%) and the electron impurity in the beam (2.4%), 2) the angular divergence of the beam (1.5%), 3) absorption of π^- mesons in the scintillators of the monitoring detectors and in the front wall of the target (2%), 4) loss due to the dead time of the apparatus (6%), 5) loss resulting from the energy selection of events (10%), and 6) the impurity of $\omega \to e^+e^-$ decays (10%).

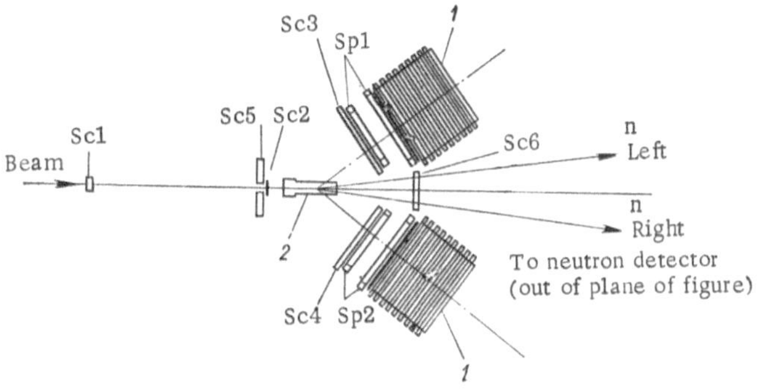

Fig. 17

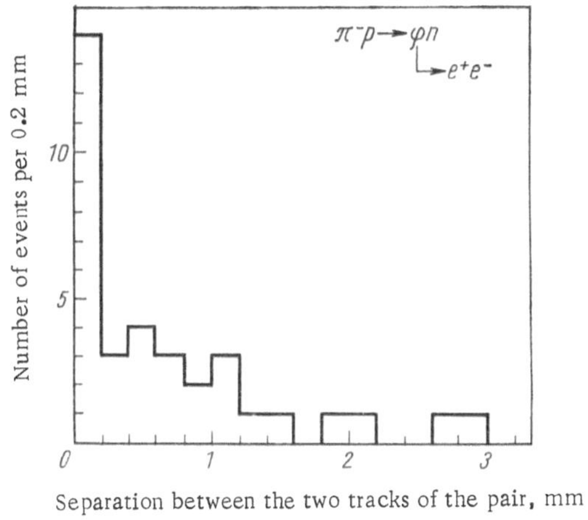

Fig. 18

Fig. 17. Diagram of the CERN experiment. Sc1-Sc4) Scintillation counters connected in coincidence; Sc5, Sc6) scintillation counters connected in anticoincidence; Sp) spark chambers; 1) electron detectors; 2) target.

Fig. 18. Distribution of the separations between tracks of the e^+e^- pairs.

3. ANALYSIS OF $V \to e^+e^-$ DECAYS BY NEUTRON TIME-OF-FLIGHT SPECTROMETERS. PROCEDURE AND RESULTS

Measurement of the Relative Probabilities for $\omega \to e^+e^-$ and $\varphi \to e^+e^-$ Decays at CERN [32, 33]

a) $\varphi \to e^+e^-$. In the experiment, φ mesons were produced by reaction (11a) with 1.93-GeV/c π^--mesons. The apparatus (Fig. 17) consists of a beam particle telescope, a liquid-hydrogen target 40 cm long, two electron detectors, and two neutron detectors. Each electron detector consists of scintillation counters Sc3 and Sc4, which eliminate events for which there are more than three charged particles; two spark chambers, Sp1 and Sp2, used to establish the electron direction; and a shower detector, which consists of nine layers of optical spark chambers, scintillators, and lead plates.

The lead plates have a total thickness equivalent to 10 radiation lengths. The pulses from the nine scintillators are summed and discriminated at the 150-MeV level. With this threshold the shower detector detects 450-MeV electrons with an efficiency of 80% and 1050-MeV electrons with an efficiency of 94%. Charged π mesons are detected at efficiencies of 5% and 15% at the respective energies. The background of charged π mesons is also suppressed by a visual analysis of the angle between the shower axis and the direction of the incident particle. The factor by which the π mesons are suppressed by the electronics and by visual selection is $9 \cdot 10^{-4}$ at 450 MeV/c and $6 \cdot 10^{-4}$ at 1050 MeV/c. The electron-detection efficiencies are 72% and 85% at the respective energies. Accordingly, coincidences between the two shower detectors can be used to suppress hadron decay by a factor of $5 \cdot 10^{-7}$.

The neutron detector consists of 24 scintillators $18 \times 18 \times 100$ cm^3 in size arranged symmetrically with respect to the beam axis in banks of 12, 4 m from the target. The flight times and coordinates in each of the 24 counters are measured by two KhR-1040 photomultipliers 12.5 cm in diameter at the two opposite ends of the scintillator.

The neutron detector detects the neutron coordinates in the scintillator within ±0.35° in polar angle and ±3° in azimuthal angle (a value governed by the 18-cm thickness of the scintillator). The flight time is measured within ±0.35 nsec. The average effective-mass resolution is ±15 MeV.

TABLE 5

Parameter	$\varphi \to e^+ e^-$	$\omega \to e^+ e^-$
Number of events	9 ± 3	11
$\sigma(\pi^- p \to Vn_{\hookrightarrow e^+ e^-})$, 10^{-33} cm²	$18{,}4 \pm 6{,}9$	67 ± 25
$\sigma(\pi^- p \to Vn)$, 10^{-27} cm²	$(30 \pm 6) \cdot 10^{-3}$	$1{,}67 \pm 0{,}07$
$B(V \to e^+ e^-)$, 10^{-5}	61 ± 26	$4{,}0 \pm 1{,}5$
$\Gamma(V \to e^+ e^-)$, keV	$2{,}1 \pm 0{,}9$	$0{,}49 \pm 0{,}19$

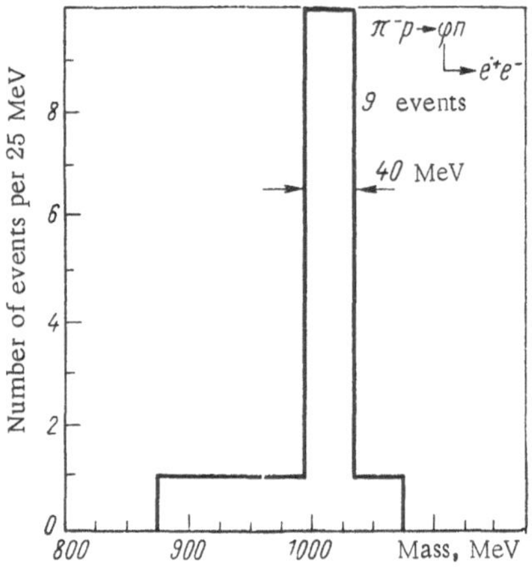

Fig. 19

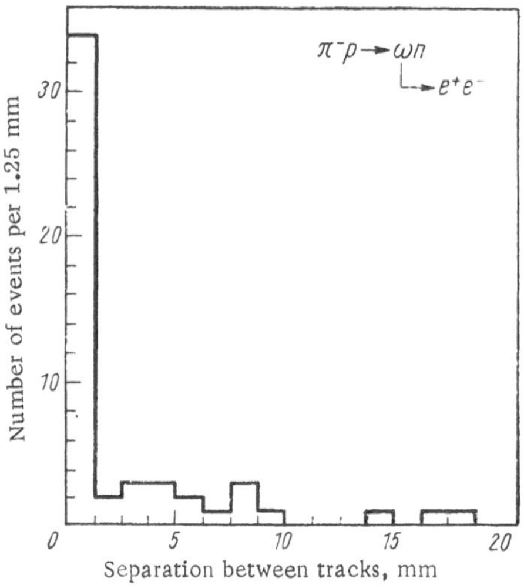

Fig. 20

Fig. 19. Effective-mass spectrum for $\varphi \to e^+e^-$ decay.

Fig. 20. Distribution of separations between tracks of the e^+e^- pairs.

The apparatus is triggered if: 1) at least one of the 24 neutron counters operates; 2) there are fewer than three charged particles; and 3) the total energy evolved in the two shower detectors is at least 750 MeV and at least 150 MeV is evolved in each detector.

Background events are caused by the superposition of charged particles and γ rays, which simulate electrons. Estimates based on an analysis of the distribution of distances between charged particles and γ rays show this effect to be negligible. The second background source is the conversion of γ rays in the target walls and in Sc3 and Sc4. The conversion pairs may not be separated in the spark chambers and may imitate electrons. The conversion background is determined by analyzing the separations between the electrons of the pairs (Fig. 18). The relative number of unseparated conversion pairs is about 25%. Figure 19 shows the mass distribution for $l = 0$ events; the maximum near 1020 MeV is due to $\varphi \to e^+e^-$ decay (9 ± 3 events). Estimates show that the number of background events does not exceed 10% of the effect.

b) $\omega \to e^+e^-$. In this experiment ω mesons were produced by 1.67-GeV/c π mesons. The effective mass of the ω meson was determined, as in the previous experiment, by measurement of the neutron direction and velocity ($\Delta M = \pm 10$ MeV) and by means of shower detectors ($\Delta M = \pm 35$ MeV).

Figure 20 illustrates the method used to evaluate the conversion background. This figure shows the distribution of events as a function of the spacing between the electrons of the conversion pair, including the cases in which the track of a charged particle is detected or when the particles of the conversion pair do not diverge. The events in Fig. 20 for $l = 0$ include $\rho \to e^+e^-$ and $\omega \to e^+e^-$ decays and a background of about 7%. Figure 21 shows the mass distribution for $l = 0$ events for the sum of $\rho \to e^+e^-$ and $\omega \to e^+e^-$ decays.

The $\omega - \rho$ interference was evaluated by identifying the events in the experimental data of [34] by means of the peripheral model with an account of absorption [35]. Constructive or destructive interference

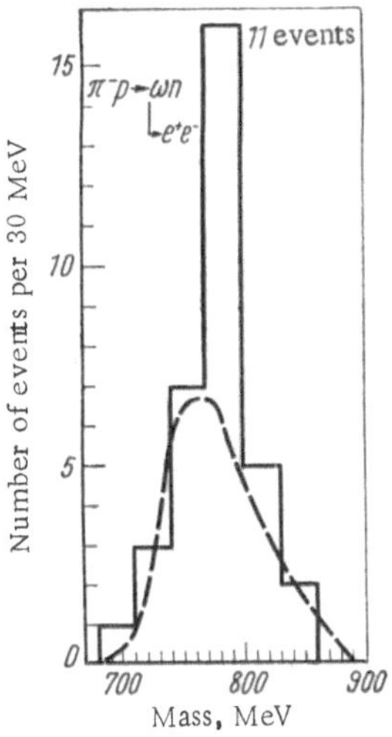

Fig. 21

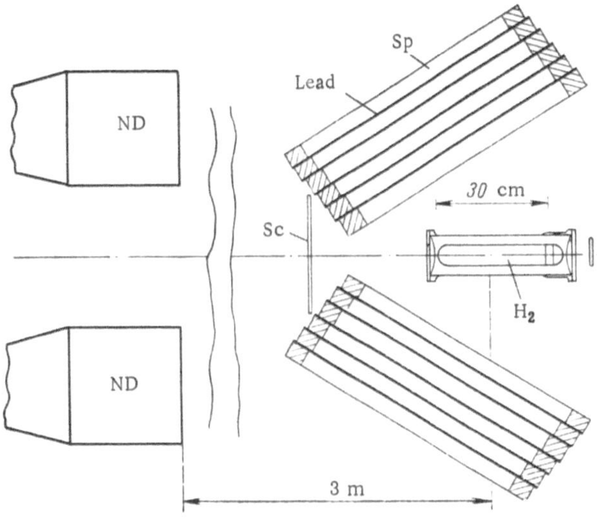

Fig. 22

Fig. 21. Effective-mass spectrum for $\omega \to e^+e^-$ decay.

Fig. 22. Nimrod experimental setup. ND) Neutron detectors; Sc) anticoincidence scintillation counter; Sp) spark chambers.

changes the number of $\omega \to e^+e^-$ events by +35% or −35%. The experimental results were analyzed under the assumption of no $\omega-\rho$ interference; the relative contributions of ω and ρ decays were evaluated by comparing the expected ω and ρ distributions with the experimental distributions. The $\omega \to e^+e^-$ and $\rho \to e^+e^-$ decays were simulated by means of the known angular distributions for ω and ρ production, the density matrices, the resonance width, the experimental geometry, and the parameters of the experimental apparatus. The broken curve in Fig. 21 shows the ρ background. Eleven events grouped at M = 780 MeV are $\omega \to e^+e^-$ decays. The $\pi^-p \to e^+e^-n$ cross section calculated on the basis of these 11 events is

$$\sigma(\pi^- p \to n + \omega,\ \omega \to e^+ e^-) = (67 \pm 25) \cdot 10^{-33}\ \text{cm}^2.$$

Estimates [36] show that the radiation corrections are small, not exceeding a few percent.

Experimental Results

The CERN experimental results are summarized in Table 5. The $B(V \to e^+e^-)$ and $\Gamma(V \to e^+e^-)$ calculations were based on data given in the literature for the total cross sections for ω and φ production. The values of $\Gamma(\omega \to \text{total})$ and $\Gamma(\varphi \to \text{total})$ were taken from the Rosenfeld tables [37].

Study of $\omega \to e^+e^-$ and $\varphi \to e^+e^-$ Decays in the Rutherford Laboratory

a) $\omega \to e^+e^-$ [38]. In the experiments carried out on the Nimrod 7-GeV proton synchrotron at the Rutherford Laboratory, ω mesons were produced in reaction (11a) by π^- mesons having energies only slightly above the threshold for ω production. The ω mesons were distinguished through measurements of the neutron time of flight by means of six scintillation counters 3 m from the liquid-hydrogen target. Electrons were detected by means of spark chambers and lead plates. Figure 22 shows the experimental setup, and Fig. 23 shows the neutron time-of-flight spectra. Analysis of the latter showed the ω mesons produced to be unpolarized.

The following considerations governed the selection of e^+e^- events: 1) The mass deficiency for the event was equal to the ω mass. 2) A single charged-particle track forming a cascade shower was observed in each spark chamber. 3) The momentum-deficiency vector lay in the plane defined by the tracks of the electron–positron pair. 4) The angles between the vectors in this plane satisfied the kinematic

TABLE 6

M, MeV/c^2	779	780	797
ΔM, MeV/c^2	±11	±14	±8

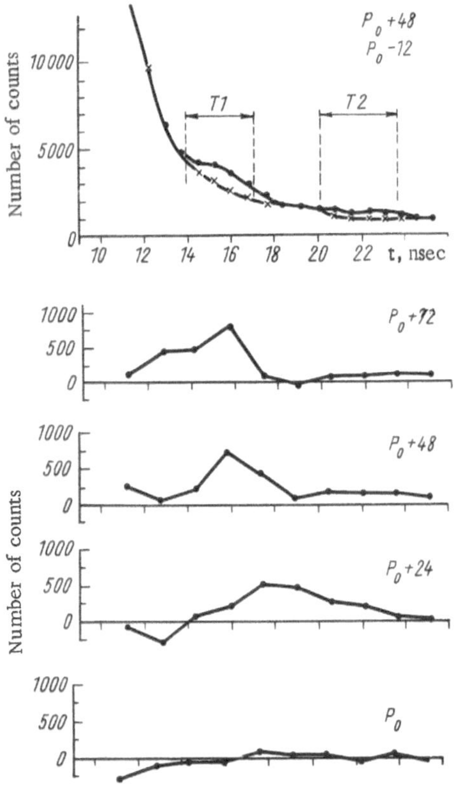

Fig. 23. Time-of-flight spectra of neutrons produced at energies above and below the threshold for ω production. The four lower curves were obtained after subtraction of background. Here we have $P_0 = 1082$ MeV/c. Neutrons emitted at 0° and 180°, respectively, in the c.m. system are detected in time gates T1 and T2.

relationships for $\omega \to e^+e^-$ decays. Analysis of 240,000 photographs revealed only three events satisfying these conditions. Table 6 shows the masses and errors for these events. The range of mass-deficiency values found experimentally was from 760 to 810 MeV/c^2, i.e., 50 MeV.

Several background sources were taken into account.

1. Simulation of e^+e^- decay by the $\pi^-p \to n + \pi^+ + \pi^-$ reaction (particularly, by $\rho^0 \to \pi^+ + \pi^-$ decay). Separation of electrons and π mesons requires the use of experimental data obtained by irradiating the spark chambers by electrons and π mesons. From the number of tracks in the shower, the number of interactions in the lead plates, and the lapses between interactions it was found that the electron-identification efficiency was greater than 90%, while the probability that an electron would be simulated by a π meson was about 0.5%.

Of the 240,000 photographs about 4000 have two π meson tracks and a mass equal to that of the ω meson. The probability that an electron is simulated by a π meson is about 0.5% per channel, so the expected number of false events is between 0.1 and 0.2.

2. Simulation of $\omega \to e^+e^-$ decay by $\rho \to e^+e^-$ decay. The effective number of ρ mesons in the experiment was estimated to be about 2000, much less than the number of ω mesons (about 14,000). Under the assumption of equal partial widths $\Gamma(\omega \to e^+e^-)$ and $\Gamma(\rho \to e^+e^-)$, the number of decays of the first type ($\omega \to e^+e^-$) was predicted to be 50 times the number of $\rho \to e^+e^-$ decays. As a control measure, another 140,000 photographs, obtained at energies below the threshold for ω production, were analyzed; about 2000 of these revealed effective ρ mesons, and there was one case of a mass of 741 ± 15 MeV/c^2 which satisfied selection conditions 1-4. However, the cascade lines in the spark chambers for this event are atypical and apparently result from the production of π^- mesons by δ electrons.

The investigators believe that the strongest argument against identifying these three events as ρ decays is found in the results of [13],* where it was shown that the relative probability for $\rho \to e^+e^-$ was less than 10^{-4}, so the expected number of $\rho \to e^+e^-$ events should not exceed 0.2 event.

3. Nonresonant production of e^+e^- pairs in the $\pi^-p \to ne^+e^-$ reaction. It was shown in [39] that the cross section for e^+e^- production with a mass equal to that of a ω meson is negigibly small. The relative probability for $\omega \to e^+e^-$ decay calculated on the basis of these three cases is $B = (21^{+39}_{-16}) \cdot 10^{-5}$.

b) $\varphi \to e^+e^-$ [40]. The second series of experiments was devoted to measuring the relative width for φ decays into e^+e^- and K^+K^- pairs. The φ mesons were produced in reaction (11a) by 1.58-GeV/c π mesons.

* The events were not identified in [13], so the corresponding results can be used only to estimate an upper limit for $\rho \to e^+e^-$ decay.

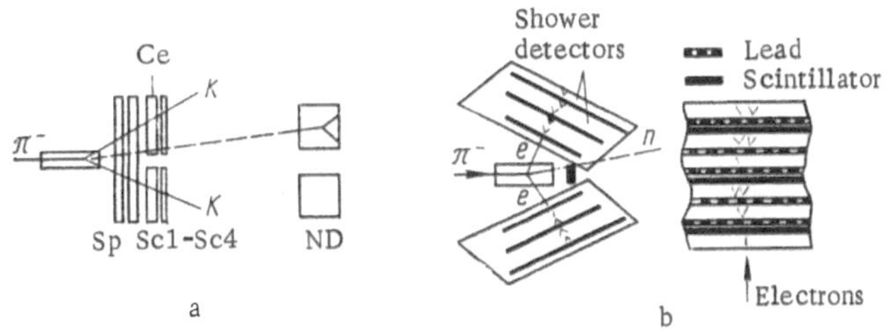

Fig. 24. Experimental setup for study of the reaction $\pi^-p \rightarrow \varphi n$. a) $\Phi \rightarrow K^+K^-$. Here Ce is a Cerenkov threshold detector. Sc1-Sc4 are scintillation counters, ND are neutron detectors, and Sp are spark chambers; b) $\Phi \rightarrow e^+e^-$. Electrons are distinguished by means of a sandwich of scintillation counters, spark chambers, and lead plates having a thickness of 1.1 radiation lengths.

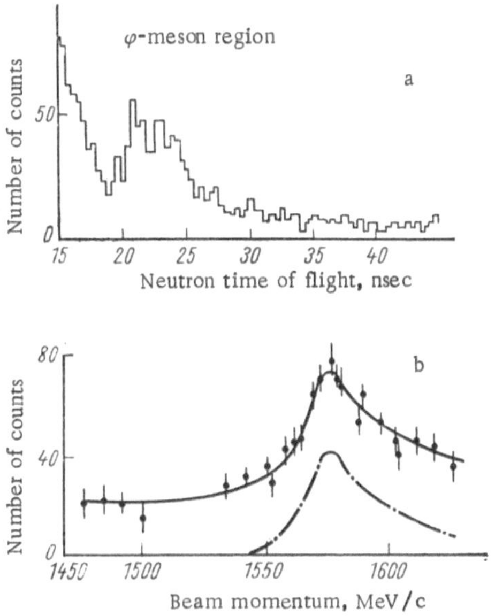

Fig. 25. a) Typical neutron time-of-flight spectrum; b) number of K^+K^- pairs as a function of the π momentum. The solid curve was calculated from $\sigma_\varphi \sim p^*$, where p^* is the π^- momentum in the c.m. system; this curve includes the background of nonresonant K^+K^- pairs and the π background. The broken curve was obtained by subtracting the background events.

The experimental setup is shown in Fig. 24. The K^+K^- pairs are distinguished by scintillation counters Sc1-Sc4 and by the Cerenkov threshold counter Ce connected in anticoincidence.

Figure 25a shows the neutron time-of-flight spectrum for a momentum above the threshold for φ production. Because of the large K^+K^- background, this spectrum is not suitable for determining the number of φ mesons. The K mesons can be distinguished from the background by measuring the beam-momentum dependence of the number of neutrons in the corresponding time region (Fig. 25b). Analysis of the K-meson events showed that the experimental results could be described by a simple S-wave model, for which the total cross section for φ production at the threshold is a linear function of the momentum p^* (in the c.m. system), and that the angular distributions for the production and decay of φ mesons are isotropic.

TABLE 7

Parameter	$\omega \to e^+ e^-$	$\varphi \to e^+ e^-$
Number of events	3	$10{,}1 \pm 4{,}8$
$B(V \to e^+e^-)$, 10^{-5}	21^{+39}_{-16}	72 ± 39
$\Gamma(V \to e^+e^-)$, keV	—	$2{,}4 \pm 1{,}5$
$f_V^2/4\pi$	—	$7{,}6^{+12}_{-2{,}8}$

The total cross section for φ production is calculated from $\sigma = (0.28 \pm 0.07)p^*$ μb, where p^* is given in MeV/c. The electron pairs are detected by thin-walled spark chambers separated by lead plates and scintillation counters (Fig. 24). The electrons are identified on the basis of showers generated in five lead plates each 1.1 radiation lengths thick. The spark chambers nearest the target are used to determine the direction of the electron emission and to eliminate some of the background events. The scintillation counters immediately behind the first chambers are connected in coincidence, while the counters in front of the third and fifth lead plates, i.e., at depths of 2 and 4 radiation lengths, respectively, give information about the magnitude of the shower. Under these experimental conditions the electron (or positron) energy spectra lie in the range 400-900 MeV, and the sum of these energies, equal to the total energy of the φ meson, is nearly constant. An electric method is used to distinguish events on the basis of the energy sum.

The ratio of the number of e^+e^- and K^+K^- pairs was measured at two π^- energies: below the threshold for φ production and at the maximum of the excitation curve, between 1574 and 1590 MeV/c. From 220,000 photographs, 4700 were selected, containing primarily background events due to the superposition of photon showers with π mesons, γ-ray conversion, etc. More stringent criteria based on analysis of the tracks in the shower, the scattering angles, the shower symmetry, etc., reduced the number of candidates to 255. Kinematic considerations further reduced this number to 80. Since only a small fraction of these 80 events can be $\varphi \to e^+e^-$ decays, two of the criteria used previously – the degree of shower symmetry and the mean free path of electrons before interactions – were reassessed and again used for the analysis. This left 27 candidates, of which 17 fell in the energy range above the φ-production threshold and ten fell in the energy range below this threshold, i.e., were background events. The number of $\varphi \to e^+e^-$ decays was found after a monitoring normalization to be $N(\varphi \to e^+e^-) = 10.1 \pm 4.8$. The results of the two experiments are given in Table 7.

The calculations of $\Gamma(\varphi \to e^+e^-)$ and $f_\varphi^2/4\pi$ were based on the value $\Gamma(\varphi \to \text{total}) = 3.4 \pm 0.8$ MeV from the tables of [41].

4. EXPERIMENTS WITH CLASHING ELECTRON – POSITRON BEAMS

States with photon quantum numbers 1^- must arise in the single-photon channel during electron–positron annihilation. The ρ, ω, and φ mesons have these quantum numbers $[1^-, M(V) \neq 0]$.

Experiments with clashing electron – positron beams have been carried out at Novosibirsk and Orsay.

Study of $e^+e^- \to V$ Reactions at Novosibirsk

a) $e^+e^- \to \rho^0 \to \pi^+\pi^-$ [42].

Description of the Apparatus. The storage device is a weakly focused racetrack with four identical gaps. The injector is the synchrotron with an extracted beam current of 500 mA at a pulse length shorter than 20 nsec; this current corresponds to about $6 \cdot 10^{10}$ particles. During positron storage, the positrons are produced in a tungsten target-converter by 250-MeV electrons. After the positrons and electrons are stored, their energy is raised to the appropriate level E, and a trigger pulse is sent to the spark-chamber system. The entire measurement cycle lasts about 2 h. A third of this time is spent on accumulating the positrons and electrons. The initial positron current averages 20 mA, and the initial electron current averages 50 mA. The particle lifetime in the storage device is longer than 3000 sec.

Description of the Measurement Apparatus. The detection apparatus consists of two symmetric systems of spark chambers and scintillation counters covering an angle of $2 \cdot 0.6$ sr near the vertical (Fig. 26). The emission angle and the interaction point are determined by spark chambers with thin plates, while the type of particle (a π meson or electron) is determined by shower and range chambers. A shower

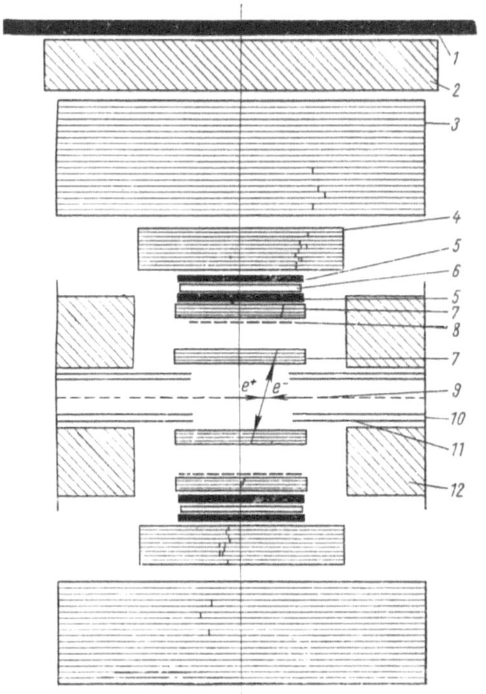

Fig. 26. Experimental setup at Novosibirsk. 1) Anticoincidence scintillation counter; 2) lead layer 200 mm thick; 3) range spark chamber; 4) shower spark chamber; 5) scintillation counters; 6) dural layer 20 mm thick; 7) spark chambers with thin plates; 8) lead radiator; 10 mm thick; 9) clashing region; 10, 11) inner and outer vacuum chambers; 12) storage magnet.

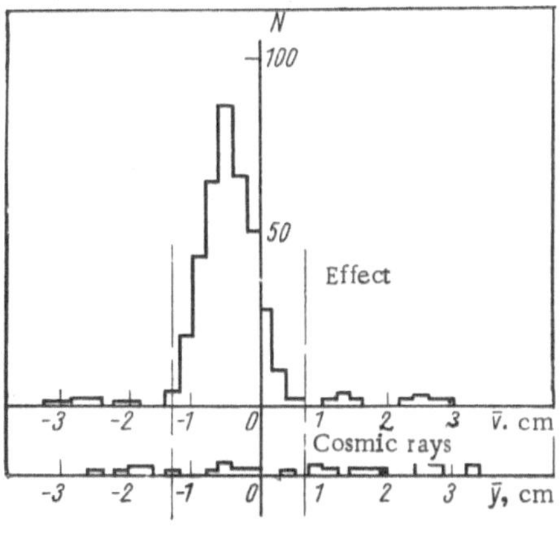

Fig. 27. Histogram of the coordinate distribution of interaction points.

chamber consists of 11 copper plates each 6 mm thick, while a range chamber consists of 21 stainless steel plates each 8 mm thick. A single camera photographs the spark chambers. The spark chambers are triggered by pulses from our scintillation counters connected in coincidence; the coincidence resolving time is 20 nsec. An anticoincidence scintillation counter 160 × 160 cm^2 in area reduces the number of background triggering events due to cosmic rays. A lead filter 20 cm thick placed between the anticoincidence counter and the spark chambers eliminates the possibility that counter 1 will be triggered by particles produced within the vacuum chamber of the accelerator. Counter 1 reduces the number of cosmic-ray triggering events by a factor of about 100. Phase synchronization of the triggering of the spark chambers and the voltage supplied to the storage resonator reduces this number by another factor of five. Under these conditions the number of triggering events due to cosmic rays does not exceed 15 h^{-1}.

Analysis of Experimental Data. An experimental study was made of the reaction

$$e^+ + e^- \rightarrow \rho^0 \rightarrow \pi^+ \pi^-. \tag{26}$$

The measurements were carried out at nine energies, from 2 × 290 to 2 × 510 MeV. About 50,000 photographs were obtained in 6 months of continuous operation.

The following criteria were used to select events. Photographs were selected with tracks in all four thin chambers, but the range of the tracks in chambers 3 could not exceed the maximum possible π^+ range. In the geometric reconstruction of the events: 1) the tracks had to be colinear within 10°; 2) the distance between intersections of tracks with the median plane of the storage device could not exceed 3 cm; and 3) the distance from the interaction point to the beam-interaction region (e^+e^-) could not exceed 1 cm.

Figure 27 shows the results of an analysis on the basis of the latter criterion after selection on the basis of the first two criteria, along with data on the cosmic-ray background, obtained with the accelerator closed off. The background due to interaction of beam particles with the residual gas was measured in the absence of one of the beams. The experimental results are shown in Table 8.

This table shows that the cosmic-ray background amounts to only 3% of the effect. The number of background events generated by interaction of beam particles with the residual gas within the vacuum chamber of the accelerator is less than 0.5%.

The events remaining after use of the criteria listed above contain events resulting from quasielastic scattering of electrons by positrons and production of π pairs. It is very difficult to distinguish between these two processes in this energy range because the electron-shower pattern is not defined clearly enough. The distinction was based on the possibility of an independent determination of the nature of the particle ($e^\pm$ or $\pi^\pm$) in the lower and upper spark-chamber systems.

TABLE 8

Type of beam	Measurement time, h	Current loss, A	No. of photographs, x1000	No. of events detected
Effect	785	29,4	31,6	371
Cosmic rays	662	0	9,6	10
Beam	40	3,7	5,2	1

TABLE 9

Parameter	Effect	Background
Measurement time, h	208	64
Discharge, C	564	167
Monitor, x1000	1216	2,2
No. of photographs, x1000	81,3	24,0
No. of elastically scattered e^+e^- pairs	528	0
No. of K^+K^- pairs	647	35

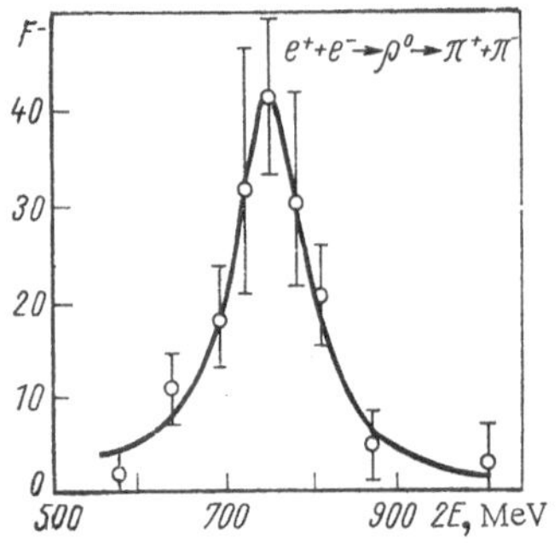

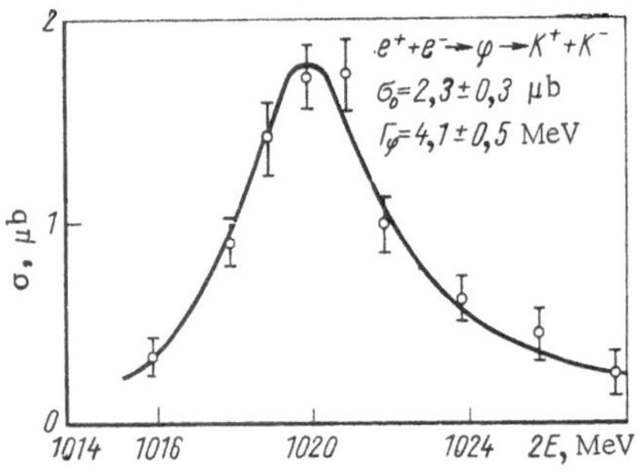

Fig. 28

Fig. 29

Fig. 28. Experimental values of $F^2(E)$, approximated by a Breit–Wigner curve.

Fig. 29. Cross section for the $e^+ + e^- \rightarrow K^+ + K^-$ reaction.

The cross section for reaction (26) can be written

$$\frac{d\sigma_{\pi\pi}}{d\Omega} = \frac{1}{32} \cdot \frac{r_0}{\gamma^2} \cdot \beta_\pi^2 \sin^2\theta \cdot F^2(E), \tag{27}$$

where r_0 is the classical radius of the electron, γ is the relativistic factor of the clashing particles, β_π is the velocity of the π meson, θ is the π emission angle, and F(E) is the modulus of the form factor for $\pi^+\pi^-$ production.

Point quantitative electrodynamics yields F = 1 in the absence of other forces. In the absence of accurate data on the beam intensity in the apparatus, use was made of the elastic scattering of electrons by positrons as a normalizing process. The cross section for this reaction is

$$\frac{d\sigma_{ee}}{d\Omega} = \frac{1}{16} \cdot \frac{r_0^2}{\gamma^2} \left(\frac{3+\cos^2\theta}{1-\cos\theta} \right). \tag{28}$$

Equations (27) and (28) give the ratio of the number of $\pi^+\pi^-$ pairs to the number of electrons scattered by positrons for a given E:

$$\frac{N_{\pi\pi}}{N_{ee}} = \frac{\beta_{\pi\pi}^2}{a} F^2(E)(1+\delta), \tag{29}$$

TABLE 10

Parameter	$e^+e^- \to \rho^0$		$e^+e^- \to \varphi$	
	effect	bckg	effect	bckg
Number of events	—	—	208	64
σ_V, 10^{-30} cm^2	(1,3±0,2)	—	(4,3±0,6)	—
$B(V \to e^+e^-)$, 10^{-5}	(5,0±1,0)	—	(34±4)	—
$\Gamma(V \to e^+e^-)$, keV	5,30±0,91	—	1,42±0,13	—
$f_V^2/4\pi$	2,56±0,44	—	12,7±1,16	—
$\Gamma(V \to \text{total})$, MeV	105±21	—	4,1±0,5	—
m_V, MeV	754±9	—	—	—

where $(1 + \delta)$ is a correction factor, and a is a constant which depends on the experimental geometry ($a = 20.3$).

The quantity δ incorporates radiation corrections [43] and corrections for the efficiency of the $\pi^+\pi^-$ and e^+e^- detection. The integral correction does not exceed 20% in the energy range studied. The experimental values of the function $F^2(E)$ are shown in Fig. 28; approximation of the experimental data by the Breit–Wigner equation

$$F^2(E) = \frac{F_0^2 m^2 \Gamma^2}{(4E^2 - m^2)^2 + m^2\Gamma^2} \tag{30}$$

yields $F_0^2 = 42 \pm 8$, $m = 754 \pm 9$ MeV, and $\Gamma = 105 \pm 20$ MeV. The indicated errors contain, in addition to the statistical errors the errors in the determination of the energy of the initial particles (0.5%) and possible systematic errors. The use of an equation expressing the form factor F(E) in terms of the phase of the $\pi\pi$ scattering in the $l = 1$ state has essentially no effect on the values of these parameters. These results can be used to evaluate the total cross section for reaction (26) at the maximum, corresponding to the formation of an intermediate ρ meson, to calculate the relative probability for $\rho \to e^+e^-$ decay, and to calculate the $\rho - \gamma$ coupling constant (Table 10).

b) $e^+e^- \to \varphi$

The same apparatus was used for a study of the annihilation of an electron and positron resulting in φ production; for this purpose the apparatus was extensively modified: the length of the interaction region was reduced, the solid angle of the spark-chamber system was increased, apertures were formed in the primary magnet to allow monitoring of the double bremsstrahlung, the vacuum in the storage device was improved so that the particle lifetime was raised to 5 h at a current of 100 mA, and the intensity of the synchrotron-injector beam was increased to 10^{11} particles/pulse.

Description of the Experimental Apparatus. The system of spark chambers and scintillation counters (Fig. 26) can simultaneously detect the three basic types of φ-meson decay

$$\varphi \to \left.\begin{matrix} K^+ + K^-; \\ K_S{}^0 + K_L{}^0, \quad K_S{}^0 \to \pi^+\pi^-; \\ \pi^+ + \pi^- + \pi^0. \end{matrix}\right\} \tag{31}$$

Minimum-ionization particles are distinguished by four scintillation counters operating in coincidence. Charged K mesons, whose energy is 16 MeV and for which the corresponding light pulse in the scintillator is much greater than that for a relativistic particle, are detected only by the counters nearest the storage vacuum chamber. Lead plates 10 mm thick are placed between the storage vacuum chamber and the scintillation counters to detect neutral types of φ decay. The absolute energy is determined within 1% by magnetic measurements.

Experimental Results. Most of the experiments were carried out over a period of 1 month at nine energies of the particles in the storage device, from 508 to 514 MeV. The beam-clashing conditions were monitored and controlled through detection of double-bremsstrahlung events by means of two counters. About 100 photographs were obtained in these experiments; the basic results are given in Table 9.

TABLE 11

Energy, MeV	2×322,0	2×352,3	2×382,0	2×412,4	2×442,9
L (10^{31} $cm^{-2} \cdot h^{-1}$)	1,3	1,7	2,5	3,4	4,5
$\int Ldt$, 10^{32} cm^{-2}	3,0	6,7	8,2	10,4	6,2
I^+, I^-, mA	6	6,7	7,3	8,5	10
τ^+, τ^-, h	11	14	18	20	20
σ_l, cm	4,8	5,2	5,6	6,0	6,5
No. of photographs, ×1000	20	30	50	45	45
S, mm^2	1,9	2,2	2,1	2,4	2,6
Pexp, 10^{-10} torr	1,15	1,5	1,7	2,0	2,15

Note: L) Maximum beam intensity; $\int Ldt$) integral number of particles; $I^{\pm}$) number of electrons (or positrons) in cycle; $\tau^{\pm}$) beam lifetime; σ_l) longitudinal spread of beam particles; S) cross-sectional area of the beam-interaction region; p_{exp}) gas pressure in the experimental section of the ring.

The detection efficiency for K^+K^- pairs (7.4%) was determined by the Monte Carlo method on a computer. The basic experimental results are shown in Fig. 29. The experimental points are approximated by a Breit–Wigner curve with $\Gamma_{\varphi} = 4.1 \pm 0.5$ MeV and $\sigma_0 = (2.3 \pm 0.3) \cdot 10^{-30}$ cm^2.

Using the tabulated value of $W(\varphi \to K^+K^-) = 47 \pm 3\%$, we find the total cross section for φ production to be $\sigma(\varphi) = 4.8 \pm 0.6$ μb. These results can be used to calculate the relative probability and width for $\varphi \to e^+e^-$ decay; the basic results of the two experiments are shown in Table 10.

Study of $e^+e^- \to V$ Reactions at Orsay [45-49]

a) $e^+e^- \to \rho^0 \to \pi^+\pi^-$

The first results of this study were published in 1967 [45]. The experimental apparatus used at Novosibirsk, consisting of a symmetric system of film-type spark chambers and scintillation counters (Fig. 30). The thin-walled spark chambers nearest the interaction region are used to establish the track direction. The large-mass spark chambers are used to identify the secondary particles formed as a result of e^+e^- interactions.

Several improvements were later incorporated in the apparatus: 1) The system for distinguishing and identifying electrons was improved. 2) The thickness of the range chambers was increased to match the range of π mesons having energies up to 440 MeV. 3) The thickness of the absorbers between the interaction region and the scintillation counters triggering the chambers was reduced. 4) The background of particles scattered by the residual gas was reduced by 35%. 5) The same intensity in the storage device was increased (by a factor of three or four). 6) The accuracy with which the electron energy in the ring was measured was improved to ±0.3%. The basic parameters of the apparatus are shown in Table 11.

Experimental Data. First, many of the background events due to scattering of beam particles or to cosmic rays were eliminated. Then events were selected for which: 1) the track continuations intersected in the beam-interaction region, 2) the angle between the projections of the two tracks on the plane perpendicular to the beam direction was less than 10°, 3) the actual angle between the two tracks did not exceed 15°, 4) the angle formed by the tracks with the normal to the surface of the spark chambers was less than 42°, and 5) there was a coincidence between the time at which scintillation counters Sc1-Sc4 (which trigger the spark chambers) operated and the time at which the electron-positron beams interacted. This latter condition reduced the background of neutral cosmic-ray particles by a factor of five, i.e., to a value on the order of a few percent.

Analysis of Experimental Data. The experimental data consist primarily of events containing e^+e^-, $\mu^+\mu^-$, and $\pi^+\pi^-$ pairs. The $\mu^+\mu^-$ events can be separated from the $\pi^+\pi^=$ events on the basis of the difference between the ranges of the μ and π mesons in the spark chambers. Since not all the secondary particles can be unambiguously identified, a statistical method, based on an independent check of each track of

TABLE 12

Energy, MeV	2×322,0	2×352,3	2×382,0	2×412,4	2×442,9
No. $N_{\pi\pi}$ of $\pi\pi$ pairs	41	157	389	182	68
No. N_{ee} of ee pairs	183	344	355	387	225
$N_{\pi\pi}/N_{ee}$	0,223±0,046	0,457±0,053	1,098±0,095	0,469±0,050	0,304±0,053
Correction $1+\lambda_{\pi}$ for π-meson absorption	1,21±0,06	1,21±0,06	1,20±0,06	1,195±0,06	1,185±0,06
Correction $1+\lambda_e$ for electron absorption	1,053±0,03	1,047±0,028	1,041±0,025	1,036±0,022	1,033±0,020
π-meson form factor F_{π}^2	13,1±2,8	26,3±3,4	58,0±6,2	19,5±2,4	11,7±2,1

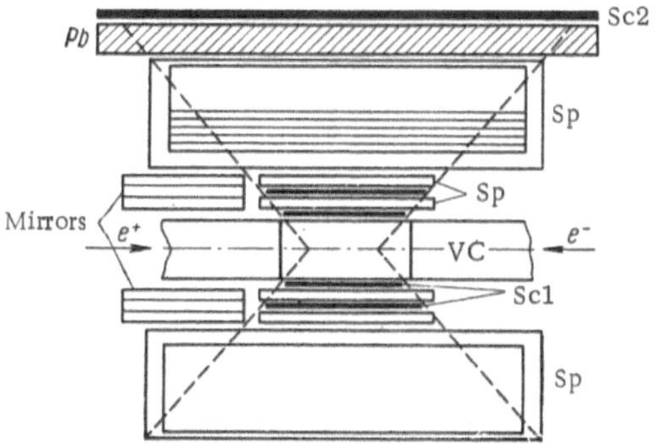

Fig. 30

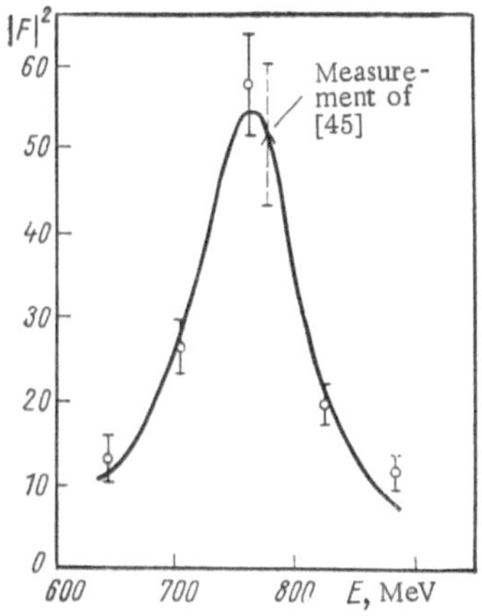

Fig. 31

Fig. 30. Experimental setup at Orsay. Sc1) Scintillation counters; Sp) spark chambers; Sc2) anticoincidence scintillation counter; VC) vacuum chamber.

Fig. 31. Form factor for the π meson ($e^+ + e^- \to \rho^0 \to \pi^+\pi^-$). The experimental points are approximated by a Breit–Wigner curve.

the event, is used to analyze them. The tracks are classified only if the spark is typical. On the basis of the number of events having two, one, or zero tracks identifiable as corresponding to a π or e, one can evaluate the probabilities for identifying the event as a π-meson track or electron shower. In this manner one can determine the ratio of the number of $\pi^+\pi^-$ and e^+e^- pairs among the unidentified events and the statistical error within which the ratio $N_{\pi^+\pi^-}/N_{e^+e^-}$ is determined. Account is also taken of the possibility of identifying a π-meson track as an electron track as a result of charge exchange. At no energy does the number of unidentified events exceed 20%. Table 12 shows the numbers $N_{\pi^+\pi^-}$ and $N_{e^+e^-}$ for each energy, along with the corresponding corrections.

The π-meson absorption λ_{π} in the second group of scintillation counters was calculated on the basis of the known cross sections for the inelastic interactions of π mesons with carbon and copper. Account was taken of the possible triggering of the second counter by inelastic events; this effect can be evaluated experimentally by observing inelastic events in the first intermediate range spark chambers. The quantity λ_e, which takes into account electron absorption, was also calculated. Here the calculation accuracy is affected by the large uncertainty (about 60%) in the bremsstrahlung spectrum and in the range of low-energy electrons. Radiation corrections including effects taking into account vacuum polarization were calculated on the basis of the results of [50, 51]. The form factor for the π meson was calculated from

$$|F_\pi|^2=\frac{N_{\pi\pi}}{N_{ee}}\cdot\frac{(1+\lambda_\pi)}{(1+\lambda_e)}\cdot\frac{(1+\delta_e)}{(1+\delta_\pi)}\cdot\frac{12\sigma_{ee}}{\varepsilon_\pi r_e^2}\left(\frac{E}{m_0c^2}\right)^2\cdot\frac{1}{\beta_\pi^2}, \tag{32}$$

where σ_{ee} is the effective cross section for the reaction $e^+e^- \to e^+e^-$, and ε_π is the number of $\pi^+\pi^-$ pairs. The errors given in Table 12 for $|F_\pi|^2$ are statistical. The experimental $|F|^2$ values are approximated by Eq. (30) (Fig. 31). The final calculation took into account the systematic error of ±6% due to the uncertainty in π-meson and electron absorption (the relative probability also changes by ± 6%), and the polarization of particles accelerated in the ring with respect to the direction of the accelerator magnetic field. This effect, calculated in [52], reduces m_ρ, Γ_ρ, and B by 0.1%, 0.7%, and 3.6%, respectively.

When these corrections and the errors, on the order of ±0.3%, in the energy of the accelerator particles are taken into account, we find m_ρ = 760 ± 16 MeV, Γ_ρ = 112 ± 12 MeV, and F_0^2 = 55 ± 9. The relative probabilities, coupling constants, and partial widths for $\rho^0 \to e^+e^-$ decay are shown in Table 14.

b) $e^+e^- \to \omega \to \pi^+\pi^-\pi^0$.

The experimental data were obtained at seven energies grouped around the mass of the ω meson. Three-particle events were identified and separated from the two-particle and background events by selecting events with two noncollinear tracks.

A possible background source is that of π mesons scattered by the wall of the vacuum chamber. When the angle between the tracks of the two π mesons is large enough, the geometric reconstruction is accurate enough to eliminate from the analysis events in which the angle θ_S between the tracks is less than 20°.

Another background source is due to two-particle events of the type

$$\left.\begin{array}{l} e^++e^-\to e^++e^-+\gamma; \\ e^++e^-\to \pi^++\pi^-+\gamma. \end{array}\right\} \tag{33}$$

In these reactions the photon is emitted parallel to the electrons, so the angle between the projections of the tracks on the plane perpendicular to the beam direction is always 0°. Accordingly, events for which the angle θ_T between the tracks in the plane perpendicular to the beam is less than 10° can be eliminated. Those events for which, in the plane perpendicular to the beam direction, there is an angle greater than 40° with respect to the normal to the spark-chamber plane and an angle greater than 55° in the beam plane are also rejected. The geometric efficiency of the apparatus was calculated by the Monte Carlo method (with an account of the selection criteria and the length of the bunch, whose total width at half-maximum is 17.7 cm). The cross sections for the production and decay of ω mesons are calculated from

$$\frac{d^3\sigma}{d\omega_+\,d\omega_-\,d(\cos\theta_N)}\sim\frac{\alpha}{(2\pi)^2}\cdot\frac{1}{64E^2}|H\times A|^2\sin^2\theta_N(\mathbf{p}_+\mathbf{p}_-)^2,$$

where

$$\begin{gathered} H=1/(4E^2-m_\omega^2+im_\omega\Gamma_\omega), \\ A=\sum_i 1/(4E^2+m_\pi^2-m_\rho^2-4E\omega_i+im_\rho\Gamma_\rho), \end{gathered} \tag{34}$$

ω_+ and ω_- are the total energies of the π^+ and π^- mesons, θ_N is the angle between the normal to the ω decay plane and the beam direction, and E is the energy of the beam electrons.

The simulation program includes corrections for nuclear absorption of π mesons in the absorber in front of the counters. These corrections reduce the number of detected events by 0.710 ± 0.078.

Account is also taken of the loss caused by the fact that in 14% of the cases one of the π mesons is not energetic enough to reach the trigger counter. The Monte Carlo method also takes into account the efficiency for events in which a charged π meson and a γ ray are detected from the decay of a π^0 meson which has undergone conversion in the wall of the vacuum chamber. The number of such events is 25 ± 8%.

The polarization of the electron–positron beam reduces the efficiency of the apparatus by 7.2 ±6.6%; after all the corrections are taken into account, this efficiency is 5.2%. The total intensity for a given en-

TABLE 13

$E-m_\omega$, MeV	−10	−5	−2,5	0	2,5	5	20
Number of events	8	17	30	122	27	21	4
Correction for hadron vacuum polarization	1,01	1,01	1,01	0,978	0,949	0,951	0,96
Radiation correction	1,255	1,28	1,29	1,255	1,14	1,01	0,645
Intensity, 10^{32} cm^{-2}	9,17	6,63	7,00	16,62	5,59	6,63	9,17
σ_0, μb	0,213 ±0,075	0,637 ±0,159	1,08 ±0,204	1,74 ±0,165	1,05 ±0,205	0,575 ±0,132	0,051 ±0,026

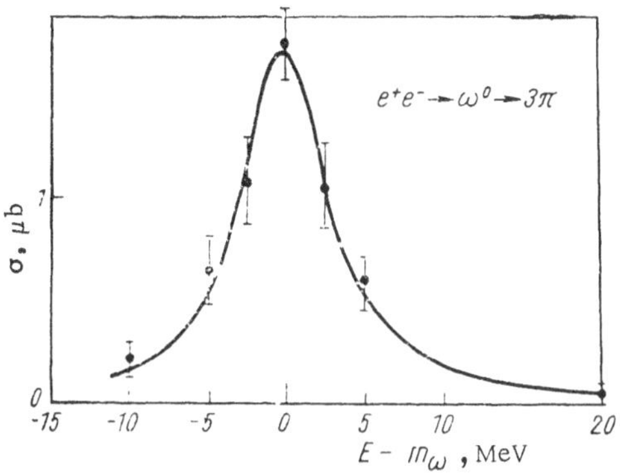

Fig. 32. Cross section for the reaction $e^+ + e^- \to \pi^+ + \pi^- + \pi^0$. The experimental points are approximated by a Breit–Wigner curve.

ergy is calculated on the basis of the two-particle events, e^+e^- and $\pi^+\pi^-$ pairs due to meson decay. The number of two-particle events is compared with the number calculated for $e^+e^- \to e^+e^-$ and $e^+e^- \to \rho^0$ reactions; the results are shown in Table 13. The radiation corrections are calculated from [53]

$$p(k) = \frac{d}{dk} e^{-\delta_s(k)}, \tag{35}$$

where p(k) is the probability for the emission of a γ ray with energy k, and

$$\delta_s(k) = b\,[\lg(E/k) - 13/12]; \quad b = \frac{4\alpha}{\pi}\left[\lg(2E/m) - \frac{1}{2}\right].$$

The corrections for vacuum polarization were taken from [51]. The experimental results are shown in Fig. 32; the curve here is calculated from the Breit–Wigner equation.

The cross section for the reaction $e^+e^- \to 3\pi$ was found to be 1.71 ± 0.165 μb, and a width of $\Gamma(\omega)$ = 14.0 ± 2.4 MeV was calculated. The error in the effective cross section includes the errors for nuclear π-meson absorption (±11%), and monitor error (±5%), the error involved in the measurement of the beam polarization (±6.6%), and the efficiency with which $\pi^\pm\gamma$ events are detected. Calculation of the total cross section for ω-meson production also requires a correction for the relative number of $\omega \to \pi^0\gamma$ decays (10 ± 1%).

The values of $B(\omega \to e^+e^-)$, $\Gamma(\omega \to e^+e^-)$, and $f_\omega^2/4\pi$ found after summation of the systematic and statistical errors are given in Table 13.

c) $e^+ + e^- \to \varphi$.

The third series of experiments was devoted to the $e^+e^- \to \varphi$ reaction. Since the width of the φ meson is small (about 4 MeV), it was crucial to know the beam energy accurately in this experiment. Absolute energy calibration was based on the known value for the φ mass: 1019.3 MeV. The excitation curve for the resonance was measured by changing the energy of each beam in 0.5-MeV steps from 502 to 525 MeV. At the beam intensity of $3 \cdot 10^{31}$ cm^{-2} · h^{-1} the number of false triggering events due to background particles did not exceed 70 counts/min. A total of 500,000 photographs were obtained in this experiment.

1. $e^+ + e^- \to \varphi \to K_S^0 + K_L^0$. This event was identified on the basis of $K_S^0 \to \pi^+\pi^-$ decay. For $K_S^0 \to \pi^+\pi^-$ the angle θ between the two π mesons is approximately 180° and with 2E = m(φ) is always greater than the minimum angle θ = 150°. This circumstance permitted an easy identification of K_S^0 decay, for

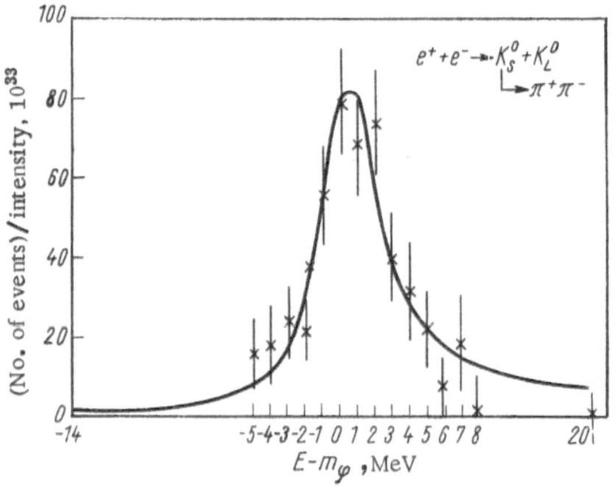

Fig. 33. Excitation curve for the reaction $e^+ + e^- \to K_S^0 + K_L^0$, $K_S \to \pi^+ + \pi^-$.

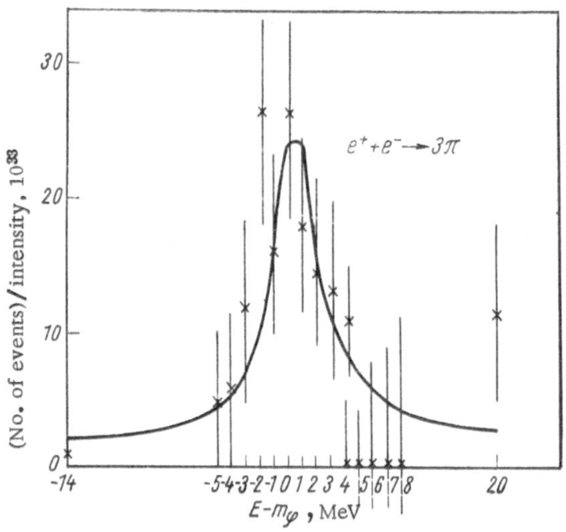

Fig. 34. Excitation curve for the reaction $e^+ + e^- \to 3\pi$.

which the calculated detection efficiency is 10.2%. Analysis of the background processes showed that: a) the $K_S^0 \to 2\pi$ decay could be simulated by decay to 3π (5 ± 1%); b) it could be simulated by the decay of regenerated K_L^0 ($K_L^0 \to K_S^0 \to 2\pi$); c) $e^+e^- \to \pi^+\pi^-$ events were partially eliminated by selection of the basis of the angle θ ($\theta < 170°$). The corresponding correction to the number of detected events was 4 + 1%. The reaction $e^+e^- \to \pi^+\pi^-\gamma$ and analogous reactions were distinguished from the effect on the basis of the projection of the angle θ on the plane normal to the beam ($\theta_T < 175°$). The corresponding correction to the number of K_S^0 decays was 10 ± 2%; d) the π meson could be simulated by conversion electrons (e.g., in $\pi^+\pi^-\pi^0$ decay). The corresponding correction was on the order of 1% for K_S^0 detection and about 5 ± 2% in the decay $\varphi \to \pi^+\pi^-\pi^0$; e) corrections had to be made for scattering of π mesons before chambers C1 (4%) and for the absorption of charged π mesons in the copper absorber between counters $S_1(S_1')$ and $S_2(S_2')$; 23 ± 7%; f) some of the charged π mesons could form showers ($\pi^+ \to \pi^0 \to 2\gamma$) as a result of charge-exchange scattering. The corresponding correction, taking into account the relative number of discarded events, was 5 ± 2%; and g) the number of K^0 mesons detected could have been reduced a few percent by polarization of the electron beam.

On the basis of $K_S^0 \to 2\pi$ decay, 150 φ-meson events were identified; the data are shown in Fig. 33.

2. $e^+ + e^- \to \varphi \to \pi^+ + \pi^- + \pi^0$. In this reaction the angle between the charged mesons (the π^0 meson is not detected) is not distinguished and there is no maximum near 180°, so the kinematics of the process cannot be reconstructed. The efficiency with which $\varphi \to 3\pi$ decays can be detected was calculated to be 4%. Experimentally, 53 $\varphi \to 3\pi$ decays were identified; their distribution is shown in Fig. 34. The experimental results were fitted on the basis of the width $\Gamma(\varphi)$ and the position of the excitation-curve maximum, both measured on the basis of K_S^0 decay.

3. $e^+ + e^- \to \varphi \to K^+K^-$. This type of experiment was stimulated to a large extent by the contradictory results obtained on the basis of the two previous types of φ decay. Since the K mesons produced in φ decay have a low energy, efforts were made to reduce the amount of matter (including that in the walls of the accelerator chamber) to 0.14 g/cm². With this modification, the efficiency with which K mesons could be detected in the scintillation counters was improved. The K^+K^- pairs were selected on the basis of a minimum number of sparks in the track and on the basis of track collinearity within 10°. Events whose tracks intersected more than 12 mm from the beam-interaction region were discarded. The number of triggering events due to cosmic rays in the experiments was 20 h^{-1}. The average number of triggering events due to $\varphi \to K^+K^-$ decay was 15 h^{-1} (at the resonance energy).

The final result incorporated corrections for the cosmic-ray background, nuclear absorption, and multiple scattering of K mesons. The integral experimental error due to all the corrections and to the statistical error (3.8%) was 6.5%. Measurements were carried out at 15 energies near m_φ (about 1500 events); the data are shown in Fig. 35. The Orsay results are summarized in Table 14.

TABLE 14

Parameter	$e^+e^- \to \rho^0$	$e^+e^- \to \omega$	$e^+e^- \to \varphi$
σ_V, 10^{-30} cm^{-2}	1,60±0,20	1,82±0,34	5,27±0,35
$B(V \to e^+e^-)$, 10^{-5}	6,4±0,8	7,7±1,4	37,3±2,5
$\Gamma(V \to e^+e^-)$, keV	7,4±0,6	0,94±0,18	1,58±0,13
$f_V^2/4\pi$	1,99±0,11	14,9±2,8	11,5±0,9
$\Gamma(V \to \text{total})$, MeV	110,7±5,3	14,0±2,4	4,24±0,28
m_V, MeV	773,5±5,4	—	—

TABLE 15

Laboratory	$B = \frac{\Gamma(V \to e^+e^-)}{\Gamma(V \to \text{total})}$		
	$B(\rho \to e^+e^-) \times 10^5$	$B(\omega \to e^+e^-) \times 10^5$	$B(\varphi \to e^+e^-) \times 10^5$
Dubna	5,1±1,0	—	66^{+44}_{-28}
CERN	—	4,0±1,5	61±26
DESY-MIT [66]	6,4±1,5	—	29±8
Novosibirsk	5,0±1,0	—	34±4
Orsay	6,4±0,8	7,7±1,4	37,3±2,5
Rutherford laboratory	—	21^{+39}_{-16}	72±39
$B(V \to e^+e^-)$, average	5,7±0,5	6,0±1,0	36,3±2,0
$\Gamma(V \to e^+e^-)$, average, keV	7,1±0,6	0,76±0,13	1,41±0,078
$f_V^2/4\pi$, average	1,92±0,018	18,4±3,1	12,8±0,7

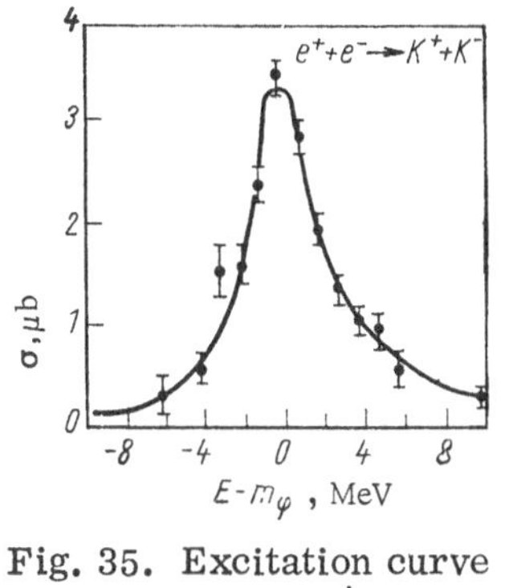

Fig. 35. Excitation curve for the reaction $e^+ + e^- \to K^+ + K^-$.

DISCUSSION OF RESULTS

1. Relative Probabilities, Partial Widths, and Coupling Constants for $V \to e^+e^-$ Decays

The results obtained in six laboratories are summarized in Table 15. The relative probabilities for $V \to e^+e^-$ decays measured by the various methods lie within a standard deviation. The average values for $B(V \to e^+e^-)$, $\Gamma(V \to e^+e^-)$, and $f_V^2/4\pi$ were calculated on the basis of the results in Table 15.

2. $\omega - \varphi$ Mixing

As we mentioned in the Introduction, the problem of $\omega - \varphi$ mixing is one of the fundamental problems of the physics of elementary particles. Its importance stems largely from the associated possibility of determining the nature of an interaction violating SU(3) symmetry. In addition to the classical mixing angle θ introduced by Sakurai [9], the literature now reveals the angles θ_Y and θ_N and the generalized mixing angle θ_α [8, 9,

TABLE 16

Experimental data	θ_Y, deg	θ_N, deg	θ_a, deg	Kf_{ρ}^{-2}	Kf_{ω}^{-2}	Kf_{φ}^{-2}	Kf_{Y}^{-2}	Ref.
Orsay	41,6±±3,0	29^{+4}_{-5}	31,2±±3	9	1,20±±0,24	1,56±±0,15	11,1±±1,2	—
SU(6)	33,3	33,3	—	9	1,0	2,00	12,00	—
Current mixing	33	21	26,5	9	0,79	1,89	10,8	[8]
Mass mixing	39	39	—	9	0,72	1,11	7,3	[8]
	47,2	32,7	39,8	9	1,17	1,0	8,7	[58]
	35	22,5	28,2	9	0,65	1,33	7,9	[59]
	—	—	—	9	0,97	1,14	8,5	[60]
	40	26,3	32,8	—	—	—	—	[62]
	—	—	—	9	1,24	1,76	—	[63]

54-59]. The angles θ_Y and θ_N and the constants f_Y and f_N relate the currents corresponding to the $\omega_\mu(x)$ and $\varphi_\mu(x)$ fields to the baryon (N) and hypercharge (Y) currents. The angles θ_Y and θ_N are introduced on the basis of the following arguments. Since φ_1 and φ_8 have the same quantum numbers, the following transition is possible:

$$\varphi_8 \rightleftarrows \varphi_1. \tag{36}$$

This transition between two states is possible because it does not violate any conservation laws except for SU(3) symmetry. Condition (36) can be used to calculate the inverse of the propagator for the mixed system:

$$D = AK^2 + BM^2, \tag{37}$$

where A and B are two-row matrices, K is the square of the momentum, and M is the mass of the two states.

The mixing effect is manifested in two ways [8, 57]: 1) mass mixing; 2) current mixing. In the former case it is assumed that condition (36) is inconsistent with diagonality of mixing B_0 (D_0, A_0, and B_0 are the values of the matrices D, A, and B before mixing). The matrix A_0 remains diagonal. Current mixing disrupts diagonality of matrix A_0, while B_0 remains diagonal. In this case mixing is described on the basis of four parameters expressed in terms of the coupling constants f_Y and f_N and the mixing angles θ_Y and θ_N. Matrices A and B_0 are symmetric (because of T invariance). Accordingly, the mixing angles θ_Y and θ_N can be related by [8]

$$\frac{\mathrm{tg}\,\theta_Y}{\mathrm{tg}\,\theta_N} = \frac{m_\varphi^2}{m_\omega^2} \tag{38}$$

or

$$\frac{m_\omega}{m_\varphi}\,\mathrm{tg}\,\theta_Y = \frac{m_\varphi}{m_\omega}\,\mathrm{tg}\,\theta_N = \mathrm{tg}\,\theta_a. \tag{39}$$

Relation (39) holds for current mixing; for mass mixing, on the other hand, we have

$$\theta_a = \theta_Y = \theta_N. \tag{40}$$

The coupling constants f_ω and f_φ are expressed in terms of f_Y and mixing angle θ_Y by

$$\left.\begin{aligned} f_\varphi &= 2f_Y\,[\cos\theta_Y]^{-1}; \\ f_\omega &= 2f_Y\,[\sin\theta_Y]^{-1} \end{aligned}\right\} \tag{41}$$

and these constants can be calculated from

$$\frac{f_Y^2}{4\pi} = \frac{\alpha^2}{12}\left[\frac{\Gamma(\omega\to e^+e^-)}{m_\omega} + \frac{\Gamma(\varphi\to e^+e^-)}{m_\varphi}\right]^{-1}. \tag{42}$$

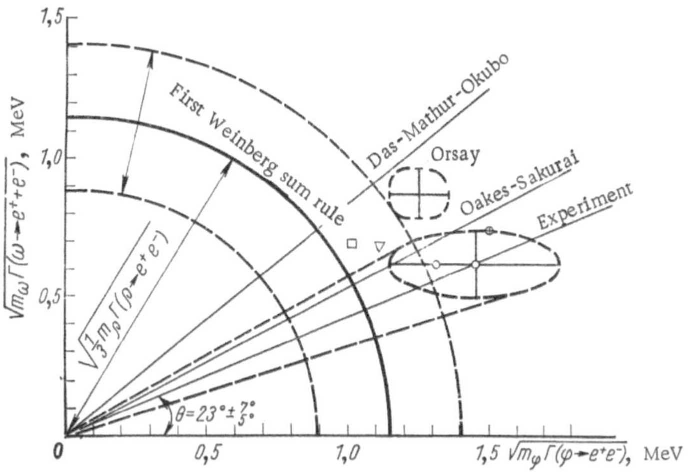

Fig. 36. Comparison of the experimental data for $V \to e^+e^-$ decays with the first Weinberg sum rule and various theoretical models. □) Quark model; ∇) mass-mixing model ($\theta_Y = \theta_N = 39°$); ○) mass-mixing model ($\theta_Y = \theta_N = 32°$); ⊕) current-mixing model ($\theta_Y = 33°$; $\theta_N = 21°$).

The relations between θ_Y and θ_N and the partial widths $\Gamma(\omega \to e^+e^-)$ and $\Gamma(\varphi \to e^+e^-)$ can be found from (43) and (44):

$$\mathrm{tg}^2\,\theta_Y = \frac{m_\varphi}{m_\omega} \cdot \frac{\Gamma(\omega \to e^+e^-)}{\Gamma(\varphi \to e^+e^-)}; \tag{43}$$

$$(1 + \mathrm{tg}\,\theta_Y \cdot \mathrm{tg}\,\theta_N)^2 = \frac{\alpha^2}{144} \cdot \frac{\beta_k^2 m_\varphi^2}{\Gamma(\varphi \to e^+e^-) \cdot \Gamma(\varphi \to K\overline{K})}. \tag{44*}$$

The experimental coupling constants in mixing angles found at Orsay are compared in Table 16 with the theoretical predictions [8, 58-60].

The CERN result for the mixing angle is $\theta_a = 23^{+7}_{-5}$. For completely constructive or completely destructive $\omega - \rho$ interference, θ_a changes by ±3°, i.e., it lies between 20° and 26°.

3. Spectral Weinberg Sum Rules

The first Weinberg sum rule [64], associated with the current-mixing model, yields a relation for the width $\Gamma(V \to e^+e^-)$ of lepton decays of V mesons:

$$\Sigma_1 = \frac{1}{3} m_\rho \Gamma(\rho \to e^+e^-) - m_\omega \Gamma(\omega \to e^+e^-) - m_\varphi \Gamma(\varphi \to e^+e^-) \approx 0. \tag{45}$$

The experimental results found at Orsay yield

$$\Sigma_1 = (-0{,}44 \pm 0{,}24)\ (\mathrm{MeV}/c)^2. \tag{46}$$

Another form of relation (45), which takes into account the violation of SU(3) symmetry [61] is

$$\Sigma_2 = \frac{1}{3}\,\frac{4m_{K^*}^2 - m_\rho^2}{3m_\rho} \Gamma(\rho \to e^+e^-) - m_\omega \Gamma(\omega \to e^+e^-) - m_\varphi \Gamma(\varphi \to e^+e^-) = 0; \tag{47}$$

$$\Sigma_2 = (+0{,}39 \pm 0{,}27)\ (\mathrm{MeV}/c)^2. \tag{48}$$

---

* Equation (44) holds in the approximation of the vector-dominance model.

TABLE 17

Curve	Confidence level, %	$[\Gamma(\omega \to \pi^+\pi^-)]^{1/2}$, $MeV^{1/2}$	α, deg	m_ρ, MeV	Γ_ρ, MeV
1	15	0	0	769,5±5	120,5±3
2	55	0,63±0,23	197±27	773,6±53	111±5,3

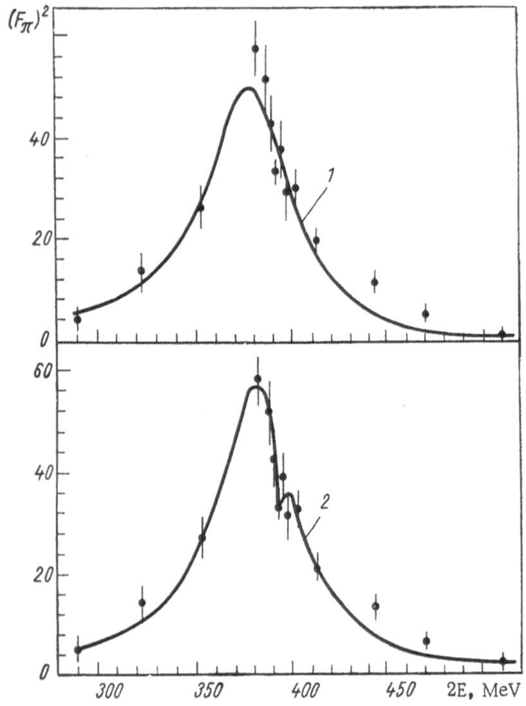

Fig. 37. Form factor for the π meson and $\rho-\omega$ interference. 1) Obtained under the assumption of no $\rho-\omega$ interference; 2) under the assumption of $\rho-\omega$ interference.

The two results agree, within the experimental errors, with relations (45) and (47). The agreement between the experimental results for $V \to e^+e^-$ decays and the first Weinberg sum rule and the various theoretical models is illustrated in Fig. 36. The radii of the circles in Fig. 36 are $1/3m_\rho\Gamma(\rho \to e^+e^-)$. The boundary values are governed by the experimental errors.

4. $\omega - \rho$ Interference

The Orsay results show that $\omega-\rho$ interference may occur. The results of the analysis are shown in Fig. 37; curve 1 was calculated from Eq. (32) with an account of only the π meson form factor, while curve 2 was calculated from

$$|F_\pi(s)|^2 = |F_{\rho\pi}(s) + \xi e^{i\alpha} F_{\omega\pi}(s)|^2, \qquad (49)$$

where $F_{\pi\rho}(s)$ and $F_{\omega\pi}(s)$ are the contributions of the ρ and φ mesons to the π meson form factor. In Eq. (49), the quantity $\xi e^{i\alpha}$ takes into account the coupling between the ω and $\pi^+\pi^-$ mesons in the final state; ξ is given in terms of the partial width $\Gamma(\omega \to \pi^+\pi^-)$ by

$$\xi = \frac{6}{\alpha} \frac{[\Gamma(\omega \to \pi^+\pi^-)\,\Gamma(\omega \to e^+e^-)]^{1/2}}{\beta_\pi^{3/2} m_\omega}. \qquad (50)$$

The parameters used to calculate the curves are shown in Table 17.

These results apparently constitute evidence for $\omega-\rho$ mixing with a quite large value for the partial width of $\omega \to \pi^+\pi^-$ decays. However the possibility is not ruled out that the observed effect is a result of a statistical fluctuation, in view of the large experimental errors.

5. Basic Results and Conclusions

1. The occurrence of $\rho^0 \to e^+e^-$, $\omega \to e^+e^-$, and $\varphi \to e^+e^-$ decays has been demonstrated.

2. Measurements have been made of the relative probabilities for the $V \to e^+e^-$ decays, and the coupling constants between the γ rays and the vector mesons have been calculated.

3. The hypothesis of $\omega-\varphi$ mixing has been supported experimentally.

4. The experimental data for $\Gamma(V \to e^+e^-)$ agree with the first Weinberg sum rule.

5. The mass of the ρ^0 meson and the total width of the ρ, ω, and φ mesons have been measured on the basis of their lepton decays.

In conclusion we should point out that the large experimental errors unfortunately rule out a definite choice of model for $\omega-\varphi$ mixing. The $\omega \to e^+e^-$ partial width has been measured less accurately (within ~20%). In view of the considerable interest in this problem, future experiments will apparently resolve this problem.

Another point which should be taken into account in future experiments is that a search for heavy vector particles with a mass greater than that of the φ meson is of fundamental importance for the theory. In this connection there is considerable interest in the predictions based on the Veneziano model [65] regarding the existence of vector mesons having masses of 1.3 GeV (φ'), 1.7 GeV (ρ''), 2.0 GeV (ρ'''), etc. However, since $1/f_{\rho'}^2$ and $1/f_{\rho''}^2$ are much smaller than $1/f_\rho^2$, the corresponding experiments will require the development of new and efficient experimental methods.

LITERATURE CITED

1. C. Yang and R. Mills, Phys. Rev., 96, 191 (1954).
2. I. Sakurai, Ann. Phys., 11, 1 (1960).
3. R. Uttijama, Phys. Rev., 101, 1597 (1956).
4. Y. Nambu, Phys. Rev., 106, 1366 (1957).
5. W. Frazer and J. Fulko, Phys. Rev. Letters, 2, 365 (1959).
6. M. Gell-Mann and F. Zachariasen, Phys. Rev., 124, 953 (1961).
7. M. Krammer and D. Schildknecht, DESY Preprint 68/33 (1968).
8. N. Kroll, T. Lee, and B. Zumino, Phys. Rev., 157, 1376 (1967).
9. I. Sakurai, Phys. Rev. Letters, 9, 472 (1962).
10. V. Barmin et al., Zh. Éksp. Teor. Fiz., 45, No. 6 (12), 2082 (1963).
11. J. Murray et al., Phys. Letters, 7, 358 (1963).
12. A. Bezaguet et al., Phys. Letters, 12, 70 (1964).
13. R. Zdanis et al., Phys. Rev. Letters, 14, 721 (1965).
14. A. Barbaro-Galtieri et al., UCRL-11428 (1964).
15. I. Nambu and I. Sakurai, Phys. Rev. Letters, 8, 79 (1962).
16. M. Gell-Mann, Phys. Rev. Letters, 8, 261 (1962).
17. R. Dachen and D. Sharp, Phys. Rev., 133, B1585 (1964).
18. Nguyen Van Hieu and Fang K'ung Ty, Preprint R-2338, JINR (1965).
19. M. N. Khachaturyan et al., Author's certificate No. 182249 (1966); M. A. Azimov et al., Preprint B-7-2070, JINR (1964).
20. M. A. Azimov et al., Nucl. Instrum. and Meth., 51, 309 (1967).
21. ABBHLM Collaboration, Nuovo Cimento, 31, 729 (1964).
22. I. Derado et al., Phys. Rev. Letters, 14, 872 (1965).
23. Eisner et al., Phys. Rev., 164, 1699 (1967).
24. Y. Lee et al., Phys. Rev., 159, 1156 (1967).
25. T. Bacon et al., Phys. Rev., 157, 1263 (1967).
26. J. Allard et al., Nuovo Cimento, 50, 106 (1967).
27. H. Cohn et al., Phys. Letters, 15, 344 (1965).
28. G. Berson, Thesis, University of Michigan (1966).
29. M. A. Azimov, M. N. Khachaturyan, et al., Yad. Fiz., 3, No. 6, 515 (1967); M. A. Azimov et al., Preprint E1-3148, JINR (1967).
30. M. N. Khachaturyan et al., Phys. Letters, 24B, 349 (1967).
31. R. G. Astvacaturov et al., Phys. Letters, 27B, 45 (1968).
32. D. Bollini et al., Nuovo Cimento, 56A, 404 (1968).
33. D. Bollini et al., Nuovo Cimento, 56A, 1173 (1968).
34. T. Bacon et al., Phys. Rev., 157, 1263 (1967).
35. M. Abolins et al., Proceedings of the Conference on Elementary Particles, Heidelberg (1967), p. 509.
36. A. Baracca, Submitted to Nuovo Cimento.
36. E. Ehlotzky and H. Mitter, Nuovo Cimento, 55A, 181 (1968).
37. A. Rosenfeld et al., Rev. Mod. Phys., 40, 77 (1968).
38. D. Binnie et al., Phys. Letters, 18, 348 (1965).
39. N. Kroll and W. Wada, Phys. Rev., 95, 1360 (1954).
40. D. Binnie et al., Phys. Letters, 27B, 106 (1968).
41. A. Rosenfeld et al., UCRL-8030 (September, 1967).
42. V. Auslander et al., Phys. Letters, 25B, 433 (1967).
43. B. Baier and B. Fadin, Phys. Letters, 27B, 223 (1968).
44. V. E. Balakin et al., Preprint 327, Institute of Nuclear Physics, Siberian Branch, Academy of Sciences of the USSR (1969).

45. J. Augustin et al., Phys. Rev. Letters, 20, 126 (1968).
46. J. Augustin et al., Phys. Letters, 28B, 508 (1969).
47. J. Augustin et al., Phys. Letters, 28B, 513 (1969).
48. J. Augustin et al., Phys. Letters, 28B, 517 (1969).
49. J. Perez-Jorba, LAL 1222 (September, 1969).
50. S. Tavernier, RI-68/7, LAL, Orsay.
51. G. Altarelli et al., Nuovo Cimento, 47, 113 (1967).
52. A. Sokolov and I. Ternov, Sov. Phys. – Dokl., 8, 1203 (1964).
53. J. Perez-Jorba, Raport Technique A.C.O. 15-67.
54. I. Sakarai, Phys. Rev., 132, 434 (1963).
55. S. Glashow, Phys. Rev. Letters, 11, 48 (1963).
56. S. Okubo, Phys. Rev. Letters, 5, 165 (1963).
57. S. Coleman and H. Schnitzer, Phys. Rev., 134B, 863 (1964).
58. T. Das, V. Mathur, and S. Okubo, Phys. Rev. Letters, 19, 470 (1967).
59. R. Oakes and I. Sakurai, Phys. Rev. Letters, 19, 1266 (1967).
60. A. Dar and V. Weisskopf, Phys. Letters, 26B, 670 (1968).
61. H. Sugawara, Phys. Rev. Letters, 21, 772 (1968).
62. Glavelli and R. Torgenson, Submitted to Phys. Rev.
63. S. B. Gerasimov, Preprint P2-4522, JINR (1969).
64. S. Weinberg, Phys. Rev. Letters, 18, 507 (1967).
65. G. Veneziano, Nuovo Cimento, 57A, 190 (1968).
66. Problemy Fiziki Élementarnykh Chastits i Atomnogo Yadra, 1, No. 1 (1970).

THREE-DIMENSIONAL FORMULATION OF THE RELATIVISTIC TWO-BODY PROBLEM

V. G. Kadyshevskii, R. M. Mir-Kasimov, and N. B. Skachkov

A quasipotential approach to the relativistic two-body problem is developed on the basis of the three-dimensional Hamiltonian formulation of quantum field theory. A relativistic configuration representation is introduced and studied on the basis of a decomposition with respect to matrix elements of irreducible unitary representations of the Lorentz group. A representation is obtained for the quasipotential scattering amplitude at high energies that generalizes the nonrelativistic eikonal formula.

INTRODUCTION

The description of a two-particle relativistic system is one of the central problems of quantum field theory. The completely covariant Bethe–Salpeter equation [1-3] was introduced to study this problem in the four-dimensional Feynman–Dyson formalism. Although it has some advantages, this equation is not entirely satisfactory. In particular, in the framework of the four-dimensional Bethe–Salpeter approach there is no clear physical interpretation of the dependence of the wave function on the relative time of the two particles.

On the other hand, the relativistic two-body problem was investigated in a three-dimensional approach [4-6] (the Tamm–Dancoff method) before the appearance of the covariant formalism of field theory. It is natural to ask whether covariant field theory allows a formalism that, being three-dimensional and admitting a probability interpretation of the wave function, has the same fundamental advantages (renormalizability, analyticity, etc.) as the completely covariant theory.

This question was answered in the affirmative some years ago by the quasipotential approach to the problem of the interaction of two relativistic particles developed by Logunov and Tavkhelidze [7].

In this approach the wave function appears as a direct generalization of a nonrelativistic wave function since it depends on a single time argument and satisfies a Schrödinger-type equation. For spinless particles of equal masses m, this equation can be written in the momentum representation in the form (all the momenta are referred to the center-of-mass system):

$$(\mathbf{p}^2-\mathbf{q}^2)\,\Psi_q(\mathbf{p})=\frac{1}{4\,(2\pi)^3}\,\frac{1}{\sqrt{m^2+\mathbf{p}^2}}\int V(\mathbf{p},\ \mathbf{k},\ E_q)\,\Psi_q(\mathbf{k})\,d\mathbf{k}. \tag{0.1}$$

The quasipotential $V(\mathbf{p}, \mathbf{k}; E_q)$, which is complex in the general case, can be constructed [7-13] by perturbation theory from either the two-time Green's function of the system or the scattering amplitude on the mass shell. Besides Eq. (0.1) for the wave function, it is also necessary to investigate the Lippmann–Schwinger equation [7-13] for the invariant scattering amplitude off the mass shell:

Joint Institute for Nuclear Research, Dubna. Translated from Problemy Fiziki Élementarnykh Chastits i Atomnogo Yadra, Vol. 2, No. 3, pp. 635-690, 1972.

$$T(\mathbf{p}, \mathbf{q}) = V(\mathbf{p}, \mathbf{q}; E_q) + \frac{1}{4(2\pi)^3} \int V(\mathbf{p}, \mathbf{k}; E_q) \frac{d\mathbf{k}}{\sqrt{m^2+\mathbf{k}^2}} \cdot \frac{T(\mathbf{k}, \mathbf{q})}{k^2 - q^2 - i\varepsilon}. \tag{0.2}$$

Recently, a modification of the quasipotential approach has been proposed [14, 15]. It is quite unrelated to the Bethe–Salpeter formalism and the Feynman–Dyson covariant formalism; instead, it employs the Hamiltonian formulation of quantum field theory [16-18]. This formulation is inherently three-dimensional, and all the particles, even in intermediate states, are physical, i.e., they are on the mass shells. It is therefore a natural point of departure for deriving equations that describe two-particle relativistic systems. It has been found that the equations obtained in the Hamiltonian field theory possess a further important property: they can, in principle, be cast in the same form as the corresponding nonrelativistic Schrödinger and Lippmann–Schwinger equations. Indeed, there is only one difference between the two cases – in the latter, the integration is over the three-dimensional Euclidean momentum space, whereas in the relativistic case all the integrations are over a three-dimensional Lobachevskii momentum space. This space is realized on the upper sheet of the hyperboloid $k_0^2 - \mathbf{k}^2 = m^2$, i.e., on the mass shell of one relativistic particle.

In this connection, a very fruitful approach has been based on a detailed study of the properties of the three-dimensional relativistic equations associated with the Lorentz group, which plays the role of the group of motions of the Lobachevskii space. This has made it possible to construct an adequate formulation of the quasipotential theory in the configuration representation [19-24].

SECTION 1. COVARIANT FORMULATION OF RELATIVISTIC HAMILTONIAN THEORY

1. Equation of Motion for the Operator $R(\lambda\varkappa)$ in the Momentum Representation

Let

$$S(\infty, -\infty) = T\exp\left\{-i\int H(x)\,d^4x\right\} \tag{1.1}$$

be the total S matrix corresponding to the interaction Hamiltonian H(x). As an example, we shall consider a nucleon–meson interaction of the form*

$$H(x) = -g : \bar{\psi}(x)\gamma_5\psi(x)\varphi(x): \tag{1.2}$$

[$\psi(x)$ is the field of nucleons and antinucleons with mass M; $\varphi(x)$ is the field of pseudoscalar neutral mesons with mass m].

The matrix $S(\infty, -\infty)$ can be understood as a limit of the form

$$S(\infty, -\infty) = \lim_{\sigma\to\infty} S(\sigma, -\infty), \tag{1.3}$$

where σ is an arbitrary space-like surface [in Eq. (1.3) it is shifted to infinitely large values of the time]. As is well known, $S(\sigma, -\infty)$ satisfies the Tomonaga–Schwinger equation:

$$i\frac{\delta S(\sigma, -\infty)}{\delta\sigma(x)} = H(x)\,S(\sigma, -\infty). \tag{1.4}$$

As the surfaces σ we shall take only space-like planes described by the equation

$$\lambda x \equiv \lambda_0 x_0 - \boldsymbol{\lambda}\mathbf{x} = \sigma. \tag{1.5}$$

Here $\lambda = (\lambda_0, \boldsymbol{\lambda})$ is the vector of the normal to the plane satisfying the conditions

$$\lambda^2 = \lambda_0^2 - \boldsymbol{\lambda}^2 = 1, \quad \lambda_0 > 0, \tag{1.6}$$

---

* All the operators are given in the interaction representation.

and σ is the pseudo-Euclidean distance of the plane from the coordinate origin. Then the functional derivative in (1.4) can be written in the form

$$\frac{\delta S(\sigma, -\infty)}{\delta\sigma(x)} = \lim_{\Delta\sigma\to 0} \frac{S(\sigma+\Delta\sigma, -\infty) - S(\sigma, -\infty)}{\int [\vartheta(\sigma+\Delta\sigma-\lambda x) - \vartheta(\sigma-\lambda x)]\, d^4x} = \frac{\partial S(\sigma, -\infty)}{\partial\sigma} \cdot \frac{1}{\int \delta(\sigma-\lambda x)\, d^4x}. \tag{1.7}$$

Then (1.4) is replaced by

$$i\frac{\partial S(\sigma, -\infty)}{\partial\sigma} = \left[\int H(x)\,\delta(\sigma-\lambda x)\, d^4x\right] S(\sigma, -\infty). \tag{1.8}$$

The quantity

$$H(\sigma, \lambda) = \int H(x)\,\delta(\sigma-\lambda x)\, d^4x \tag{1.9}$$

is related to the four-dimensional Fourier transform of this operator,

$$\tilde{H}(p) = \int e^{-ipx} H(x)\, d^4x, \tag{1.10}$$

by the equation

$$H(\sigma, \lambda) = \frac{1}{2\pi}\int e^{-i\sigma\varkappa}\tilde{H}(\lambda\varkappa)\, d\varkappa, \tag{1.11}$$

where $\varkappa$ is a one-dimensional invariant parameter.

The solution of Eq. (1.11) is a T exponential function in which, instead of purely time ϑ functions, invariant functions of the following form are used:

$$\vartheta(\lambda(x-y)) = \frac{1}{2\pi i}\int_{-\infty}^{\infty} e^{i\varkappa\lambda(x-y)}\frac{d\varkappa}{\varkappa - i\varepsilon}. \tag{1.12}$$

The total S matrix $S(\infty, -\infty)$ does not depend on the direction of the vector λ, because for time-like intervals $(x-y)^2 > 0$ we always have $\vartheta(\lambda(x-y)) = \vartheta(x^0 - y^0) = \text{inv}$, and for $(x-y)^2 < 0$ the function $\vartheta(\lambda(x-y))$ gives no contribution because of the "locality" of the Hamiltonian H(x):

$$[H(x), H(y)] = 0, \quad \text{if} \quad (x-y)^2 < 0. \tag{1.13}$$

We now set

$$S(\sigma, -\infty) = 1 + \frac{1}{2\pi}\int_{-\infty}^{\infty}\frac{R(\lambda\varkappa)\, e^{i\varkappa\sigma}}{\varkappa - i\varepsilon}\, d\varkappa \tag{1.14}$$

and substitute (1.10) and (1.14) into (1.8). We obtain an integral equation [16]:

$$R(\lambda\varkappa) = -\tilde{H}(\lambda\varkappa) - \frac{1}{2\pi}\int \tilde{H}(\lambda\varkappa - \lambda\varkappa_1)\frac{d\varkappa_1}{\varkappa_1 - i\varepsilon} R(\lambda\varkappa_1). \tag{1.15}$$

Since

$$\lim_{\sigma\to\infty}\frac{1}{2\pi i}\frac{e^{i\varkappa\sigma}}{\varkappa - i\varepsilon} = \delta(\varkappa),$$

it follows from (1.14) that

$$S(\infty, -\infty) = 1 + iR(0). \tag{1.16}$$

Thus, to find the S matrix it is sufficient to solve Eq. (1.15) and then set $\varkappa = 0$. At the same time, R(0) is a completely invariant quantity that does not depend on the direction of the vector λ. Consequently, if $\varkappa \neq 0$, this vector can be taken collinear to any time-like vector encountered in a given problem. Every such choice corresponds to a definite extrapolation of the matrix elements off the surface $\varkappa = 0$. We shall refer to the latter as the energy-momentum surface.

In particular, for a system of two interacting particles with four-momenta p_1 and p_2 we can set

$$\lambda_n \sim \mathcal{P}_n, \tag{1.17}$$

where $\mathcal{P} = p_1 + p_2$.

In the following constructions we shall frequently employ an equation of a more general form than (1.15):

$$R(\lambda\varkappa, \lambda\varkappa') = -\widetilde{H}(\lambda\varkappa - \lambda\varkappa') - \frac{1}{2\pi}\int \widetilde{H}(\lambda\varkappa - \lambda\varkappa_1)\frac{d\varkappa_1}{\varkappa_1 - i\varepsilon} R(\lambda\varkappa_1, \lambda\varkappa'), \tag{1.18}$$

assuming by definition that

$$\left.\begin{aligned} R(\lambda\varkappa) &= R(\lambda\varkappa, 0); \\ S(\infty, -\infty) &= 1 + iR(0, 0). \end{aligned}\right\} \tag{1.19}$$

In contrast to (1.15), Eq. (1.18) is not related in a simple manner to the Tomonaga–Schwinger equation (1.14).

2. Diagram Technique

Using perturbation theory* [17], let us investigate Eq. (1.18) for the operators $P(\lambda\varkappa, \lambda\varkappa')$ with the Hamiltonian (1.2). We introduce standard Fourier expansions of the operators $\bar\psi$, ψ, and φ in (1.2)†:

$$\begin{aligned}
\psi(x) &= \frac{1}{(2\pi)^{3/2}}\int e^{iqx}\psi(q)\,dq = \frac{1}{(2\pi)^{3/2}}\sum_{\nu=1,2}\int\frac{dq}{\sqrt{2q_0}}e^{iqx}b_\nu^+(\mathbf{q})v^\nu(\mathbf{q}) \\
&+ \frac{1}{(2\pi)^{3/2}}\sum_{\nu=1,2}\int\frac{dq}{\sqrt{2q_0}}e^{-iqx}a_\nu(\mathbf{q})u^\nu(\mathbf{q}) \equiv \psi^{(+)}(x) + \psi^{(-)}(x); \\
\bar\psi(x) = \psi^+(x)\gamma^0 &= \frac{1}{(2\pi)^{3/2}}\int e^{ipx}\psi^+(-p)\gamma^0\,dp = \frac{1}{(2\pi)^{3/2}}\sum_{\nu=1,2}\int\frac{dp}{\sqrt{2p_0}}e^{ipx}a_\nu^+(\mathbf{p})\bar u^\nu(\mathbf{p}) \\
&+ \frac{1}{(2\pi)^{3/2}}\sum_{\nu=1,2}\int\frac{dp}{\sqrt{2p_0}}e^{-ipx}b_\nu(\mathbf{p})\bar v^\nu(\mathbf{p}) \equiv \bar\psi^{(+)}(x) + \bar\psi^{(+)}(x); \\
\varphi(x) &= \frac{1}{(2\pi)^{3/2}}\int\varphi(k)e^{ikx}d^4k = \frac{1}{(2\pi)^{3/2}}\int e^{ikx}\frac{dk}{\sqrt{2k_0}}\alpha^{(+)}(\mathbf{k}) + \frac{1}{(2\pi)^{3/2}}\int e^{-ikx}\frac{dk}{\sqrt{2k_0}}\alpha(\mathbf{k}) \\
&\equiv \varphi^{(+)}(x) + \varphi^{(-)}(x).
\end{aligned} \tag{1.20}$$

We adopt the rules of graphical representation of particles in the initial and final states given in Table 1.

It follows from Eq. (1.18) that in the first approximation in the coupling constant the operator $R(\lambda\varkappa, \lambda\varkappa')$ is given by

---

* We recall that the corresponding results for the operator $R(\lambda\varkappa)$ can be obtained by setting $\varkappa' = 0$ [see Eq. (1.19)].

† The conditions of orthonormality and completeness for the spinors u and v in (1.20) have the invariant form

$$\sum_{\alpha=1}^{4}\bar u_\alpha^\mu(\mathbf{q})u_\alpha^\nu(\mathbf{q}) = -\sum_{\alpha=1}^{4}\bar v_\alpha^\mu(\mathbf{q})v_\alpha^\nu(\mathbf{q}) = 2M\delta^{\mu\nu}; \tag{1.21}$$

$$\begin{aligned}
&\sum_{\mu=1,2}\bar u_\alpha^\mu(\mathbf{q})u_\beta^\mu(\mathbf{q}) = (\hat q + M)_{\beta\alpha}; \\
&\sum_{\mu=1,2}v_\alpha^\mu(\mathbf{q})\bar v_\beta^\mu(\mathbf{q}) = (\hat q - M)_{\alpha\beta} \\
&(\hat q = q_0\gamma_0 - \mathbf{q}\boldsymbol{\gamma})
\end{aligned} \tag{1.22}$$

TABLE 1

Line	Particle	State	Factor in the matrix element
	Nucleon	in	$\frac{(2\pi)^{3/2}}{\sqrt{2q_0}}\, u^{\nu}(q)$
	Antinucleon	in	$(2\pi)^{3/2}/\sqrt{2q_0}\,\bar{v}^{\nu}(q)$
	Meson	in	$\frac{(2\pi)^{3/2}}{\sqrt{2k_0}}$
	Nucleon	out	$\frac{(2\pi)^{3/2}}{\sqrt{2q_0}}\, \bar{u}^{\mu}(p)$
	Antinucleon	out	$\frac{(2\pi)^{3/2}}{\sqrt{2q_0}}\, v^{\mu}(p)$
	Meson	out	$\frac{(2\pi)^{3/2}}{\sqrt{2k_0}}$

$$R_1(\lambda\varkappa,\ \lambda\varkappa') = -\tilde{H}(\lambda\varkappa - \lambda\varkappa') = g\int e^{-i\lambda(\varkappa-\varkappa')x} : \bar{\psi}(x)\,\gamma_5\,\psi(x)\,\varphi(x) : d^4x, \tag{1.23}$$

or, with allowance for (1.20),

$$R_1(\lambda\varkappa,\ \lambda\varkappa') = \frac{g}{\sqrt{2\pi}}\int \delta(\lambda\varkappa - \lambda\varkappa' - p - q - k) \\ : \bar{\psi}(p)\,\gamma_5\,\psi(q)\,\varphi(k) : dp\,dq\,dk. \tag{1.24}$$

One can readily obtain an expression for R_1 as a sum of normal products of the operators $\psi^{(\pm)}(x)$, $\bar{\psi}^{(\pm)}(x)$, $\varphi^{(\pm)}(x)$. For example, the operator

$$-g\int e^{-i\lambda(\varkappa-\varkappa')x}\,d^4x : \bar{\psi}^{(+)}(x)\,\gamma_5\,\psi^{(+)}(x)\,\varphi^{(+)}(x) : \ = \int e^{-i\lambda(\varkappa-\varkappa')x} H_1(x)\,d^4x$$

has a nonvanishing matrix element for the transition vacuum → nucleon + antinucleon + meson:

$$\left\langle p_1\,\mu_1;\ p_2,\ \mu_2;\ k \left| \int e^{-i\lambda(\varkappa-\varkappa')x} H_1(x)\,dx \right| 0 \right\rangle \\ = -(2\pi)^{9/2}\frac{g}{\sqrt{2\pi}}\,\delta(\lambda\varkappa - \lambda\varkappa' - p_1 - p_2 - k)\,\frac{\bar{u}^{\mu}(p_1)\,\gamma_5\,v^{\mu_2}(p_2)}{\sqrt{2p_{10}\,2p_{20}\,2k_0}}. \tag{1.25}$$

The process (1.25) can be associated with the diagram

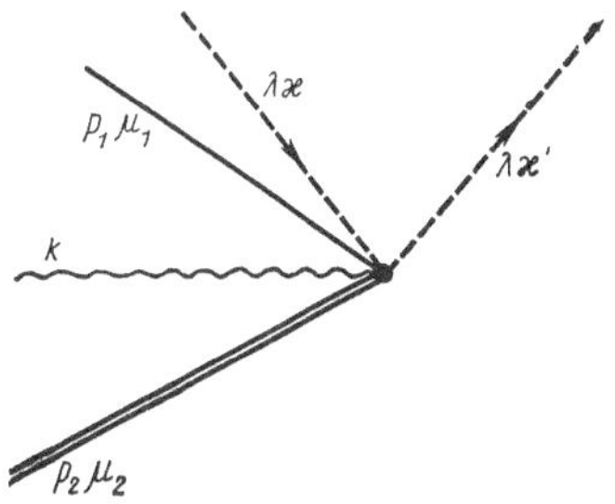

The dashed spurion lines – quasiparticles – with four-momenta $\lambda\varkappa$ and $\lambda\varkappa'$ are introduced to ensure that the conservation law $\lambda\varkappa - \lambda\varkappa' - p_1 - p_2 - k = 0$ is satisfied at the vertex.*

* It will become clear in this subsection why the surface $\varkappa = \varkappa' = 0$ may be called the energy-momentum surface.

TABLE 2

Line	Particle	Pairing	Factor in the matrix element
	Nucleon	$\underbrace{\psi_\beta(q_{j'})\,\overline{\psi}_\alpha(q_j)}_{}{}_{j'<j}$	$S^{(+)}_{\alpha\beta}(p_j,\ M)$
	Anti-nucleon	$\underbrace{\overline{\psi}_\alpha(p_{j'})\,\psi_\beta(q_{j'})}_{}{}_{j'<j}$	$S^{(+)}_{\beta\alpha}(q_j-M)$
	Meson	$\underbrace{\psi(k_{j'})\,\varphi(k_j)}_{}{}_{j'<j}$	$\Delta^{(+)}(k_j)$

In the x representation [(1.23) and (1.25)], plane waves of the form $\exp(-i\lambda\varkappa x)$ correspond to the quasiparticles. Consequently, the operator $\widetilde{H}(\lambda\varkappa-\lambda\varkappa')$ can be interpreted as an interaction of the fields ψ, $\overline{\psi}$, and φ with a plane wave. When Eq. (1.18) is iterated, we obtain operator expressions of the form

$$R_n(\lambda\varkappa,\ \lambda\varkappa')=\frac{(-1)^n}{(2\pi)^{n-1}}\int\widetilde{H}(\lambda\varkappa-\lambda\varkappa_1)\frac{d\varkappa_1}{\varkappa_1-i\varepsilon}\cdots$$
$$\cdots\frac{d\varkappa_{j-1}}{\varkappa_{j-1}-i\varepsilon}\widetilde{H}(\lambda\varkappa_{j-1}-\lambda\varkappa_j)\frac{d\varkappa_j}{\varkappa_j-i\varepsilon}\cdots\frac{d\varkappa_{n-1}}{\varkappa_{n-1}-i\varepsilon}\widetilde{H}(\lambda\varkappa_{n-1}-\lambda\varkappa'). \tag{1.26}$$

We reduce them to normal form, assuming that the Hamiltonians in (1.26) are labeled in such a way that the serial number of $\widetilde{H}(\lambda\varkappa-\lambda\varkappa_1)$ is equal to unity, that of $\widetilde{H}(\lambda\varkappa_1-\lambda\varkappa_2)$ to two, etc. this continuing up to n, which corresponds to the operator $\widetilde{H}(\lambda\varkappa_{n-1}-\lambda\varkappa')$. With each of the operators $\overline{\psi}$, ψ, and φ we associate the serial number of the Hamiltonian $\widetilde{H}$ to which these operators belong. If $\widetilde{H}$ is already given in normal form, only operators with different serial numbers need be paired to reduce $R_n(\lambda\varkappa,\lambda\varkappa')$ to normal form. In the given case, the absence of the chronological product in $R_n(\lambda\varkappa,\lambda\varkappa')$ means that the pairings have the form

$$\underbrace{\psi_\beta(q)\,\overline{\psi}_\alpha(p)}=\delta(q+p)\,\vartheta(p_0)(\hat{p}+M)_{\beta\alpha}\,\delta(p^2-M^2)\equiv\delta(q+p)\,S^{(+)}_{\beta\alpha}(p,\ M); \tag{1.27}$$

$$\underbrace{\overline{\psi}_\alpha(p)\,\psi_\beta(q)}=\delta(p+q)\,\vartheta(q_0)(\hat{q}-M)_{\beta\alpha}\,\delta(q^2-M^2)\equiv\delta(p+q)\,S^{(+)}_{\beta\alpha}(q_1-M); \tag{1.28}$$

$$\underbrace{\varphi(k)\,\varphi(k')}=\delta(k+k')\,\vartheta(k_0')\,\delta(k'^2-m^2)\equiv\delta(k+k')\,\Delta^{(+)}(k'). \tag{1.29}$$

It is readily seen that the arguments of the $S^{(+)}$ and $\Delta^{(+)}$ functions in (1.27)-(1.29) are the four-momenta of operators that stand on the right in the pairings, i.e., have higher serial numbers. This determines the rule for orienting the lines in the graphical representation of the pairings (1.27)-(1.29) (Table 2).

The first pairing in Table 2 is associated with a nucleon and the second with an antinucleon because these pairings correspond exactly to the contribution of the nucleon and antinucleon states to the total system of vectors of the u-th state.

As can be seen from (1.26), the matrix elements beginning with the second order in g also contain the factors

$$g_0(\varkappa_j)=(1/2\pi)\cdot(1/\varkappa_j-i\varepsilon)\ \ (j=1,\ 2,\ldots,\ n-1), \tag{1.30}$$

which correspond to a "virtual" quasiparticle with four-momentum $\lambda\varkappa_j$ emanating from the vertex with number j and entering the vertex with number j + 1. Graphically, such a quasiparticle will be represented as follows:

Now, taking into account Tables 1 and 2, we can formulate general rules for writing down matrix elements in the formalism.

1. Draw the Feynman diagram (or set of diagrams) that corresponds to the process, representing the free nucleon and antinucleon states in accordance with Table 1. Label the vertices in an arbitrary manner and orient each internal line in the direction from higher to lower serial number. Then, without changing

the orientations, replace certain of the unary (nucleon) internal lines by double (antinucleon) lines in such a manner that the nucleon charge is conserved at each vertex of the diagram. Ascribe a certain four-momentum p to each internal line.

2. Join the first vertex to the second, the second to the third, the third to the fourth, etc., by dashed lines oriented in the direction of increasing serial numbers, and ascribe each of them the four-momentum $\lambda \varkappa_j$, where $j = 1, 2, \ldots, n-1$ is the serial number of the vertex from which the given dashed line emanates. In addition, attach to the first vertex an ingoing external dashed line with momentum $\lambda\varkappa$ and to the last vertex (with serial number n) an outgoing dashed line with momentum $\lambda\varkappa'$.

3. With each internal dashed line with four-momentum $\lambda\varkappa_j$ associate the propagator (1.30), and with each internal physical line with momentum p one of the functions $S^+(p, M)$, $S^+(p, -M)$, or $\Delta^+(p)$ (in accordance with Table 2).

4. With each vertex of the diagram associate a factor $g\gamma_5/\sqrt{2\pi}$ * and a four-dimensional δ function that takes into account the conservation of the total four-momentum of the particles and quasiparticles that enter and leave a given vertex.

5. Integrate over infinite limits with respect to all variables $\varkappa_j$ and the independent momenta in the set of vectors p.

6. Repeat the steps indicated in 1°-5° for all n! numberings of the vertices of the given diagram, add the resulting expressions, and multiply the result by δ_p/η, where η is the number of permutations of the external vertices that occur in the diagram in a symmetric manner; δ_p is a known sign factor associated with the parity of the permutation of the external nucleon and antinucleon lines.

We now illustrate the procedure with an example: the scattering of nucleons by antinucleons in the second order in g:

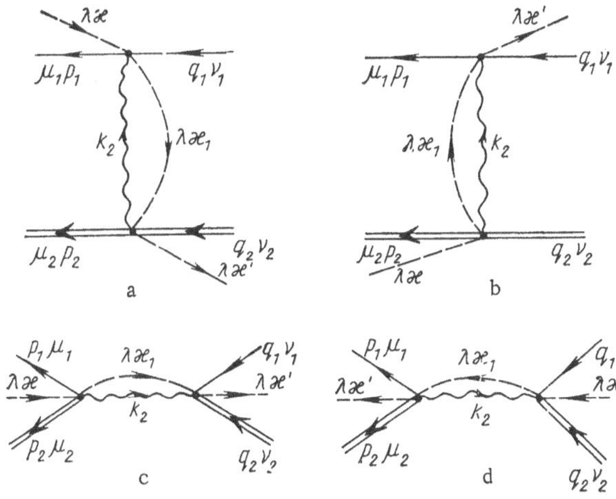

We now take the vector λ as in (1.17)†:

$$\lambda = \frac{p_1 + p_2}{\sqrt{(p_1+p_2)^2}} = \frac{q_1 + q_2}{\sqrt{(q_1+q_2)^2}}. \tag{1.31}$$

---

* All matrices that act on spinor indices must be arranged in the sequence, read from left to right, in which they are encountered if one moves along a spinor line passing through the antinucleon sections in the direction of the orientation and the nucleon sections in the opposite direction.

† It is readily seen that, because of the conservation of the four-momentum, collinearity of the vectors λ and $p_1 + p_2/[(p_1 + p_2)^2]^{1/2}$ automatically leads to collinearity of $p_1 + p_2/[(p_1 + p_2)^2]^{1/2}$ and $q_1 + q_2/[(q_1 + q_2)^2]^{1/2}$. In other words, with our choice, λ, which is the four-velocity of the system, is even conserved off the surface $\varkappa = \varkappa' = 0$.

Then some simple calculations yield

$$(T_2)^{\mu_1\nu_1}_{\mu_2\nu_2}(\varkappa',\ p_1 p_2 \mid q_1,\ q_2,\ \varkappa) = -\frac{g^2}{\sqrt{m^2-t+\frac{1}{4}(\varkappa-\varkappa')^2}}\cdot\frac{1}{\frac{1}{2}(\varkappa+\varkappa')^2+\sqrt{m^2-t+\frac{1}{4}(\varkappa-\varkappa')^2-i\varepsilon}}$$

$$\times\bar{u}^{\mu_1}(\mathbf{p}_1)\gamma_5 u^{\nu_1}(\mathbf{q}_1)\bar{v}^{\nu_2}(\mathbf{q}_2)\gamma_5 v^{\mu_2}(\mathbf{p}_2)$$

$$+\frac{g^2}{2m}\left(\frac{1}{\varkappa'+\sqrt{s_p}+m-i\varepsilon}+\frac{1}{\varkappa-\sqrt{s_p}+m-i\varepsilon}\right)\bar{u}^{\mu_1}(\mathbf{p}_1)\gamma_5 v^{\mu_2}(\mathbf{p}^2)\bar{v}^{\nu_2}(\mathbf{q}_2)\gamma_5 u^{\nu_1}(\mathbf{q}_1), \tag{1.32}$$

where

$$t=(p_1-q_1)^2,\quad s_p=(p_1+p_2)^2,\quad s_q=(q_1+q_2)^2, \tag{1.33}$$

and

$$\varkappa'+\sqrt{s_p}=\varkappa+\sqrt{s_q}. \tag{1.34}$$

It is obvious that the formula gives the same result as the Feynman technique on the energy-momentum surface.

We conclude this section by noting an important property of the formalism. The well-known ultraviolet divergences arise in this formalism only in the one-dimensional invariant integrals with respect to $\varkappa$. One can therefore develop an appropriate technique that makes it possible to eliminate the ultraviolet divergences in a relativistically invariant manner [17].

SECTION 2. EQUATION OF QUASIPOTENTIAL TYPE FOR THE RELATIVISTIC SCATTERING AMPLITUDE

3. Equation of Quasipotential Type for the Relativistic Scattering Amplitude

Using the diagram technique developed in Sec. 2, we obtain an equation analogous to (0.2) for the scattering amplitude. To be specific we shall consider the elastic scattering of nucleons by antinucleons

$$N+\bar{N}\to N+\bar{N} \tag{2.1}$$

in the scheme with the interaction Hamiltonian (1.2).

We shall take the time axis in our diagrams from the right to the left and orient the free ends of the diagrams that describe the process (2.1) off the surface $\varkappa = \varkappa' = 0$ accordingly. We shall say that a connected diagram in this class is irreducible if one cannot distinguish it in two connected subdiagrams that are joined by two spinor lines (nucleon and antinucleon) oriented from right to left and a single dashed line oriented in the opposite direction.

Otherwise the diagram is said to be reducible. Clearly, all connected diagrams can be constructed from irreducible components. We shall use this circumstance in writing down equations for the scattering amplitude (compare this with the corresponding procedure in the Bethe–Salpeter formalism [2, 3]).

Suppose

$$V=\frac{(2\pi)^4\delta(\lambda\varkappa'+p_1+p_2-q_1-q_2-\lambda\varkappa)}{\sqrt{2p_{10}2p_{20}2q_{10}2q_{20}}}V^{\mu_1\nu_1}_{\mu_2\nu_2}(\lambda\varkappa,\ p_1,\ p_2\mid\lambda\varkappa',\ q_1,\ q_2)$$

$$=\frac{(2\pi)^4\delta(\lambda\varkappa'+p_1+p_2-q_1-q_2-\lambda\varkappa)}{\sqrt{2p_{10}2p_{20}2q_{10}2q_{20}}}\bar{u}^{\mu_1}_{\alpha_1}(\mathbf{p}_1)\bar{v}^{\nu_2}_{\alpha_2}(\mathbf{q}_2)V_{\alpha_1\beta_1;\,\alpha_2\beta_2}(\lambda\varkappa,\ p_1,\ p_2\mid\lambda\varkappa',\ q_1,\ q_2)\cdot u^{\nu_1}_{\beta_1}(\mathbf{q}_1)v^{\mu_2}_{\beta_2}(\mathbf{p}_2) \tag{2.2}$$

is the matrix element corresponding to the set of all irreducible diagrams that describe the process (2.1) [the variables and the indices have the same meaning as in the second-order matrix element (1.32)]. Graphically, V can be represented as follows:

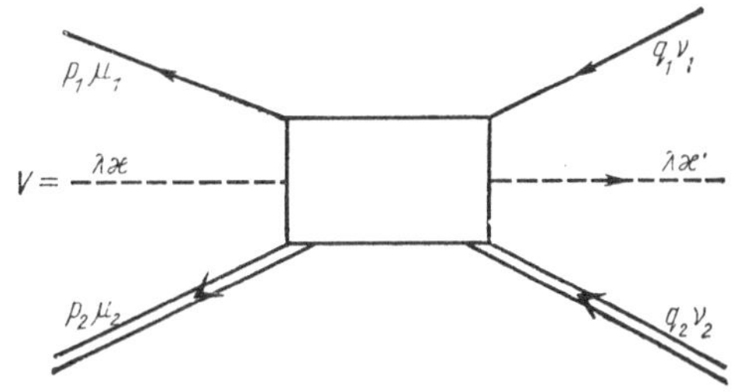

(2.3)

Suppose further that the amplitude T, which we represent graphically in the form

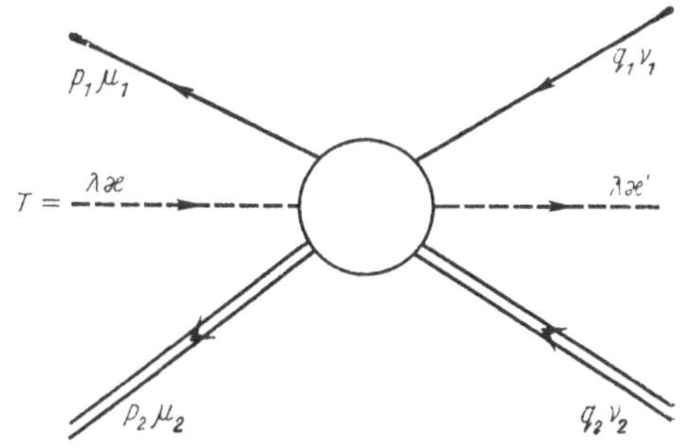

$$= \frac{(2\pi)^4\,\delta(-\lambda\varkappa+p_1+p_2-q_1-q_2+\lambda\varkappa')}{\sqrt{2p_{10}\,2p_{20}\,2q_{10}\,2q_{20}}}\,\bar{u}^{\mu_1}_{\alpha_1}(\mathrm{p}_1)\,\bar{v}^{\mu_2}_{\alpha_2}(\mathrm{q}_2)$$
$$\times T_{\alpha_1\beta_1;\,\alpha_2\beta_2}(\lambda\varkappa;\ p_1,\ p_2\,|\,\lambda\varkappa';\ q_1,\ q_2)\,u^{\nu_1}_{\beta_1}(\mathrm{q}_1)\,u^{\mu_2}_{\beta_2}(\mathrm{p}_2) \tag{2.4}$$

is the $\overline{\mathrm{N}}\mathrm{N}$ scattering amplitude of the surface $\varkappa' = \varkappa = 0$, i.e., the set of all connected diagrams that correspond to the process (2.1). Then, taking into account the definition of irreducibility, we can write the following graphical equation:

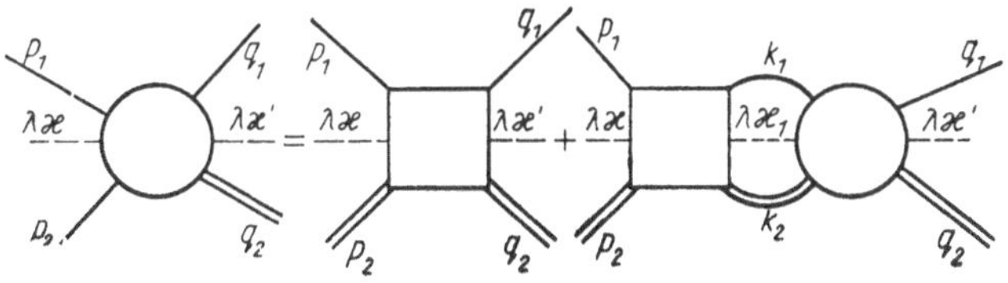

(2.5)

which, in accordance with the rules of our diagram technique (see Sec. 2), is equivalent to the integral equation

$$T^{\mu_1\nu_1}_{\mu_2\nu_2}(\lambda\varkappa,\ p_1,\ p_2\,|\,\lambda\varkappa',\ q_1,\ q_2) = V^{\mu_1\nu_1}_{\mu_2\nu_2}(\lambda\varkappa,\ p_1\,p_2\,|\,\lambda\varkappa',\ q_1,\ q_2)$$
$$+\sum_{\rho_1\rho_2}\int V^{\mu_1\rho_1}_{\mu_2\rho_2}(\lambda\varkappa,\ p_1,\ p_2\,|\,\lambda\varkappa_1,\ k_1,\ k_2)\,\vartheta(k_1^0)\,\delta(k_1^2-M^2)\,d^4k_1$$
$$\times\left\{\frac{1}{(2\pi)^3}\,\frac{\vartheta(k_2^0)\,\delta(k_2^2-M^2)}{\varkappa_1-i\varepsilon}\,\delta(-\lambda\varkappa_1+k_1+k_2-q_1-q_2+\lambda\varkappa')\right\}$$
$$\times d^4k_2\,d\varkappa_1\,T^{\rho_1\nu_1}_{\rho_2\nu_2}(\lambda\varkappa_1,\ k_1,\ k_2\,|\,\lambda\varkappa',\ q_1,\ q_2). \tag{2.6}$$

In deriving Eq. (2.6) we have used the completeness relation (1.22) and omitted the δ function, which expresses the conservation law $p_1 + p_2 - \lambda\varkappa = q_1 + q_2 - \lambda\varkappa'$, common to both sides of (2.4).

If [cf. with (1.17) and (1.31)]

$$\lambda = \frac{p_1+p_2}{\sqrt{(p_1+p_2)^2}} = \frac{k_1+k_2}{\sqrt{(k_1+k_2)^2}} = \frac{q_1+q_2}{\sqrt{(q_1+q_2)^2}}, \tag{2.7}$$

the standard invariant variables

$$\left.\begin{aligned} s_p &= (p_1+p_2)^2;\quad s_k = (k_1+k_2)^2;\quad s_q = (q_1+q_2)^2;\\ t_{pq} &= (p_1-q_1)^2;\quad u_{pq} = (p_2-q_1)^2;\quad t_{pk} = (p_1-k_1)^2 \end{aligned}\right\} \tag{2.8}$$

can be introduced in Eq. (2.6). It is readily seen that the quantities (2.8) are related by equations of the form

$$\sqrt{s_k s_q} + t_{kq} + u_{kq} = 4M^2, \tag{2.9}$$

which go over on the energy-momentum surface into the well-known equation $s + t + u = 4M^2$.

If we now integrate with respect to k_2 and $\varkappa_1$ in (2.6), the expression in the braces becomes

$$\frac{1}{(2\pi)^3} \cdot \frac{1}{\sqrt{s_k}} \cdot \frac{1}{\varkappa' + \sqrt{s_k} - \sqrt{s_q} - i\varepsilon}. \tag{2.10}$$

The parameter $\varkappa'$, which was introduced in Sec. 1 to formulate the procedure more generally, can now be set equal to zero. Then the expression (2.10) becomes

$$\frac{1}{(2\pi)^3} \cdot \frac{1}{\sqrt{s_k}(\sqrt{s_k} - \sqrt{s_q} - i\varepsilon)} \equiv G_q^{(0)}(k) \tag{2.11}$$

or, with allowance for (2.9),

$$G_q^{(0)}(k) = \frac{1}{(2\pi)^3} \cdot \frac{1}{s_k + t_{kq} + u_{kq} - 4M^2 - i\varepsilon}. \tag{2.12}$$

Comparing the structures of Eqs. (2.6) and (0.2), we see that $G_q^{(0)}(k)\delta_{\rho_1\sigma_1}\delta_{\rho_2\sigma_2}$ in Eq. (2.6) is the free Green's function of the two-particle system and V is the quasipotential. The intimate analogy between Eqs. (2.6) and (0.2) becomes especially pronounced if we go over in (2.6) to the center-of-mass system. Introducing the notation

$$\begin{gathered} \mathbf{p}_1 = -\mathbf{p}_2 = \mathbf{p};\ \mathbf{q}_1 = -\mathbf{q}_2 = \mathbf{q};\ \mathbf{k}_1 = -\mathbf{k}_2 = \mathbf{k}; \\ E_p = \sqrt{\mathbf{p}^2 + M^2};\ E_k = \sqrt{\mathbf{k}^2 + M^2},\ E_q = \sqrt{\mathbf{q}^2 + M^2} \end{gathered} \tag{2.13}$$

and recalling that

$$2E_q + \varkappa = 2E_p;\ 2E_k + \varkappa_1 = 2E_p + \varkappa',$$

we can replace (2.6) by

$$T^{\mu_1\nu_1}_{\mu_2\nu_2}(\mathbf{p}, \mathbf{q}) = V^{\mu_1\nu_1}_{\mu_2\nu_2}(\mathbf{p}, \mathbf{q}; E_q) + \frac{1}{(4\pi)^3} \sum_{\rho_1\rho_2} \int V^{\mu_1\rho_1}_{\mu_2\rho_2}(\mathbf{p}, \mathbf{k}; E_q) \frac{d\mathbf{k}}{\sqrt{\mathbf{k}^2 + M^2}} \frac{T^{\rho_1\nu_1}_{\rho_2\nu_2}(\mathbf{k}, \mathbf{q})}{E_k(E_k - E_q - i\varepsilon)}. \tag{2.14}$$

As can be seen from (2.14), the Green's function, or the "energy" denominator, does not depend on the spin indices. All the features that are introduced into the scattering amplitude by the spin are a consequence of the spin dependence of the quasipotential.* Thus, the situation here is exactly the same as in the nonrelativistic theory, in which the free Hamiltonian is a scalar in the spin space, and only the interaction terms can depend on the spin.

In view of this remark, we shall concentrate our attention in what follows on the scalar equation

$$T(\mathbf{p}, \mathbf{q}) = V(\mathbf{p}, \mathbf{q}; E_q) + \frac{1}{(4\pi)^3} \int V(\mathbf{p}, \mathbf{k}; E_q) \frac{d\mathbf{k}}{\sqrt{\mathbf{k}^2 + m^2}} \frac{T(\mathbf{k}, \mathbf{q})}{E_k(E_k - E_q - i\varepsilon)}. \tag{2.15}$$

4. Quasipotential Formalism with a "Nonrelativistic" Normalization of the Scattering Amplitude

As we have mentioned above (see the end of Sec. 3), our problem is now to investigate the scalar quasipotential equation (2.15). In this equation, T(p, q) must be interpreted as the scattering amplitude off the

---

* We emphasize that the matrix $V^{\mu_1\rho_1}_{\mu_2\rho_2}$ depends parametrically on the energy E_q and is not Hermitian in the general case.

energy-momentum surface,* and $V(\mathbf{p}, \mathbf{k}, E_q)$ as the quasipotential that corresponds to the sum of the corresponding irreducible diagrams. In the g^2 approximation

$$V^{(2)}(\mathbf{p}, \mathbf{k}, E_q) = \frac{g^2}{\sqrt{\mu^2+(\mathbf{p}-\mathbf{k})^2}} \frac{1}{E_p+E_k-2E_q-i\varepsilon+\sqrt{\mu^2+(\mathbf{p}-\mathbf{k})^2}}, \tag{2.16}$$

where

$$E_p = \sqrt{\mathbf{p}^2+m^2}, \quad E_k = \sqrt{\mathbf{k}^2+m^2}, \quad E_q = \sqrt{\mathbf{q}^2+m^2}.$$

On the surface $E_p = E_q$, the amplitude $T(\mathbf{p}, \mathbf{q})$ is equal to the invariant amplitude $T(s, t)$ related to the differential cross section of elastic scattering by the equation

$$\frac{d\sigma}{d\omega} = \frac{|T(s, t)|^2}{(8\pi)^2 s}. \tag{2.17}$$

We replace T by a new amplitude A, setting

$$A(s, t) = \frac{T(s, t)}{8\pi\sqrt{s}}. \tag{2.18}$$

Then (2.17) becomes

$$\frac{d\sigma}{d\omega} = |A(s, t)|^2, \tag{2.19}$$

which is identical with the condition of normalization of the nonrelativistic elastic scattering amplitude.

To obtain from Eq. (2.15) the corresponding equation for A, we must extrapolate (2.18) off the energy-momentum surface. To this end we rewrite (2.18) in the form

$$A(s, t) = \frac{T(s, t)}{8\pi\sqrt{4m^2-t-u}}$$

and note that, in accordance with (2.9), the following relation holds when we leave the energy-momentum surface:

$$\sqrt{s_p s_q} + t + u = 4m^2,$$

where

$$s_p = (p_1+p_2)^2 = 4E_p^2, \quad s_q = (q_1+q_2)^2 = 4E_q^2.$$

Consequently, the desired extrapolation of $A(s, t)$ has the form

$$A(s, t) = A(\mathbf{p}, \mathbf{q})|_{E_p=E_q} \to A(\mathbf{p}, \mathbf{q})\Big|_{E_p \neq E_q} = \frac{T(\mathbf{p}, \mathbf{q})}{8\pi\sqrt{4E_p E_q}}\Big|_{E_p \neq E_q}. \tag{2.20}$$

Then, making the substitution

$$T(\mathbf{p}, \mathbf{q}) = 8\pi\sqrt{4E_p E_q}\, A(\mathbf{p}, \mathbf{q}); \tag{2.21}$$

$$V(\mathbf{p}, \mathbf{k}; E_q) = -2m\sqrt{4E_p E_k}\, \widetilde{V}(\mathbf{p}, \mathbf{k}; E_q), \tag{2.22}$$

in (2.15), we finally obtain

$$A(\mathbf{p}, \mathbf{q}) = -\frac{m}{4\pi}\widetilde{V}(\mathbf{p}, \mathbf{q}; E_q) + \frac{1}{(2\pi)^3}\int \widetilde{V}(\mathbf{p}, \mathbf{k}; E_q) \frac{d\mathbf{k}}{\sqrt{1+\frac{\mathbf{k}^2}{m^2}}} \frac{A(\mathbf{k}, \mathbf{q})}{2E_q-2E_k+i\varepsilon}. \tag{2.23}$$

---

* It is obvious that with our definition of the λ vector the energy momentum surface in the center-of-mass system is simply the energy surface $E_p = E_q$.

In form, Eq. (2.23) is obviously a direct relativistic generalization of the Lippmann–Schwinger equation.

If the quasipotential is real, Eq. (2.23) yields a relation that has exactly the same form as the nonrelativistic unitarity condition:

$$\operatorname{Im} A(\mathbf{p},\mathbf{q}) = \frac{|\mathbf{q}|}{4\pi}\int A(\mathbf{p},\mathbf{k})\, A^*(\mathbf{k},\mathbf{q})\, d\omega_k; \quad E_p = E_k = E_q = \sqrt{\mathbf{q}^2+m^2}. \tag{2.24}$$

Writing this equation in terms of the amplitude $T(\mathbf{p}, \mathbf{q})$ [see Eq. (2.21)], we readily see that it is equivalent to the relativistic condition of two-particle unitarity*

$$\operatorname{Im} T(\mathbf{p},\mathbf{q}) = \frac{1}{(8\pi)^2}\sqrt{\frac{E_q{}^2-m^2}{E_q{}^2}}\int T(\mathbf{p},\mathbf{k})\, T^*(\mathbf{k},\mathbf{q})\, d\omega_{\mathbf{k}}. \tag{2.25}$$

We now set

$$\Psi_q(\mathbf{p}) = \left(2\pi\right)^3\sqrt{1+\frac{\mathbf{p}^2}{m^2}}\,\delta(\mathbf{p}-\mathbf{q}) - \frac{4\pi}{m}\cdot\frac{A(\mathbf{p},\mathbf{q})}{2E_q-2E_p+i\varepsilon} \quad \left(E_p=\sqrt{\mathbf{p}^2+m^2},\ E_q=\sqrt{\mathbf{q}^2+m^2}\right). \tag{2.26}$$

From Eq. (2.23), we then find an equation for the function

$$\Psi_q(\mathbf{p}) = (2\pi)^3\sqrt{1+\frac{\mathbf{p}^2}{m^2}}\,\delta(\mathbf{p}-\mathbf{q}) + \frac{1}{(2\pi)^3}\frac{\int \tilde{V}(\mathbf{p},\mathbf{k};E_q)\,\Psi_q(\mathbf{k})\, d\Omega_k}{2E_q-2E_p+i\varepsilon}, \tag{2.27}$$

or

$$(2E_q-2E_p)\,\Psi_q(\mathbf{p}) = \frac{1}{(2\pi)^3}\int \tilde{V}(\mathbf{p},\mathbf{k};E_q)\,\Psi_q(\mathbf{k})\, d\Omega_k, \tag{2.28}$$

where

$$d\Omega_k = \frac{d\mathbf{k}}{\sqrt{1+\frac{\mathbf{k}^2}{m^2}}}. \tag{2.29}$$

Consequently, $\Psi_q(\mathbf{p})$ can be regarded as the relativistic wave function of the two-particle system in the $\mathbf{p}$ representation and Eqs. (2.27) and (2.28) as two forms (integral and differential) of the relativistic Schrödinger equation. Our next problem is to construct an equally natural relativistic generalization. In this problem fundamental importance attaches to the fact that all the four-momenta in our equations belong to the upper sheet of the four-dimensional hyperboloid

$$p_0{}^2-\mathbf{p}^2=m^2 \tag{2.30}$$

which is, of course, one of the realizations of a space of constant negative curvature (Lobachevskii space).

5. The Case of Unequal Masses

Hitherto we have considered relativistic equations that describe the interactions of particles with equal masses. We now set ourselves the task of finding a relativistic generalization of the Lipmann–Schwinger equation in the case of unequal masses [23] that is as simple as (2.23).

To make the transition to unequal masses in the nonrelativistic Lippmann–Schwinger equation, it is sufficient to replace the parameter m in that equation by twice the reduced mass:

---

* Equation (2.25) with Im V = 0 is also satisfied by an amplitude $T(\mathbf{p}, \mathbf{q})$ that satisfies the quasipotential equation (0.2). This circumstance reflects one of the principal ideas of the quasipotential approach.

$$m \to 2\mu = \frac{2m_1 m_2}{m_1 + m_2}. \tag{2.31}$$

However, in the relativistic case it would be quite unjustified to apply the procedure (2.31) to Eq. (2.23) to obtain the desired equation. To demonstrate this more clearly, let us recall how the very concept of a reduced mass arises in the nonrelativistic theory.

Let

$$E = \frac{\mathbf{k}_1^2}{2m_1} + \frac{\mathbf{k}_2^2}{2m_2} = \frac{m\mathbf{v}_1^2}{2} + \frac{m_2 \mathbf{v}_2^2}{2} \tag{2.32}$$

be the total energy of two free nonrelativistic particles in an arbitrary frame of reference. Consider the total-momentum vector

$$\mathbf{K} = \mathbf{k}_1 + \mathbf{k}_2 \tag{2.33}$$

and the relative-momentum vector

$$\mathbf{k} = \frac{m_2 \mathbf{k}_1 - m_1 \mathbf{k}_2}{m_1 + m_2} = \mu(\mathbf{v}_1 - \mathbf{v}_2). \tag{2.34}$$

Then, making the Galilean transformation

$$\mathbf{v}_1 = \frac{\mathbf{k}}{m_1} + \frac{\mathbf{K}}{m_1 + m_2}, \quad \mathbf{v}_2 = -\frac{\mathbf{k}}{m_2} + \frac{\mathbf{K}}{m_1 + m_2}, \tag{2.35}$$

we replace (2.32) by

$$E = \frac{\mathbf{K}^2}{2(m_1 + m_2)} + \frac{\mathbf{k}^2}{2\mu}. \tag{2.36}$$

In the center-of-mass system, $\mathbf{K} = 0$, and therefore

$$E = \frac{\mathbf{k}^2}{2\mu}. \tag{2.37}$$

The difference of the two single-particle energies of the type (2.37) gives the energy denominator of the Lippmann–Schwinger equation. It must be emphasized that the momentum of the effective particle that replaces the two-particle system is the relative momentum **k**, whose length is, by virtue of the formula

$$\mathbf{k}^2 = \mu^2 (\mathbf{v}_1 - \mathbf{v}_2)^2 \equiv \mu^2 \mathbf{v}^2, \tag{2.38}$$

an invariant under Galilean transformations.

In the relativistic treatment the cms four-momenta of two free particles can be written in the form

$$k_1 = \left(\sqrt{m_1^2 + \mathbf{k}^2},\ \mathbf{k}\right),\ k_2 = \left(\sqrt{m_2^2 + \mathbf{k}^2}, -\mathbf{k}\right). \tag{2.39}$$

Therefore the total energy of the particles in this frame of reference is

$$\sqrt{m_1^2 + \mathbf{k}^2} + \sqrt{m_2^2 + \mathbf{k}^2} \equiv \sqrt{s_k}. \tag{2.40}$$

If $m_1 = m_2 = m$,

$$2\sqrt{m^2 + \mathbf{k}^2} = \sqrt{s_k}. \tag{2.41}$$

The energy denominator in Eq. (2.23) is obviously the difference of expressions of the type (2.41).

It follows from (2.40) and (2.41), that, irrespective of the relationship between the masses, the total energy of two free relativistic particles in the center-of-mass system cannot be interpreted as the energy of some effective relativistic particle with momentum **k**.

However, the fact that in the case of equal masses the total energy (2.41) is proportional to the energy of one particle in fact enables one to regard the energy denominator of the relativistic Lippmann–Schwinger equation (2.23) as a single-particle denominator and, as a result, to reduce the relativistic problem of two particles with equal masses to the problem of describing the motion of particles of mass m in a quasi-potential field. It is therefore clear that if we could succeed in representing (2.40) as an expression proportional to the energy of a relativistic particle with some mass m' (which goes over into the mass m when $m_1 = m_2 = m$) then also in the case of unequal masses the relativistic two-body problem would be equivalent to a single-particle problem. Under such conditions it would be possible to apply the mathematical method employed in the problem of particles with equal masses.

Denoting by **k'** the momentum of the particle with mass m', we have, in accordance with the foregoing,

$$\sqrt{s_k} = \sqrt{m_1^2 + \mathbf{k}^2} + \sqrt{m_2^2 + \mathbf{k}^2} = \text{const}\,\sqrt{m'^2 + \mathbf{k}'^2}. \tag{2.42}$$

It is shown in [23] that the four-vector k must be taken in the form

$$k_\mu' = (k_0', \mathbf{k}') = \sqrt{\frac{\mathcal{K}^2}{\mathcal{K}_\perp}}\,(\mathcal{K}_\perp)_\mu, \tag{2.43}$$

where*

$$\left.\begin{aligned} \mathcal{K} &= \frac{m_2 k_1 - m_1 k_2}{m_1 + m_2};\\ \mathcal{K}_\perp &= \mathcal{K} - \lambda(\lambda\mathcal{K}) = \frac{k_1 - k_2}{2} - \frac{m_1^2 - m_2^2}{2\sqrt{s_k}}\lambda;\\ (\lambda\mathcal{K}_\perp) &= 0; \qquad \lambda = \frac{k_1 + k_2}{\sqrt{s_k}}. \end{aligned}\right\} \tag{2.44}$$

As can be seen from (2.43),

$$\mathbf{k}'^2 = -\mathcal{K}^2 \tag{2.45}$$

in all coordinate systems. In the center-of-mass system, for which $\lambda = (1, 0)$, we have $k_0' = 0$, and the square of the three-dimensional relative momentum is given by

$$\mathbf{k}'^2 = -\left(\frac{m_2 k_1 - m_1 k_2}{m_1 + m_2}\right)^2 = 4\mu^2 \operatorname{sh}^2 \frac{s}{2} = -\mu^2 (u_1 - u_2)^2, \tag{2.46}$$

where u_1 and u_2 are the relativistic four-velocities, which, in the general case, have the form

$$\begin{aligned} u_1 &= \frac{k_1}{m_1} = \left(\frac{1}{\sqrt{1-\mathbf{v}_1^2}}, \frac{\mathbf{v}_1}{\sqrt{1-\mathbf{v}_1^2}}\right);\\ u_2 &= \frac{k_2}{m_2} = \left(\frac{1}{\sqrt{1-\mathbf{v}_2^2}}, \frac{\mathbf{v}_2}{\sqrt{1-\mathbf{v}_2^2}}\right), \end{aligned} \tag{2.47}$$

and the parameter s is related to the relativistic relative velocity $\mathbf{v}_{rel}$ by

$$\frac{1}{\sqrt{1-\mathbf{v}_{rel}^2}} = \operatorname{ch} s = u_1 u_2 = u_{10} u_{20} - \mathbf{u}_1 \mathbf{u}_2. \tag{2.48}$$

Combining the relations (2.46) and (2.48), we obtain

$$\frac{\mathbf{k}'^2}{2\mu} = \frac{\mu}{\sqrt{1-\mathbf{v}_{rel}^2}} - \mu, \tag{2.49}$$

from which it follows that in the relativistic case the kinetic energy of the relative motion can be expressed in terms of the relative momentum **k'** and the reduced mass μ in the nonrelativistic manner. Then, substituting (2.46) into (2.42), we find

$$\sqrt{s_k} = \sqrt{m_1^2 + \mathbf{k}^2} + \sqrt{m_2^2 + \mathbf{k}^2} = \frac{m_1 + m_2}{\sqrt{m_1 m_2}}\sqrt{m_1 m_2 + \mathbf{k}'^2}. \tag{2.50}$$

---

* The vector $\mathcal{K}_\perp$ that occurs here is known as the Wightman–Gårding relative momentum [25].

Consequently, the mass m' of the "effective" particle in the relation (2.42) is equal to the geometric mean of the masses m_1 and m_2:

$$m' = \sqrt{m_1 m_2}. \tag{2.51}$$

It is also readily seen that in the nonrelativistic limit the denominator of the quasipotential equation in the case of unequal masses, which has the form

$$\frac{1}{\sqrt{s_k} - \sqrt{s_q} - i\varepsilon} = \frac{1}{\sqrt{m_1^2 + k} + \sqrt{m_2^2 + k} - \sqrt{m_1^2 + q^2} - \sqrt{m_2^2 + q^2} - i\varepsilon}, \tag{2.52}$$

goes over exactly into the denominator of the corresponding nonrelativistic Lippmann–Schwinger equation, which depends on the reduced mass μ:

$$\frac{1}{\frac{m_1 + m_2}{\sqrt{m_1 m_2}}\left(\sqrt{m_1 m_2 + k'^2} - \sqrt{m_1 m_2 + q'^2} - i\varepsilon\right)} \to \frac{1}{\frac{k'^2}{2\mu} - \frac{q'^2}{2m} - i\varepsilon}. \tag{2.53}$$

Making some simple calculations, we arrive at a relativistic equation for the scattering amplitude of particles with unequal masses:

$$A(p', q') = -\frac{\mu}{2\pi} V(p', q'; E_{q'}) + \frac{1}{(2\pi)^3} \frac{\sqrt{m_1 m_2}}{m_1 + m_2} \int \frac{V(p,' k'; E_{q'})\, d\Omega_{k'}\, A(k', q')}{E_{q'} - E_{k'} + i\varepsilon}, \tag{2.54}$$

where

$$E_{k'} = \sqrt{m_1 m_2 + k'^2}; \quad E_{q'} = \sqrt{m_1 m_2 + q'^2}. \tag{2.55}$$

Equation (2.54) can be regarded as the relativistic generalization of the Lippmann–Schwinger equation in the spirit of the Lobachevskii geometry that is realized on the upper sheet of the hyperboloid $p'^2 = m_1 m_2$.

We should like to emphasize that Eq. (2.54) describes the scattering of an effective relativistic particle by the quasipotential $V(p', q'; E_{q'})$.

SECTION 3. RELATIVISTIC CONFIGURATION REPRESENTATION

6. The Momentum Space as a Lobachevskii Space and the Shapiro Transformation

We can consider the quasipotential equations (2.23), (2.27), and (2.28). An important feature of the quasipotential formalism is the presence of three-dimensional kinematic entities having a relativistic origin. For example, in the above equations the integration is extended over the three-dimensional momentum space with volume element

$$d\Omega_p = \frac{dp}{\sqrt{1 + \frac{p^2}{m^2}}}, \tag{3.1}$$

which serves as a Lorentz invariant measure on the hyperboloid (2.30).

Equations (2.23), (2.27), and (2.28) can be regarded as direct relativistic generalizations of the Lippmann–Schwinger and Schrödinger equations; for these equations have an "absolute" nature with respect to the geometry of the momentum space, i.e., they can be written in a form in which, superficially, they do not differ from their nonrelativistic analogs. A difference is that in this case the nonrelativistic (Euclidean) expressions for the energy and the volume element are replaced by their relativistic (non-Euclidean) analogs:

$$
\left.\begin{aligned}
E_p &= \frac{\mathbf{p}^2}{2m} \to E_p = \sqrt{\mathbf{p}^2+m^2};\\
d\Omega_p &= d\mathbf{p} \to d\Omega_p = \frac{d\mathbf{p}}{\sqrt{1+\frac{\mathbf{p}^2}{m^2}}}.
\end{aligned}\right\} \tag{3.2}
$$

The upper sheet of the hyperboloid (2.30), which is embedded in the four-dimensional momentum space, serves as a model of a relativistic non-Euclidean momentum space. On this surface one can obviously introduce different coordinate systems and, thus, obtain different expressions for the element of length, element of volume, Laplacian, etc. As coordinates on the surface (2.30), we introduce the components of the momentum vector **p**. These are Cartesian coordinates on the hyperplane $p_0 = 0$, into which the hyperboloid is mapped under projection from the point $(\infty, 0)$. As a model of the Lobachevskii space, we now have the complete **p** space with metric

$$
ds^2 = d\mathbf{p}^2 - \frac{(\mathbf{p}d\mathbf{p})^2}{m^2+\mathbf{p}^2} \equiv g_{ik}\, dp_i\, dp_k. \tag{3.3}
$$

Although these remarks are almost self-evident, they cast light on the meaning of the subsequent constructions. Our intention is to adopt consistently the point of view that in the relativistic case the momentum space of a particle with mass m is the Lobachevskii space with the metric (3.3). Since our problem of the interaction of two relativistic particles reduces to the problem of the behavior of a single relativistic particle in a quasipotential field, this enables us to employ effectively some group-theoretical and geometrical arguments and thus render the procedure more rigorous and more closely related to the nonrelativistic theory.

Let $\psi(\mathbf{p})$ be the wave function of a particle with spin zero, mass m, and momentum **p** belonging to a Lobachevskii space. The question arises of how, in this formalism, one defines the angular momentum and coordinate operators, **L** and **x**. Since the group of motions of Lobachevskii space – the Lorentz group – contains the subgroup O(3), the operator **L** has the usual form:

$$
L_k = \frac{1}{i}\, \varepsilon_{klm}\, p_l\, \frac{\partial}{\partial p_m}. \tag{3.4}
$$

The coordinate **x** in the nonrelativistic theory is the generator of translations of the Euclidean **p** space. Since the proper Lorentz transformations $\Lambda_{\mathbf{k}}$ [$\mathbf{k} = (k_1, k_2, k_3)$ are the parameters] play the role of translations in the curved **p** space, introduce the relativistic operator **x** as the generator of these transformations.*

It is easily shown [31] that

$$
\mathbf{p}' = \overrightarrow{(\Lambda_{\mathbf{k}}\,\mathbf{p})} = \mathbf{p}(+)\mathbf{k} = \mathbf{p} + \mathbf{k}\left[\sqrt{1+\frac{\mathbf{p}^2}{m^2}} + \frac{\mathbf{p}\mathbf{k}}{m^2\left(1+\sqrt{1+\frac{\mathbf{k}^2}{m^2}}\right)}\right]. \tag{3.5}
$$

Therefore,

$$
\mathbf{x} = i\sqrt{1+\frac{\mathbf{p}^2}{m^2}} \cdot \frac{\partial}{\partial \mathbf{p}}. \tag{3.6}
$$

The operator (3.6) is Hermitian in the metric

$$
(\Psi, \Phi) = \int \Psi^*(\mathbf{p})\, \Phi(\mathbf{p})\, d\Omega_p. \tag{3.7}
$$

However, its components do not commute with one another and cannot be reduced simultaneously to diagonal form. Thus, (3.6) is not suited to the construction of the configuration representation, and a different approach must therefore be adopted if we wish to introduce a relativistic coordinate.

---

* The coordinate operators are defined similarly in the theory of quantized space–time [26-29] (in this connection see also [30]). We take the notation for the operation of a shift in a space of constant curvature (formula (3.5) from [27]).

In this connection we note that the identity of the nonrelativistic coordinate **x** and the generator of shifts of the **p** space is not the only group-theoretical property of **x**. Indeed, it is obvious that the scalar square of **x**, i.e., $\left(i\frac{\partial}{\partial \mathbf{p}}\right)^2 = r^2$ is the Casimir operator of the group of motions of the Euclidean **p** space. At the same time, the exponential functions $e^{i\mathbf{pk}}$ which realize the transformation from the **p** representation to the **r** representation are matrix elements of unitary representations of the group that correspond to pure translations and values $r^2 > 0$.

We now turn to the Lorentz group and ask which of its Casimir operators C goes over in the nonrelativistic limit $m \to \infty$ into $\left(i\frac{\partial}{\partial \mathbf{p}}\right)^2$.* It is

$$C = \mathbf{x}^2 - \frac{1}{m^2}\mathbf{L}^2. \tag{3.8}$$

We set this quantity equal to

$$\mathbf{x}^2 - \frac{1}{m^2}\mathbf{L}^2 = \frac{1}{m^2} + r^2 > 0, \tag{3.9}$$

where

$$0 < r < \infty. \tag{3.10}$$

The relations (3.8)-(3.10) single out the so-called principal series of unitary representations of the Lorentz group (see, for example, [32]).

The role of the plane waves that correspond to the "translations" (3.5) is played by the functions

$$\xi(\mathbf{p}, \mathbf{r}) = \left(\frac{p_0 - \mathbf{pn}}{m}\right)^{-1-irm}, \tag{3.11}$$

where

$$p_0 = \sqrt{m^2 + \mathbf{p}^2}, \quad \mathbf{r} = r\mathbf{n}\,(\mathbf{n}^2 = 1).$$

In the nonrelativistic limit

$$\lim_{m\ \infty} \xi(\mathbf{p}, \mathbf{r}) = e^{i\mathbf{pr}}. \tag{3.12}$$

The analogy between these two systems of functions becomes quite clear if one writes down the orthogonality and completeness conditions for ξ (**p**, **r**):

$$\frac{1}{(2\pi)^3}\int \xi(\mathbf{p}, \mathbf{r})\,\xi^*(\mathbf{p}, \mathbf{r}')\,d\Omega_p = \delta^{(3)}(\mathbf{r} - \mathbf{r}'); \tag{3.13}$$

$$\frac{1}{(2\pi)^3}\int \xi(\mathbf{p}, \mathbf{r})\,\xi^*(\mathbf{p}', \mathbf{r})\,d\mathbf{r} = \delta^{(3)}(\mathbf{p} - \mathbf{p}')\sqrt{1 + \frac{\mathbf{p}^2}{m^2}}. \tag{3.14}$$

The expansion in the functions (3.11) was first introduced by Shapiro [33] and has thus become known as the Shapiro transformation.

Obviously the Shapiro transformation plays the role of a Fourier transformation in the Lobachevskii **p** space, mapping the latter onto some relativistic **r** space. Now the new **r** space has very specific properties with respect to the Lorentz group [33]: in accordance with (3.9), the modulus of the radius vector **r** is a relativistic invariant and its direction transforms as the three-dimensional part of the isotropic vector

$$n_\mu = (1, \mathbf{n}). \tag{3.15}$$

---

* In the given spinless case, the second Casimir operator, **Lx**, vanishes.

Thus, we have the following conclusion: if the quasipotential theory of scattering is formulated in terms of Lobachevskii geometry (see Sec. 4), a Shapiro transformation, i.e., an expansion in matrix elements of the principal series of the unitary representations of the Lorentz group, must be used for the transition to the configuration representation.

Let us consider in more detail the transformation of a "translation" of the curved **p** space defined by (3.5). Clearly, in contrast to Euclidean shifts, the set of transformations (3.5) does not form a group. However, being elements of the Lorentz group, these "shifts" have a number of group properties. In particular,

$$\left.\begin{aligned} &\mathbf{p}(+)0=\mathbf{p};\\ &\mathbf{p}(-)\mathbf{p}=0;\\ &0(+)\mathbf{k}=\mathbf{k};\\ &(\mathbf{p}(+)\mathbf{k})(-)\mathbf{k}=\mathbf{p}.\end{aligned}\right\} \tag{3.16}$$

Obviously, the volume element (3.1) is invariant under the transformation (3.5):

$$d\Omega_{\mathbf{p}(+)\mathbf{k}}=d\Omega_{\mathbf{p}}. \tag{3.17}$$

This property of $d\Omega_{\mathbf{p}}$ makes it possible to introduce the operation of convolution of functions in the Lobachevskii space [31]:

$$\Psi_1(\mathbf{p})*\Psi_2(\mathbf{p})=\int d\Omega_k\,\Psi_1(\mathbf{k})\,\Psi_2(-\mathbf{k}(+)\mathbf{p}). \tag{3.18}$$

Further, we define a δ function in the Lobachevskii space in the form

$$\delta(\mathbf{k}(-)\mathbf{p})=\frac{\delta(\mathbf{k}-\mathbf{p})}{\sqrt{g(k)}}=\sqrt{1+\frac{\mathbf{k}^2}{m^2}}\,\delta(\mathbf{k}-\mathbf{p}). \tag{3.19}$$

Note also the useful formula

$$\sqrt{1+\frac{(\mathbf{p}(-)\mathbf{k})^2}{m^2}}=\sqrt{1+\frac{\mathbf{p}^2}{m^2}}\sqrt{1+\frac{\mathbf{k}^2}{m^2}}-\frac{1}{m^2}\mathbf{pk}. \tag{3.20}$$

In the spherical coordinates

$$\left.\begin{aligned} &p_0=m\operatorname{ch}\chi_p; \quad \mathbf{p}=m\operatorname{sh}\chi_p\,\boldsymbol{\mu};\\ &k_0=m\operatorname{ch}\chi_k; \quad \mathbf{k}=m\operatorname{sh}\chi_k\,\boldsymbol{\nu}\end{aligned}\right\} \tag{3.21}$$

this equation takes the form

$$\sqrt{1+\frac{(\mathbf{p}(-)\mathbf{k})^2}{m^2}}=\operatorname{ch}\chi_p\cdot\operatorname{ch}\chi_k-\operatorname{sh}\chi_p\cdot\operatorname{sh}\chi_k\,\boldsymbol{\mu\nu}. \tag{3.22}$$

7. Quasipotential Theory in the Relativistic Configuration Representation

We now use the Shapiro transformation (see Sec. 6) to formulate the variant of quasipotential theory developed earlier in the relativistic **r** representation.

Setting (compare this with the corresponding nonrelativistic formulas)

$$\Psi_q(\mathbf{r})=\frac{1}{(2\pi)^3}\int\xi(\mathbf{p},\mathbf{r})\,\Psi_q(\mathbf{p})\,d\Omega_p; \tag{3.23}$$

$$\widetilde{V}(\mathbf{p},\mathbf{k};E_q)=\int\xi^*(\mathbf{p},\mathbf{r})\,V(\mathbf{r},\mathbf{r}';E_q)\,\xi(\mathbf{k},\mathbf{r}')\,d\mathbf{r}\,d\mathbf{r}'; \tag{3.24}$$

$$G_q(\mathbf{r},\mathbf{r}')=\frac{1}{(2\pi)^3}\int\xi(\mathbf{k},\mathbf{r})\,\frac{d\Omega_k}{2E_q-2E_k+i\varepsilon}\,\xi^*(\mathbf{k},\mathbf{r}'), \tag{3.25}$$

we obtain from (2.27)

$$\Psi_q(\mathbf{r}) = \xi(\mathbf{q}, \mathbf{r}) + \int G_q(\mathbf{r}, \mathbf{r}') V_\iota(\mathbf{r}', \mathbf{r}''; E_q) \Psi_q^\iota(\mathbf{r}'') \, d\mathbf{r}' \, d\mathbf{r}'', \tag{3.26}$$

which is the exact analog of the Schrödinger equation in integral form. The nonrelativistic expression for the amplitude can also be generalized in an obvious manner:

$$A(\mathbf{p}, \mathbf{q}) = -\frac{1}{4\pi} \cdot \frac{1}{(2\pi)^3} \int \tilde{V}(\mathbf{p}, \mathbf{k}; E_q) \Psi_q(\mathbf{k}) \, d\Omega_k = -\frac{m}{4\pi} \int \xi^*(\mathbf{p}, \mathbf{r}) V(\mathbf{r}, \mathbf{r}'; E_q) \Psi_q(\mathbf{r}') \, d\mathbf{r} d\mathbf{r}'. \tag{3.27}$$

Further, to simplify the formulas, it is convenient to go over to a system of units in which $\hbar = c = m = 1$. Then the "plane wave" $\xi(\mathbf{q}, \mathbf{r})$ is written as

$$\xi(\mathbf{q}, \mathbf{r}) = (q_0 - \mathbf{q}\mathbf{n})^{-1-ir},$$

and the passage to the limit of the nonrelativistic theory corresponds to the approximation

$$|\mathbf{q}| \ll 1; \quad |\mathbf{r}| \gg 1; \quad |\mathbf{r}| \cdot |\mathbf{q}| = \text{fixed}. \tag{3.28}$$

We now turn to an expansion in partial waves. We note first (see, for example, [34]) that

$$\xi(\mathbf{q}, \mathbf{r}) = \sqrt{\frac{\pi}{2 \operatorname{sh} \chi_q}} \sum_{l=0}^{\infty} (2l+1) \frac{\Gamma(ir+l+1)}{\Gamma(ir+1)} P^{-1/2-l}_{-1/2+ir}(\operatorname{ch} \chi_q) \, P_l\left(\frac{\mathbf{q}\mathbf{k}}{qr}\right) \equiv \sum_{l=0}^{\infty} (2l+1) \, i^l p_l(\operatorname{ch} \chi_q, r) \, P_l\left(\frac{\mathbf{q}\mathbf{r}}{qr}\right);$$

$$\xi^*(\mathbf{q}, \mathbf{r}) = \sum_{l=0}^{\infty} (2l+1)(-i)^l p_l^*(\operatorname{ch} \chi_q, r) \, P_l\left(\frac{\mathbf{q}\mathbf{r}}{qr}\right) \quad (q = \operatorname{sh} \chi_q, \quad q_0 = \operatorname{ch} \chi_q). \tag{3.29}$$

It is readily seen that in the nonrelativistic region (3.28) the functions*

$$\left.\begin{aligned} p_l(\operatorname{ch} \chi_q, r) &= \frac{(-1)^{l+1}}{r} \sqrt{\frac{\pi}{2 \operatorname{sh} \chi_q}} (-r)^{(l+1)} P^{-1/2-l}_{-1/2+ir}(\operatorname{ch} \chi_q); \\ p_l^*(\operatorname{ch} \chi_q, r) &= \frac{1}{r} \sqrt{\frac{\pi}{2 \operatorname{sh} \chi_q}} (r)^{(l+1)} P^{-1/2-l}_{-1/2+ir}(\operatorname{ch} \chi_q) \end{aligned}\right\} \tag{3.31}$$

go over to one and the same function $j_l(qr) = \sqrt{\frac{\pi}{2qr}} J_{l+1/2}(qr)$. In addition, the functions (3.31) can, like $j_l(qr)$, be expressed in terms of elementary functions:

$$\left.\begin{aligned} r p_0(\operatorname{ch} \chi_q, r) &= \frac{\sin r\chi_q}{\operatorname{sh} \chi_q}; \\ r^{(l+1)} p_l(\operatorname{ch} \chi_q, r) &= (-1)^l (\operatorname{sh} \chi_q)^l \left(\frac{d}{d \operatorname{ch} \chi_q}\right)^l r p_0(\operatorname{ch} \chi_q, r). \end{aligned}\right\} \tag{3.32}$$

It is also readily verified that

$$\frac{2 \operatorname{sh} \chi \cdot \operatorname{sh} \chi'}{\pi} \int_0^\infty r^2 dr^2 \, p_l(\operatorname{ch} \chi, r) \, p_l^*(\operatorname{ch} \chi', r) = \delta(\chi - \chi'); \tag{3.33}$$

$$\frac{2rr'}{\pi} \int_0^\infty \operatorname{sh}^2 \chi d\chi \, p_l(\operatorname{ch} \chi, r) \, p_l^*(\operatorname{ch} \chi, r') = \delta(r - r'). \tag{3.34}$$

---

* Here

$$i^\lambda \frac{\Gamma(-ir+\lambda)}{\Gamma(-ir)} \equiv r^{(\lambda)}. \tag{3.30}$$

Further,

$$\left.\begin{aligned} G_q(\mathbf{r}, \mathbf{r}') &= \frac{1}{4\pi} \sum_l (2l+1) G_{lq}(r, r') P_l\left(\frac{\mathbf{r}, \mathbf{r}'}{rr'}\right); \\ G_{lq}(r, r') &= \frac{2}{\pi} \int_0^\infty \frac{p_l(\operatorname{ch}\chi_k, r)\, p_l^*(\operatorname{ch}\chi_k, r') \operatorname{sh}^2\chi_k\, d\chi_k}{2\operatorname{ch}\chi_q - 2\operatorname{ch}\chi_k + i\varepsilon}. \end{aligned}\right\} \tag{3.35}$$

Hitherto we have been concerned with the generalization of the integral form of nonrelativistic scattering theory. We shall now find the analogs of the differential equations that the corresponding nonrelativistic functions satisfy. Using the recursion relations for the spherical functions,

$$\begin{gathered} (2\nu+1) z P_\nu^\mu(z) = (\nu-\mu+1) P_{\nu+1}^\mu(z) + (\nu+\mu) P_{\nu-1}^\mu(z), \\ P_\nu^\mu(z) = P_{-\nu-1}^\mu(z) \end{gathered} \tag{3.36}$$

it is easy to show that the functions $p_l(\cosh \chi_q, r)$ satisfy the "differential" equation

$$\left[2\operatorname{ch}\chi_q - 2\operatorname{ch}\left(i\frac{d}{dr}\right) - \frac{2i}{r}\operatorname{sh}\left(i\frac{d}{dr}\right) - \frac{l(l+1)}{r^2} e^{i\frac{d}{dr}}\right] p_l(\operatorname{ch}\chi_q, r) = 0. \tag{3.37}$$

Hence, taking into account (3.29), we obtain the analog of the differential Schrödinger equation in the relativistic region for the free case:

$$(2E_q - H_0)\xi(\mathbf{q}, \mathbf{r}) \equiv \left[2\operatorname{ch}\chi_q - 2\operatorname{ch}\left(i\frac{d}{dr}\right) - \frac{2i}{r}\operatorname{sh}\left(i\frac{d}{dr}\right) + \frac{\Delta_{\vartheta,\varphi}}{r^2} e^{i\frac{d}{dr}}\right]\xi(\mathbf{q}, \mathbf{r}) = 0, \tag{3.38}$$

where $\Delta_{\vartheta,\varphi}$ is the angular part of the Laplace operator. In the nonrelativistic limit, (3.38) goes over exactly into the free differential Schrödinger equation. Thus, Eqs. (3.37) and (3.38) can be regarded as equations in finite differences and investigated by the corresponding methods. This point of view will be discussed in detail in Sec. 4.

If we apply the Shapiro transformation to Eq. (2.28), we obtain the relativistic analog of the differential Schrödinger equation with the interaction

$$(2E_q - H_0)\Psi_q(\mathbf{r}) = \int V(\mathbf{r}, \mathbf{r}'; E_q)\Psi_q(\mathbf{r}')\, d\mathbf{r}' \tag{3.39}$$

[the free Hamiltonian H_0 is given by (3.38)].

Finally, we write down an equation for the partial Green's function (3.35):

$$\left[-2\operatorname{ch}\left(i\frac{d}{dr}\right) - \frac{2i}{r}\operatorname{sh}\left(i\frac{d}{dr}\right) - \frac{l(l+1)}{r^2} e^{i\frac{d}{dr}}\right] G_{lq}(r, r') + 2E_q G_{lq}(r, r') = \frac{1}{r^2}\delta(r - r'). \tag{3.40}$$

Thus, we have shown that the relativistic scattering theory can be formulated as a consistent geometric generalization of nonrelativistic scattering theory with a nonlocal interaction. It is natural to ask whether an analogous generalization of the theory with a local interaction can be constructed. This question will be investigated in Sec. 9.

In conclusion, let us establish the geometry inherent in the new relativistic **r** space.

If we answer this question on the basis of the relations (3.13) and (3.14) and Eqs. (3.36) and (3.39), the metric of the space must be Euclidean. On the other hand, the shift generators cannot be merely equal to $-i d/dr$ since the eigenvalues of these operators form a Euclidean **p** space and certainly not a **p** space of constant curvature.

Now the following assertion is true: the relativistic **r** space is Euclidean but the generators of shift transformations $[\mathbf{p} = (\hat{p}_1, \hat{p}_2, \hat{p}_3)]$ in this space are realized in the form of difference-differential operators whose joint spectrum is the Lobachevskii **p** space and joint eigenfunction the plane wave $\xi(\mathbf{p}, \mathbf{r})$.

Explicitly, the momentum operators $\hat{p}_1$, $\hat{p}_2$, and $\hat{p}_3$ in the **r** representation have the form

$$\hat{p}_1 = -\sin\theta\cos\varphi\left(e^{i\frac{d}{dr}} - \frac{H_0}{2}\right) - i\left(\frac{\cos\varphi\cos\theta}{r}\cdot\frac{\partial}{\partial\theta} - \frac{\sin\varphi}{r\sin\theta}\cdot\frac{\partial}{\partial\varphi}\right)e^{i\frac{d}{dr}};$$

$$\hat{p}_2 = -\sin\theta\sin\varphi\left(e^{i\frac{d}{dr}} - \frac{H_0}{2}\right) - i\left(\frac{\sin\varphi\cos\theta}{r}\cdot\frac{\partial}{\partial\theta} + \frac{\cos\varphi}{r\sin\theta}\cdot\frac{\partial}{\partial\varphi}\right)e^{i\frac{d}{dr}}; \quad (3.41)$$

$$\hat{p}_3 = -\cos\theta\left(e^{i\frac{d}{dr}} - \frac{H_0}{2}\right) + \frac{i\sin\theta}{r}\cdot\frac{\partial}{\partial\theta}e^{i\frac{d}{dr}},$$

where the operator H_0 is defined in (3.38) and θ and φ are the angular coordinates of the vector **r**.

It is easily proved directly from (3.41) that

$$[\hat{p}_2, \hat{p}_2] = [\hat{p}_2, \hat{p}_3] = [\hat{p}_3, \hat{p}_1] = 0, \quad (3.42)$$

where

$$\hat{p}\xi(p, r) = p\xi(p, r). \quad (3.43)$$

In the nonrelativistic limit, the operators (3.41) go over into the usual generators of translations, $-id/dr$, written down in a spherical coordinate system.

Of course, besides (3.42), the remaining commutation relations of the Euclidean group of motions remain unchanged in the relativistic **r** space. In particular,

$$[L_i \hat{p}_k] = i\varepsilon_{ikl} L_l, \quad (3.44)$$

where **L** is the angular momentum operator.

8. Lorentz Group, Orispherical Coordinates, and Operator Fourier Transformation

The Fourier transformation can be generalized to the relativistic case in various ways. This is because a three-parametric set of transformations that serves as the relativistic analog of the Euclidean group of transformations can be distinguished in the group of motions of the Lobachevskii space in different ways. Here we shall consider the relativistic generalization of the Fourier transformation associated with the so-called orispherical shifts of Lobachevskii space. In contrast to the shifts (3.5), orispherical shifts form a group. As a result, these transformations are quite different from Shapiro transformations (see Sec. 6). In Sec. 14 we shall employ the technique developed here to derive a high-energy representation for the relativistic scattering amplitude.

We consider a two-dimensional spinor representation of the Lorentz group realized by complex 2×2 matrices with determinant unity:

$$a = \begin{pmatrix}\alpha & \beta\\ \gamma & \delta\end{pmatrix}; \quad \det a = a\delta - \gamma\beta = 1. \quad (3.45)$$

We use the Pauli matrices σ_i (i = 1, 2, 3), the identity matrix $\sigma_0 = \begin{pmatrix}1&0\\0&1\end{pmatrix}$, and the four-vector Δ_μ to construct the spin tensor

$$\tilde{\Delta} = \Delta_\mu \sigma^\mu = \begin{pmatrix}\Delta_0 + \Delta_3 & \Delta_1 - i\Delta_2\\ \Delta_1 + i\Delta_2 & \Delta_0 - \Delta_3\end{pmatrix}$$

which transforms under Lorentz rotations $\Delta'_\mu = L^\nu_\mu \Delta_\nu$ in accordance with the law

$$\tilde{\Delta}' = a\tilde{\Delta}a^+ = \Delta'_\mu \sigma^\mu. \quad (3.46)$$

In the group (3.45) we take the three-parametric subgroup consisting of triangular matrices of the form

$$K=\begin{pmatrix} e^{\frac{a}{2}} & 0 \\ e^{\frac{a}{2}}\gamma & e^{-\frac{a}{2}} \end{pmatrix} \begin{matrix} a\text{—real;} \\ \gamma=\gamma_1+i\gamma_2. \end{matrix} \tag{3.47}$$

If we assume that the four-vector Δ belongs to the hyperboloid $\Delta^2 = 1$, and we set $\Delta_\mu = (1, 0)$ and $a = k$ in (3.43), we obtain a relation that maps the space of the group parameters (3.47) onto the upper sheet of the hyperboloid $\Delta_0^2 - \Delta^2 = 1$:

$$\begin{pmatrix} \Delta_0+\Delta_3 & \Delta_1-i\Delta_2 \\ \Delta_1+i\Delta_2 & \Delta_0-\Delta_3 \end{pmatrix} = KK^+ = \begin{pmatrix} e^a & e^a(\gamma_1-i\gamma_2) \\ e^a(\gamma_1+i\gamma_2) & e^{-a}+e^a\tilde{\gamma}^2 \end{pmatrix}. \tag{3.48}$$

The two-vector (γ_1, γ_2) is here denoted by $\tilde{\gamma}$.

Equating the corresponding elements of the matrices in (3.48), we obtain

$$\left.\begin{aligned} &\Delta_0+\Delta_3=e^a; \\ &\Delta_0-\Delta_3=e^{-a}+e^a\tilde{\gamma}^2; \\ &\Delta=(\Delta_1, \Delta_2)=e^a\tilde{\gamma}. \end{aligned}\right\} \tag{3.49}$$

The relations (3.49) define an orispherical* system of coordinates on the surface $\Delta^2 = 1$. Since the matrices (3.47) form a group, a group operation is also induced in the space of parameters $(a, \tilde{\gamma})$ and, as a result of (3.49), on the upper sheet of the hyperboloid $\Delta^2 = 1$.

When applied to four-vectors of the form $\Delta_\mu = (\sqrt{1+\Delta^2}, \Delta)$ we shall denote this operation by the symbol $\oplus$ and write

$$\Delta' = \Delta \oplus q. \tag{3.51}$$

This equation is equivalent to the relations

$$\left.\begin{aligned} &a'=a+c; \\ &\tilde{\gamma}'=e^{-c}\tilde{\gamma}+\tilde{\mu}, \end{aligned}\right\} \tag{3.52}$$

where $(a', \tilde{\gamma}')$, $(a, \tilde{\gamma})$ and $(c, \tilde{\mu})$ are the orispherical coordinates that correspond to the four-vectors Δ_μ', Δ_μ and $\tilde{q}_\mu$, respectively.

It follows directly from (3.52) that the operation $\oplus$ has all the group properties. The inverse transformation is

$$q^{-1} \oplus q = 0, \tag{3.53}$$

and hence

$$q^{-1} \equiv (-c, \; -e^c\tilde{\mu}). \tag{3.54}$$

In what follows the set of $\oplus$ transformations will be called the group of orispherical shifts and denoted by T(3). It is readily seen from formulas (3.49) and (3.52) that in the nonrelativistic limit $(a, |\gamma| \ll 1)$

---

* An orisphere in the Lobachevskii space that is realized on the upper sheet of the hyperboloid $\Delta^2 = 1$ is defined as the two-dimensional surface specified by the equation

$$\Delta\xi = \text{const}, \quad \xi^2 = \xi_0^2 - \xi^2 = 0.$$

In particular, the equation

$$a = 0 \tag{3.50}$$

defines the orisphere $\Delta_0 + \Delta_3 = 1$ by virtue of (3.49).

A remarkable property of an orisphere is the fact that its internal geometry is Euclidean.

the orispherical coordinates go over into the Cartesian coordinates of the three-dimensional Euclidean momentum space and the group T(3) into the Abelian group of shifts of this space.

In the orispherical coordinates (3.49), the volume element of the hyperboloid, $d\Omega_{\Delta}=\frac{d\Delta}{\sqrt{1+\Delta^2}}$ has the form

$$d\Omega_{\Delta}=e^{2a}\,da\,d^2\tilde{\gamma}. \tag{3.55}$$

Since T(3) is a subgroup of the Lorentz group,

$$d\Omega_{\Delta\oplus q}=d\Omega_{\Delta}, \tag{3.56}$$

[cf with (3.17)]. The property (3.56) makes it possible to interpret (3.55) as a right-invariant volume element on the group T(3) itself.

Using (3.49) and (3.52)-(3.54), we can also readily show that

$$\Delta_0\lambda_0-\boldsymbol{\Delta}\boldsymbol{\lambda}=\sqrt{1+(\boldsymbol{\Delta}\oplus\boldsymbol{\lambda}^{-1})^2}. \tag{3.57}$$

Comparing (3.57) and (3.20), we obtain an important relation:

$$(\boldsymbol{\Delta}(-)\boldsymbol{\lambda})^2=(\boldsymbol{\Delta}\oplus\boldsymbol{\lambda}^{-1})^2. \tag{3.58}$$

We now turn to the description of a Fourier transformation on T(3).

An ordinary three-dimensional Fourier transformation is an expansion in plane waves, which are one-dimensional unitary representations of the Abelian group of translations of Euclidean space – that is, unitary solutions of the functional equation

$$U(\boldsymbol{\Delta}_1+\boldsymbol{\Delta}_2)=U(\boldsymbol{\Delta}_1)\,U(\boldsymbol{\Delta}_2).$$

Since T(3) is non-Abelian, it does not have nontrivial one-dimensional representations, i.e., the functional equation

$$U(\boldsymbol{\Delta}_1\oplus\boldsymbol{\Delta}_2)=U(\boldsymbol{\Delta}_1)\,U(\boldsymbol{\Delta}_2) \tag{3.59}$$

admits only operator solutions. The matrix elements of these operators form a basis with respect to which functions on the group T(3) can be expanded. The corresponding expansions play the role of Fourier transformations on the given group. The Fourier transforms that result are operators in the same space as the operators U in Eq. (3.59).

Let us now suppose that we dispose of a complete orthogonal system of "state vectors" $|\tilde{\rho}>$:

$$\left.\begin{aligned}&\int d\tilde{\rho}\,|\tilde{\rho}><\tilde{\rho}|=I;\\&<\tilde{\rho}|\tilde{\rho}'>=\delta^{(3)}(\tilde{\rho}-\tilde{\rho}');\quad\tilde{\rho}=(\rho_1,\rho_2).\end{aligned}\right\} \tag{3.60}$$

In the space of these vectors, we define an operator $\hat{U}_z(\boldsymbol{\Delta})=\hat{U}_z(a,\tilde{\gamma})$ by setting

$$<\tilde{\rho}_1|\hat{U}_z(a,\tilde{\gamma})|\tilde{\rho}_2>e^{iaz+i\tilde{\gamma}\tilde{\rho}_1-a}\,\delta(\tilde{\rho}_1-e^{-a}\tilde{\rho}_2). \tag{3.61}$$

It follows readily from (3.61) that

$$\hat{U}_z(a_1+a_2,\,e^{-a_2}\tilde{\gamma}_1+\tilde{\gamma}_2)=\hat{U}_z(a_1,\tilde{\gamma}_1)\,\hat{U}_z(a_2,\tilde{\gamma}_2); \tag{3.62}$$

$$\hat{U}_z^{+}(a,\tilde{\gamma})=\hat{U}_z(-a,\,-e^{a}\tilde{\gamma})=\hat{U}_z^{-1}(a,\tilde{\gamma}). \tag{3.63}$$

Thus, the operators $\tilde{U}_z(a,\gamma)$ realize a unitary representation of T(3).

Further, suppose $f(\boldsymbol{\Delta})=f(a,\tilde{\gamma})$ is a function on T(3). In accordance with what we have said above, the Fourier transform on this group is defined by the expression

$$\frac{1}{(2\pi)^3}\int f(\boldsymbol{\Delta})\,d\Omega_{\Delta}<\tilde{\rho}_1|\hat{U}_z(\boldsymbol{\Delta})|\tilde{\rho}_2>\equiv<\tilde{\rho}_1|\hat{f}(z)|\tilde{\rho}_2>. \tag{3.64}$$

It can be shown that there also exists an inversion formula, so that $f(\Delta)$ can be calculated from known matrix elements $<\tilde{\rho}_1|\hat{f}(z)|\tilde{\rho}_2>$:

$$f(\Delta)=\int dz d\tilde{\rho}_1 d\tilde{\rho}_2 <\tilde{\rho}_1|\hat{f}(z)\hat{U}_z^+(\Delta)|\tilde{\rho}_2>. \tag{3.65}$$

As can be seen from (3.65), to find $f(\Delta)$ it is in fact necessary to specify only the integral of $<\tilde{\rho}_1|\hat{f}(z)|\tilde{\rho}_2>$ with respect to the parameter $\tilde{\rho}_1$:

$$\int d\tilde{\rho}_1 <\tilde{\rho}_1|\hat{f}(z)|\tilde{\rho}_2> \equiv f(z, \tilde{\rho}_2), \tag{3.66}$$

and it is not necessary to know the matrix element $<\tilde{\rho}_1|\hat{f}(z)|\tilde{\rho}_2>$ itself.

This is because (3.66) can also be inverted, i.e., given $f(z, \tilde{\rho})$ one can establish the matrix $<\tilde{\rho}_1|\hat{f}(z)|\tilde{\rho}_2>$:

$$<\tilde{\rho}_1|\hat{f}(z)|\tilde{\rho}_2> = \frac{1}{(2\pi)}\int e^{ia(z-z')}dz'\, da\, \delta^{(2)}(\tilde{\rho}_1-e^{-a}\tilde{\rho}_2) f(z', \tilde{\rho}_2). \tag{3.67}$$

Using (3.59), we can readily obtain a "convolution theorem" for the transformations (3.64) and (3.65). Namely, if

$$\hat{f}_1(z)=\frac{1}{(2\pi)^3}\int f_1(\Delta)\hat{U}_z(\Delta)\, d\Omega_\Delta$$

and

$$\hat{f}_2(z)=\frac{1}{(2\pi)^3}\int f_2(\Delta)\hat{U}_z(\Delta)\, d\Omega_\Delta,$$

then

$$\left.\begin{aligned}\hat{f}_1(z)\hat{f}_2(z)&=\frac{1}{(2\pi)^3}\int d\Omega_\Delta\{f_1(\Delta)*f_2(\Delta)\}\hat{U}_z(\Delta);\\ f_1(\Delta)*f_2(\Delta)&=\frac{1}{(2\pi)^3}\int f_1(\Delta\oplus\lambda^{-1})f_2(\lambda)\, d\Omega_\lambda.\end{aligned}\right\} \tag{3.68}$$

It is readily seen that in the nonrelativistic limit all the relations associated with the Fourier decomposition on T(3) go over into the corresponding formulas of ordinary Fourier analysis.

9. Operations of Finite-Difference Differentiation and Analogs of the Most Important Functions

In Sec. 7 we have obtained the relativistic Schrödinger equation (3.39), in which the radial part of the free Hamiltonian is a combination of the finite-shift operators exp(id/dr) and exp(−id/dr) [see (3.38)]. It is therefore clear that the properties of these equations are most expediently investigated by the methods employed in the calculus of finite differences (see [20]).

We define the operation of finite-difference differentiation Δ by setting (cf. with [35])

$$\Delta=i\left(e^{-i\frac{d}{dr}}-1\right). \tag{3.69}$$

If Δ is applied to a function $f(r)$, (3.69) shows that

$$\Delta f(r)=\frac{f(r-i)-f(r)}{-i}. \tag{3.70}$$

Obviously, the relativistic Schrödinger equations can be rewritten in terms of Δ. It is clear that in the nonrelativistic limit*

$$\Delta f(r) \to \frac{d}{dr_4} f(r). \tag{3.71}$$

It is readily seen that the Δ derivative of the product of two functions $f(r)$ and $\varphi(r)$ is given by

$$\Delta [f(r)\varphi(r)] = [\Delta f(r)]\varphi(r) + f(r)[\Delta\varphi(r)] + \frac{1}{i}[\Delta f(r)][\Delta\varphi(r)]. \tag{3.72}$$

For effective use of the operation Δ it is desirable to have a function that behaves with respect to Δ differentiation like an ordinary power function. Now the function $r^{(\lambda)}$ defined by (3.30) is just such a generalized power; for it is readily shown that

$$\Delta r^{(\lambda)} = \lambda r^{(\lambda-1)}. \tag{3.73}$$

Further, it is obvious that

$$r^{(0)} = 1. \tag{3.74}$$

If $\lambda = n$, where n is a positive integer, then, with allowance for (3.30),

$$r^{(n)} = r(r+i)(r+2i)\cdots(r+i(n-1)). \tag{3.75}$$

On the transition to the nonrelativistic theory

$$r^{(\lambda)} \to r^{\lambda}. \tag{3.76}$$

This definition of a generalized power is more general than the corresponding definition in [35]. In contrast to the ordinary analysis, the product of two generalized powers is not a generalized power. However, it can be shown that for arbitrary λ and μ

$$r^{(\lambda)} r^{(\mu)} = \frac{r^{(\lambda+\mu)}}{F(-\lambda, -\mu; -ir; 1)}, \tag{3.77}$$

where F is the hypergeometric function. Now suppose we know the Taylor expansion of some function $f(r)$:

$$f(r+a) = \sum_{n=0}^{\infty} \frac{a^n}{n!} \cdot \frac{d^n}{dr^n} f(r). \tag{3.78}$$

Let us find the analogous expansion of f in terms of the Δ operation. First, it is clear that

$$f(r+a) = e^{a\frac{d}{dr}} f(r) = \left(e^{-i\frac{d}{dr}}\right)^{ia} f(r) = (1-i\Delta)^{ia} f(r). \tag{3.79}$$

Expanding the binomial $(1 - i\Delta)^{ia}$ in a series in powers of Δ, we obtain

$$(1-i\Delta)^{ia} = 1 + ia(-i\Delta) + \frac{ia(ia-1)}{2!}(-i\Delta)^2 + \frac{ia(ia-1)(ia-2)}{3!}(-i\Delta)^3 + \dots$$

Hence, and from (3.79) and using (3.75), we finally obtain

$$f(r+a) = \sum_{n=0}^{\infty} \frac{a^{(n)}}{n!} \Delta^n f(r). \tag{3.80}$$

---

* We recall that in the ordinary units (3.70) has the form

$$\Delta f(r) = \frac{f\left(r - \frac{i\hbar}{mc}\right) - f(r)}{-\frac{i\hbar}{mc}}$$

Comparison of the expansions (3.80) and (3.78) shows that these series have exactly the same structure.

Having at our disposal a generalization of a power function, we can, using corresponding series, introduce the analogs of the most important functions employed in continuous analysis [20].

We define a generalized exponential function by an expression of the form

$$\sum_{n=0}^{\infty} \frac{a^n r^{(n)}}{n!} \equiv \exp[a; r]. \tag{3.81}$$

With allowance for (3.73) it is obvious that

$$\Delta \exp[a; r] = a \exp[a; r]. \tag{3.82}$$

It is easy to show that

$$\exp[a; r] = F(-ir, 1; 1; ia), \tag{3.83}$$

or

$$\exp[a; r] = (1 - ia)^{ir} = e^{ir \ln(1 - ia)}. \tag{3.84}$$

It is here appropriate to emphasize that the plane waves $\xi(\mathbf{p}, \mathbf{r})$ introduced in Sec. 6 have a structure that can be expressed in terms of the generalized exponential function (3.84):

$$\xi(\mathbf{p}, \mathbf{r}) = (p_0 - \mathbf{pn})^{-1-ir} = (p_0 - \mathbf{pn})^{-1} \exp[i(p_0 - 1 - \mathbf{pn}); -r] \quad (\mathbf{r} = r\mathbf{n}). \tag{3.85}$$

It is also interesting to note that, with allowance for (3.84), formula (3.79) can be written in the form

$$f(r + a) = e^{a\frac{d}{dr}} f(r) = \exp[\Delta; a] f(r).$$

Therefore, to within the imaginary unit, the operation Δ is the "generator" of translations on the straight line r in the representation given by the generalized exponential functions $\exp[\Delta, a]$. Using the concept of a generalized exponential function, we can also define generalized hyperbolic and trigonometric functions (see [20]).

Let us also consider the so-called logarithmic derivative of the Γ function:

$$\Psi(z) = \frac{\Gamma'(z)}{\Gamma(z)}. \tag{3.86}$$

Using the well-known functional relation

$$\Psi(z + 1) - \Psi(z) = \frac{1}{z}, \tag{3.87}$$

we obtain

$$\Delta\Psi(ir + 1) = \frac{1}{r - i} = r^{(-1)}. \tag{3.88}$$

Therefore, the function $\Psi(ir + 1)$ occupies the same place in the calculus of finite differences as ln r in continuous analysis. This conclusion is also confirmed by the similarity of the expansions of these functions in the Taylor series (3.80) and (3.78), respectively:

$$\Psi(i(r + a) + 1) = \Psi(ir + 1) + \frac{a}{r - i} - \frac{1}{2} \cdot \frac{a(a + i)}{(r - i)(r - 2i)} + \frac{1}{3} \frac{a(a + i)(a + 2i)}{(r - i)(r - 2i)(r - 3i)} + \ldots = \Psi(ir + 1) + \sum_{n=1}^{\infty} (-1)^{n+1} \frac{r^{(-n)} a^{(n)}}{n}; \tag{3.89}$$

$$\ln(r + a) = \ln r + \sum_{r=1}^{\infty} (-1)^{n+1} \frac{r^{-n} a^a}{n}. \tag{3.90}$$

We now construct the analog of the hypergeometric function in this formalism [22]. It is natural to assume that this function (let it be ${}_2F_1[\alpha, \beta; \gamma; a; r]$) has the following expansion in the generalized powers $r^{(n)}$:

$$ {}_2F_1[\alpha, \beta; \gamma; a; r] = 1 + \frac{\alpha\beta}{\gamma} ar + \frac{\alpha(\alpha+1)\beta(\beta+1)}{\gamma(\gamma+1)\,2!} a^2 r(r+i) + \ldots = \sum_{n=0}^{\infty} \frac{(i\alpha)^{(n)}(i\beta)^{(n)}}{(i\gamma)^{(n)}} \cdot \frac{(-ia)^n r^{(n)}}{n!}. \tag{3.91} $$

Recalling the definition of the many-parametric hypergeometric function

$$ {}_pF_q(\alpha_1, \ldots, \alpha_p; \beta_1, \ldots, \beta_q; az) = \sum_{n=0}^{\infty} \frac{(i\alpha_1)^{(n)}(i\alpha_2)^{(n)} \ldots (i\alpha_p)^{(n)}}{(i\beta_1)^{(n)} \ldots (i\beta_q)^{(n)}} \, \frac{(i^{p-q}a)^n z^n}{n!} \tag{3.92} $$

we conclude that

$$ {}_2F_1[\alpha, \beta; \gamma; a; r] = {}_3F_1(\alpha, \beta, -ir; \gamma; ia). \tag{3.93} $$

It is readily verified that

$$ \Delta\, {}_2F_1[\alpha, \beta; \gamma; a; r] = a\frac{\alpha\beta}{\gamma}\, {}_2F_1[\alpha+1, \beta+1; \gamma+1; a; r]. $$

Using this equation and the recursion relations for ${}_3F_1$, we obtain a difference equation for the function (3.91):

$$ \{[r - i(\gamma+2)]\Delta^3 + i[r - 2i(\gamma+1) + a(\alpha+ir+2)(\beta+ir+2)]\Delta^2 $$
$$ + [i\gamma - 2a\alpha\beta - a(ir+2)(\alpha+\beta+1)]\Delta - ia\alpha\beta\}\, {}_2F_1[\alpha, \beta; \gamma; a; r] = 0. \tag{3.94} $$

It can be seen that this is a third-order equation in the operation Δ. However, in the nonrelativistic limit, the term containing Δ^3 vanishes and (3.94) goes over into the hypergeometric differential equation for ${}_2F_1(\alpha, \beta; \gamma; ar)$:

$$ \left\{ r(1-ar)\frac{d^2}{dr^2} + [\gamma - (\alpha+\beta+1)ar]\frac{d}{dr} - a\alpha\beta \right\} {}_2F_1(\alpha, \beta; \gamma; ar) = 0. $$

We also define a generalized confluent hypergeometric function $\Phi[\alpha; \gamma; a; r]$ by setting

$$ \Phi[\alpha; \gamma; a; r] \equiv {}_1F_1[\alpha; \gamma; a; r] = \lim_{\xi\to\infty} {}_2F_1\left[\alpha, \xi; \gamma; \frac{a}{\xi}; r\right] = \sum_{n=0}^{\infty} \frac{(i\alpha)^{(n)}}{(i\gamma)^{(n)}} \cdot \frac{a^n r^{(n)}}{n!}. \tag{3.95} $$

It is easily shown that

$$ \Phi[\alpha; \gamma; a; r] = {}_2F_1(\alpha, -ir; \gamma; ia), \tag{3.96} $$

where ${}_2F_1(\alpha, -ir; \gamma; ia)$ is the ordinary hypergeometric function.

Applying the recursion relations for the ordinary hypergeometric function ${}_2F_1$, we can easily show that $\Phi[\alpha; \gamma; a; r]$ satisfies the difference equation

$$ \{[\gamma + i(r-i)]\Delta^2 + i[\gamma - a(r-i) + ia\alpha]\Delta - ia\alpha\}\,\Phi[\alpha; \gamma; a; r] = 0. \tag{3.97} $$

As a second-order difference equation, Eq. (3.97) must have two linearly independent solutions. One can show that besides (3.96) [we denote this expression by $y_1(r, a)$], the function

$$ y_2(r, a) = a^{1-\gamma} r^{(1-\gamma)}\,\Phi[\alpha - \beta + 1;\ 2-\gamma;\ a;\ r + i(1-\gamma)] $$
$$ = a^{1-\gamma} r^{(1-\gamma)}\, {}_2F_1(\alpha - \gamma + 1,\ -ir + 1 - \gamma;\ 2-\gamma;\ ia) $$

is also a solution of (3.97). It is remarkable that the two functions $y_1(r, a)$ and $y_2(r, a)$ satisfy one and the same hypergeometric differential equation in the argument a. This "dualism" is characteristic of both the solutions of Eq. (3.94) and also of the solutions of the difference equations considered in this and subsequent chapters. In the nonrelativistic region, (3.97) obviously goes over into the differential equation for the confluent hypergeometric function:

$$\left[r\frac{d^2}{dr^2}+(\gamma-ar)\frac{d}{dr}-\alpha a\right]\Phi(\alpha;\gamma;ar)=0.$$

Note that Eqs. (3.94) and (3.97) can serve as the basis for the construction of the theory of special functions in the calculus of finite differences.

We now find a generalization of the step function $\vartheta(r)$:

$$\vartheta(r)=\begin{cases}1 & \text{for } r>0;\\ 0 & \text{for } r<0.\end{cases} \tag{3.98}$$

Of course, the most important property of $\vartheta(r)$ is expressed by the equation

$$\frac{d\vartheta(r)}{dr}=\delta(r), \tag{3.99}$$

whose validity can be proved most simply by proceeding from the integral representation

$$\vartheta(r)=\frac{1}{2\pi i}\int_{-\infty}^{\infty}\frac{e^{i\varkappa r}}{\varkappa-i\varepsilon}\,d\varkappa. \tag{3.100}$$

Being guided by (3.99), we define the generalized ϑ function as a function $\hat{\vartheta}(r)$ for which

$$\Delta\hat{\vartheta}(r)=\delta(r). \tag{3.101}$$

In complete analogy with (3.100)

$$\hat{\vartheta}(r)=\frac{1}{2\pi i}\int_{-\infty}^{\infty}\frac{e^{i\varkappa r}\,d\varkappa}{e^{\varkappa}-1-i\varepsilon}, \tag{3.102}$$

from which, calculating the integral, we obtain

$$\hat{\vartheta}(r)=\lim_{\varepsilon\to 0}\frac{(1+i\varepsilon)^{ir-1}}{1-e^{-2\pi r}}=\frac{1}{1-e^{-2\pi r}}. \tag{3.103}$$

Thus, $\hat{\vartheta}(r)$ for $r \neq 0$ is a function with period i and is therefore to be regarded as a constant under Δ differentiation [Eq. (3.70)]. At the point $r = 0$ the expression (3.103) has a pole, which is responsible for the existence of Eq. (3.101). In the nonrelativistic limit, $|r| \gg 1$, the function (3.103) obviously goes over into (3.98).

Note also that the identity $\vartheta(r) + \vartheta(-r) = 1$, which holds for the function (3.98), also holds for the $\hat{\vartheta}$ functions:

$$\hat{\vartheta}(r)+\hat{\vartheta}(-r)=1. \tag{3.104}$$

In calculating the Δ derivative of the δ function, the latter must be interpreted as a function of a complex variable since the Δ operation is accompanied by entry into the complex r plane. We therefore introduce the following representation of $\delta(r)$:

$$\delta(r)=\lim_{\mu\to 0}\frac{1}{2\pi i}\left(\frac{1}{r-i\mu}-\frac{1}{r+i\mu}\right). \tag{3.105}$$

With allowance for (3.105) and (3.70):

$$\Delta\delta(r)=\frac{1}{2\pi i}\lim_{\mu\to 0} i\left[\frac{1}{r-i-i\mu}-\frac{1}{r-i+i\mu}-\frac{1}{r-i\mu}+\frac{1}{r+i\mu}\right]=\frac{1}{i}\,\delta(r)+\frac{i}{\pi}\lim_{\mu\to 0}\frac{\mu}{(r-i)^2+\mu^2}. \tag{3.106}$$

The term $\frac{i\mu}{\pi} \cdot \frac{1}{(r-i)^2+\mu^2}$ in (3.106) in integrals along the real axis in the r space can be assumed to be a continuous function of μ, and therefore (3.106) is in fact equivalent to

$$\Delta\delta(r) = -i\delta(r). \tag{3.107}$$

The same equation is obtained by formal Δ differentiation of the identity $r\delta(r) = 0$ in accordance with the rule (3.72). Indeed,

$$\Delta[r\delta(r)] = \delta(r) + r\Delta\delta(r) + \frac{1}{i}\Delta\delta(r) = 0,$$

and hence

$$\Delta\delta(r) - \frac{-1}{r-i}\delta(r) = -i\delta(r).$$

Note that the relation (3.105) in conjunction with (3.87) and (3.70) can be written in the form

$$\delta(r) = \frac{1}{2\pi i}\lim_{\mu\to 0}\Delta[\Psi(ir+\mu) - \Psi(ir-\mu)]. \tag{3.108}$$

Hence, noting (3.107) and (3.103), we obtain an expression for the "smeared" $\hat{\vartheta}$ function:

$$\vartheta_\mu = \frac{1}{2\pi i}[\Psi(ir+\mu) - \Psi(ir-\mu)] - \frac{\operatorname{cth}\pi r}{2}. \tag{3.109}$$

10. Local Quasipotential. Relativistic Generalization of the Yukawa and Coulomb Potentials

It is shown in [7] that if it is required of Eq. (0.2) only that it give the correct physical scattering amplitude, the quasipotential $V(\mathbf{p}, \mathbf{k}, E_q)$ can be taken to be local:

$$V(\mathbf{p}, \mathbf{k}; E_q) = V\left((\mathbf{p}-\mathbf{k})^2; E_q\right), \tag{3.110}$$

and the function (3.110) then has the spectral representation [8]*

$$V\left((\mathbf{p}-\mathbf{k})^2; E_q\right) = \int_{\nu_0}^{\infty} \frac{\rho(\nu; E_q)\,d\nu}{\nu + (\mathbf{p}-\mathbf{k})^2}. \tag{3.111}$$

In other words, the quasipotential in such an approach is a superposition of nonrelativistic Yukawa potentials with an energy-dependent spectral function.

A slight modification of the arguments given in [7] in the derivation of (3.110) enables us to prove, under the same assumptions, that Eq. (2.23) leads to the correct physical amplitude $A(\mathbf{p}, \mathbf{q})$ in the case of a quasipotential of the form [24]

$$\tilde{V}(\mathbf{p}, \mathbf{k}; E_q) = \int_{\sigma_0}^{\infty} \frac{\rho(\sigma; E_q)\,d\sigma}{\sigma - (E_p - E_q)^2 + (\mathbf{p}-\mathbf{k})^2} \tag{3.112}$$

(see the foregoing remark).

If we take into account (3.20), we can rewrite (3.112) in the form

---

* Strictly speaking, the relation (3.111) must be written out separately for the even and odd parts of the quasipotential, which correspond to the even and odd angular momenta l in the expansion of V in partial waves. In addition, subtractions must be made in this relation. However, so as not to complicate the situation, we shall operate with representations of the type (3.111) and (3.112).

$$\widetilde{V}(\mathbf{p}, \mathbf{k}; E_q) = \int_{\sigma_0}^{\infty} \frac{\rho(\sigma; E_q)\, d\sigma}{\sigma - 2 + 2\sqrt{1 + (\mathbf{p}(-)\mathbf{k})^2}} \equiv V(\mathbf{p}(-)\mathbf{k})^2;\ E_q). \tag{3.113}$$

Thus, we have constructed a quasipotential which is local in the sense of Lobachevskii geometry. Setting

$$V(r, E_q) = \frac{1}{(2\pi)^3} \int \xi(\mathbf{p}, \mathbf{r})\, V(\mathbf{p}^2; E_q)\, d\Omega_p, \tag{3.114}$$

and taking into account (3.113), we obtain

$$V(r, E_q) = -\int_{\sigma_0}^{\infty} \rho(\sigma, E_q)\, V(\sigma, r)\, d\sigma, \tag{3.115}$$

where

$$V(\sigma, r) = \begin{cases} \dfrac{1}{4\pi r} \cdot \dfrac{\operatorname{ch}\left(r \arccos\left(-1 + \frac{\sigma}{2}\right)\right)}{\operatorname{sh} \pi r} & \text{for } \sigma < 4; \\[2ex] \dfrac{1}{4\pi r} \cdot \dfrac{\cos\left(r \operatorname{Ar\,ch}\left(-1 + \frac{\sigma}{2}\right)\right)}{\operatorname{sh} \pi r} & \text{for } \sigma > 4. \end{cases} \tag{3.116}$$

Clearly, the function $V(\sigma, r)$ can be regarded as the relativistic analog of the Yukawa potential. It is readily seen that for $r \gg 1$, $\sigma < 4$

$$V(\sigma, r) \to \frac{e^{-\sqrt{\sigma} r}}{4\pi r}. \tag{3.117}$$

If we set $\sigma = 0$ in (3.116), we obviously arrive at the relativistic generalization of the Coulomb potential:

$$V(0, r) = \frac{1}{4\pi r} \operatorname{cth} \pi r. \tag{3.118}$$

We emphasize that the potentials (3.116) and (3.118) correspond to interactions that propagate the finite velocity since they are associated in the p representation with the Feynman "propagators"

$$\frac{1}{\sigma - (p_0 - k_0)^2 + (\mathbf{p} - \mathbf{k})^2} \quad \text{and} \quad \frac{1}{(\mathbf{p} - \mathbf{k})^2 - (p_0 - k_0)^2}. \tag{3.119}$$

Substituting (3.113) into (2.28), we obtain

$$(2E_q - 2E_p)\, \Psi_q(\mathbf{p}) = \frac{1}{(2\pi)^3} \int V((\mathbf{p}(-)\mathbf{k})^2;\ E_q)\, \Psi_q(\mathbf{k})\, d\Omega_k \tag{3.120}$$

or, taking into account (3.18), the definition of convolution in Lobachevskii space,

$$(2E_q - 2E_p)\, \Psi_q(\mathbf{p}) = \frac{1}{(2\pi)^3} V(\mathbf{p}^2; E_q) * \Psi_q(\mathbf{p}). \tag{3.121}$$

Because of the spherical symmetry of the function $V((\mathbf{p}(-)\mathbf{k})^2;\ E_q)$, its expansion in the "plane waves" $\xi(\mathbf{p}, \mathbf{r})$ can be reduced to the form [see Eq. (3.114)]

$$V((\mathbf{p}(-)\mathbf{k})^2;\ E_q) = \int \xi^*((\mathbf{p}(-)\mathbf{k}), \mathbf{r})\, V(r, E_q)\, d\mathbf{r} = \int_0^{\infty} r^2\, dr\, V(r, E_q) \int d\omega_n\, \xi^*((\mathbf{p}(-)\mathbf{k}), \mathbf{r}), \tag{3.122}$$

and therefore, applying the "composition theorem" for the ξ functions (see [19]),

$$\int \xi^*(\mathbf{p}(-)\mathbf{k}, \mathbf{r})\, d\omega_n = \int \xi^*(\mathbf{p}, \mathbf{r})\, \xi(\mathbf{k}, \mathbf{r})\, d\omega_n, \tag{3.123}$$

we find

$$V((\mathbf{p}(-)\mathbf{k})^2; E_q) = \int \xi^*(\mathbf{p}, \mathbf{r}) V(r, E_q) \xi(\mathbf{k}, \mathbf{r}) d\mathbf{r}. \tag{3.124}$$

Then, substituting (3.124) into (3.120) and (3.121) and introducing the notation

$$\Psi_q(\mathbf{r}) = \frac{1}{(2\pi)^3} \int \xi(\mathbf{p}, \mathbf{r}) \Psi_q(\mathbf{p}) d\Omega_p, \tag{3.125}$$

we obtain*

$$[H_0 - 2E_q + V(r, E_q)] \Psi_q(\mathbf{r}) = 0. \tag{3.126}$$

Thus, in the relativistic **r** representation, we have obtained a difference-differential equation of Schrödinger type with a local, energy-dependent quasipotential. This quasipotential is the superposition (3.115) of the relativistic Yukawa potentials (3.116), and the actual form of the spectral function $\rho(\sigma, E_q)$ can in principle be determined from field theory. It is therefore clear that Eq. (3.126), like the Schrödinger equation (0.2) in the orthodox quasipotential approach, can be used to calculate relativistic corrections to the energy levels of bound states and also to find the asymptotic behavior of the relativistic scattering amplitude at high energies.

In addition, it is of interest to use Eq. (3.126) for a phenomenological description of a system of two interacting relativistic particles. If this is done, the choice of the quasipotential in a local form means, in the light of what we have said above, that to a certain extent allowance is made for the existence of a spectral representation of the relativistic scattering amplitude or, ultimately, the properties of locality and causality in the sense of field theory. Some problems with simple local quasipotentials are considered in the next section.

SECTION 4. APPLICATIONS

11. Solutions of the Free Difference Schrödinger Equation

In accordance with (3.126), we have the following free radial equation in our formalism when the interaction is switched off:

$$\left(H_0^{\text{rad}} - 2E_q\right) \Psi_{lq}^{(0)}(r) = 0, \tag{4.1}$$

where the operator H_0^{rad} is given by the formula [cf with (3.38)]

$$H_0^{\text{rad}} = 2\,\mathrm{ch}\, i\frac{d}{dr} + \frac{2i}{r}\,\mathrm{sh}\, i\frac{d}{dr} + \frac{l(l+1)}{r^2} e^{i\frac{d}{dr}}. \tag{4.2}$$

Writing (4.1) in terms of the Δ operation, we readily see that this equation is of second order in Δ. In accordance with the general theory of difference equations [35], it follows that Eq. (4.1) has two linearly independent solutions.

Let us find the solutions of (4.1). One of them is actually already known: it is the function $p_l(\cosh \chi_q, r)$ defined by (3.31) and satisfying the normalization condition (3.33). However, as in the nonrelativistic theory, it is convenient in this case to use a solution with a different normalization. Namely, we set

$$s_l(r, \chi) = r\,\mathrm{sh}\,\chi\, p_l(\mathrm{ch}\,\chi, r). \tag{4.3}$$

Then, using (3.33) and (3.34), we obtain orthonormality and completeness conditions for the functions $s_l(r, \chi)$:

$$\frac{2}{\pi} \int_0^\infty dr s_l(r\chi) s_l^*(r, \chi') = \delta(\chi - \chi'); \tag{4.4}$$

---

* In the case $m_1 \neq m_2$ it is also possible to obtain a difference equation in the **r** representation with a local quasipotential [23].

$$\frac{2}{\pi}\int_0^\infty d\chi s_l(r,\chi)\, s_l^*(r',\chi)=\delta(r-r'). \tag{4.5}$$

Substituting $p_l(\cosh\chi, r)$, expressed in terms of $s_l(r,\chi)$, into (4.1), we obtain an equation for $s_l(r,\chi)$:

$$\hat{H}_0^{\rm rad} s_l(r,\chi)\equiv\left(2\,\mathrm{ch}\, i\frac{d}{dr}+\frac{l(l+1)}{r^{(2)}}e^{i\frac{d}{dr}}\right)s_l(r,\chi)=2Es_l(r,\chi). \tag{4.6}$$

Thus, if the substitution (4.3) is made, the term with $\frac{1}{i}\,\mathrm{sh}\, i\frac{d}{dr}$ disappears from Eq. (4.1). For comparison we recall that in the nonrelativistic theory a substitution of the type (4.3) eliminates the term with the first derivative from the Schrödinger equation.

It follows from (3.69) that the expression $\frac{1}{i}\,\mathrm{sh}\, i\frac{d}{dr}$ is the real part of the operator Δ:

$$\frac{\Delta+\Delta^*}{2}=\frac{1}{i}\,\mathrm{sh}\, i\frac{d}{dr}. \tag{4.7}$$

Applying the recursion relations between the spherical functions P_ν^μ, we can derive a recursion relation for $s_l(r,\chi)$ that uses the operator (4.7) [cf. (3.32)]:

$$s_l(r,\chi)=-\frac{(-r)^{(l+1)}}{(\mathrm{sh}\chi)^{(l)}}\left[\frac{1}{ir}\,\mathrm{sh}\, i\frac{d}{dr}\right]^l\frac{s_0(r,\chi)}{r}. \tag{4.8}$$

In complete analogy with the nonrelativistic theory, we can assert that the function

$$c_l=(-1)^l s_{-l-1} \tag{4.9}$$

is also a solution of Eq. (4.6), since this equation is unaffected by the substitution

$$l\to -l-1. \tag{4.10}$$

Using (4.9), (4.3), and (3.31), we obtain

$$c_l=\sqrt{\frac{\pi\,\mathrm{sh}\,\chi}{2}}(-r)^{(-l)}P^{1/2+l}_{-1/2+ir}(\mathrm{ch}\,\chi). \tag{4.11}$$

Using the recursion relation (3.32), we can readily find the asymptotic behavior as $r\chi\to\infty$ of the function $p_l(\cosh\chi, r)$ and consequently of the functions $s_l(r,\chi)$ and $c_l(r,\chi)$. Simple calculations give

$$\left.\begin{aligned} s_l(r,\chi)&\approx\sin\left(r\chi-\frac{l\pi}{2}\right) \quad \text{for } r\chi\gg1;\\ c_l(r,\chi)&\approx\cos\left(r\chi-\frac{l\pi}{2}\right) \quad \text{for } r\chi\gg1.\end{aligned}\right\} \tag{4.12}$$

Thus, the solutions $s_l(r,\chi)$ and $c_l(r,\chi)$ of Eq. (4.6) correspond to standing waves.

We now introduce the functions (see [36])

$$e_l^{(1)}(r,\chi)=c_l(r,\chi)+i\,s_l(r,\chi); \tag{4.13}$$

$$e_l^{(2)}(r,\chi)=c_l(r,\chi)-i\,s_l(r,\chi). \tag{4.14}$$

The free solutions of these functions can be expressed in terms of the associated Legendre functions of the second kind in the following manner:

$$e_l^{(1,2)}(r,\chi)=\frac{1}{i}\sqrt{\frac{2\,\mathrm{sh}\,\chi}{\pi}}(-r)^{(-l)}Q^{\frac{1}{2}+l}_{-\frac{1}{2}\pm ir}(\mathrm{ch}\,\chi). \tag{4.15}$$

It follows from (4.12) that

$$e_l^{(1,2)}(r,\chi) \approx e^{\pm i\left(r\chi - \frac{l\pi}{2}\right)} \quad \text{for} \quad r\chi \gg 1. \tag{4.16}$$

Clearly the choice of the solutions of Eq. (4.6) in the form (4.15) corresponds to distinguishing outgoing and ingoing spherical waves.

In the nonrelativistic limit

$$\chi \ll 1, \ r \gg 1, \chi r = \text{ fixed.} \tag{4.17}$$

The functions $s_l(r,\chi), c_l(r,\chi), e_l^{(1)}$ and $e_l^{(2)}$ go over to within a factor into the corresponding cylindrical functions:

$$\left.\begin{aligned} &s_l(r,\chi_q) \to \sqrt{\frac{\pi r q}{2}}\, J_{l+1/2}(rq);\\ &c_l(r,\chi_q) \to -\sqrt{\frac{\pi r q}{2}}\, N_{l+1/2}(rq);\\ &e^{(1,2)}(r,\chi_q) \to \pm i \sqrt{\frac{\pi r q}{2}}\, H_{l+1/2}^{(1,2)}(rq). \end{aligned}\right\} \tag{4.18}$$

According to the general theory of finite-difference equations [35], any two solutions $\varphi_1(r)$ and $\varphi_2(r)$ of the difference equation (4.6) are linearly independent if the determinant

$$\begin{vmatrix} \varphi_1(r) & \varphi_2(r) \\ \Delta\varphi_1(r) & \Delta\varphi_2(r) \end{vmatrix} \equiv W(\varphi_1, \varphi_2) \tag{4.19}$$

is nonvanishing. Clearly, (4.19) must be regarded as the analog of a Wronskian in finite-difference analysis. In the nonrelativistic limit, we obviously have

$$W(\varphi_1,\varphi_2) \to \begin{vmatrix} \varphi_1(r) & \varphi_2(r) \\ \frac{d}{dr}\varphi_1(r) & \frac{d}{dr}\varphi_2(r) \end{vmatrix}. \tag{4.20}$$

On the basis of (4.13) and (4.14),

$$\left.\begin{aligned} &W(s_l, c_l) = \frac{1}{2i} W\left(e_l^{(1)}, e_l^{(2)}\right);\\ &W(s_l, c_l) = W\left(s_l, e_l^{(1)}\right) = W\left(s_l, e_l^{(2)}\right). \end{aligned}\right\} \tag{4.21}$$

As in the continuous case, the Wronskians (4.21) can be calculated exactly. We give only the final expression for generalized Wronskian [20]:

$$W\left(e_l^{(1)}, e_l^{(2)}\right) = \frac{2i(-r)^{(l+1)}}{(r)^{(l+1)}}(-1)^l \operatorname{sh}\chi. \tag{4.22}$$

It is interesting to note the relation

$$\frac{W\left(e_l^{(1)}, e_l^{(2)}\right)}{\begin{vmatrix} e_l^{(1)}(r,\chi) & e_l^{(2)}(r,\chi) \\ \frac{\partial}{\partial\chi} e_l^{(1)}(r,\chi) & \frac{\partial}{\partial\chi} e_l^{(2)}(r,\chi) \end{vmatrix}} \equiv \frac{W\left(e_l^{(1)}, e_l^{(2)}\right)}{W_\chi\left(e_l^{(1)}, e_l^{(2)}\right)} = \frac{\operatorname{sh}\chi}{r}. \tag{4.23}$$

If it is recalled that sinh χ is the modulus of a three-dimensional momentum, the similarity of (4.23) and the analogous ratio of Wronskians in the nonrelativistic theory becomes obvious. In some calculations it is useful to remember the equation

$$W\left(e_l^{(1)}, e_l^{(2)}\right) = \frac{ir - l - 1}{ir} \operatorname{sh} \chi \begin{vmatrix} e_l^{(1)}(r, \chi) & e_l^{(2)}(r, \chi) \\ e_{l+1}^{(1)}(r, \chi) & e_{l+1}^{(2)}(r, \chi) \end{vmatrix}. \tag{4.24}$$

An analog of this equation also exists in the nonrelativistic case.

Our next problem is to use the free solutions of Eq. (4.6) to construct its Green's function $G_l(r, r'; \chi_q)$. Clearly, $G_l(r, r'; \chi_q)$ satisfies the equation [cf. (3.40)]

$$\left[2 \operatorname{ch} \chi_q - 2 \operatorname{ch} i \frac{d}{dr} - \frac{l(l+1)}{r^{(2)}} e^{i \frac{d}{dr}}\right] G_l(r, r'; \chi_q) = \delta(r - r'). \tag{4.25}$$

One can then argue as in the nonrelativistic formalism. Indeed, with allowance for (4.25), the Green's function for $r < r'$ and $r > r'$ satisfies the free equation and must therefore be a linear combination of the functions s_l, c_l, and $e_l^{(1,2)}$ with i-periodic coefficients. At the point $r = r'$, these coefficients must have a singularity so that the function $\delta(r - r')$ arises after differentiation with respect to r. Taking into account the correspondence principle and the definition of $\hat{\vartheta}(r)$ [see (3.101)-(3.103)], we readily see that it is precisely this generalized ϑ function that must occur in the expressions for the coefficients.

The exact form of $G_l(r, r'; \chi_q)$ can easily be found from the integral representation [see (3.35)]:

$$G_l(r, r'; \chi_q) = \frac{2}{\pi} \int_0^\infty \frac{d\chi s_l(r, \chi) s_l^*(r', \chi)}{2 \operatorname{ch} \chi_q - 2 \operatorname{ch} \chi + i\varepsilon}. \tag{4.26}$$

The integral (4.26) can be calculated by residue theory. Omitting the fairly long calculations, we content ourselves with the results:

$$\begin{aligned} G_l(r, r'; \chi_q) \equiv & -\frac{1}{W\left(e_l^{(1)}(r, \chi_q), e_l^{(2)}(r', \chi_q)\right)} \\ & \times \{\hat{\vartheta}(r - r') e_l^{(1)}(r, \chi_q) e_l^{(2)}(r', \chi_q) + \hat{\vartheta}(r' - r) e_l^{(1)}(r', \chi_q) e_l^{(2)}(r, \chi_q) \\ & - \hat{\vartheta}(r + r') e_l^{(1)}(r, \chi_q) e_l^{(1)}(r', \chi_q) - \hat{\vartheta}(-r - r') e_l^{(2)}(r, \chi_q) e_l^{(2)}(r', \chi_q)\}. \end{aligned} \tag{4.27}$$

The only difference between (4.27) and the corresponding expression of the ordinary theory [36] is that our Green's function contains an "acausal" term proportional to $\hat{\vartheta}(-r - r')$. In the nonrelativistic limit this term obviously vanishes.

12. Potential Well

Exactly solvable examples provide an important illustration of the technique. Let us consider the case of a spherically symmetric potential well of finite depth [21, 24]:

$$V(r) = -V_0 \quad \text{for } r \leqslant a; \tag{4.28}$$

$$V(r) = 0 \quad \text{for } r > a. \tag{4.29}$$

We shall first study the discrete spectrum, assuming for simplicity that $l = 0$.

If $r \le a$, Eq. (3.126) has the form

$$\left(2 \operatorname{ch} i \frac{d}{dr} - 2E_q - V_0\right) \Psi_{q_0}^{\mathrm{I}}(r) = 0. \tag{4.30}$$

In this case $E_q \le 1$, and it is convenient to use the parametrization

$$E_q = \cos \chi_q. \tag{4.31}$$

We seek $\Psi_{q_0}^{\mathrm{I}}(r)$ in the form

$$\Psi_{q_0}^{\mathrm{I}}(r) = A^{\mathrm{I}}(r) \sin \alpha r, \tag{4.32}$$

where $A^{\mathrm{I}}(r)$ is an i-periodic function.

The function $\Phi^{\mathrm{I}}_{q_0}(r)$ satisfies the boundary condition

$$\Psi^{\mathrm{I}}_{q_0}(0)=0. \tag{4.33}$$

Substituting $\Psi^{\mathrm{I}}_{q_0}(r)$ into (4.30), we obtain an equation for α:

$$\operatorname{ch}\alpha=E_q+\frac{V_0}{2}. \tag{4.34}$$

In the region $r > 0$, Eq. (3.126) takes the form

$$\left(2\operatorname{ch} i\frac{d}{dr}-2E_q\right)\Psi^{\mathrm{I}}_{q_0}(r)=0. \tag{4.35}$$

The solution $\Psi^{\mathrm{II}}_{q_0}$, which satisfies the condition

$$\Psi^{\mathrm{II}}_{q_0}(\infty)=0, \tag{4.36}$$

must be sought in the form

$$\Psi^{\mathrm{II}}_{q_0}(r)=A^{\mathrm{II}}(r)\,e^{-r\chi_q}. \tag{4.37}$$

We now determine the energy levels by requiring that the solutions $\Psi^{\mathrm{I}}_{q_0}(r)$ and $\Psi^{\mathrm{II}}_{q_0}(r)$ satisfy a fitting condition at the point $r = a$ [24]:

$$W\left(\Psi^{\mathrm{I}}_{q_0}(r),\ \Psi^{\mathrm{II}}_{q_0}(r)\right)\big|_{r=0}=0. \tag{4.38}$$

Substituting (4.32) and (4.37) into (4.38) and ignoring the i-periodic constants A^{I} and A^{II}, we obtain an equation for the bound states:

$$\operatorname{ctg}\alpha a=-\frac{\sin\chi_q}{\operatorname{sh}\alpha}. \tag{4.39}$$

Let us consider the case when $V_0 = 2m$ and the energy of the bound state $2E_q$ is small compared with m:

$$\mu=2E_q=2m-|W|\ll m. \tag{4.40}$$

In the language of the composite models of elementary particles (see, for example, [37-39]), the condition (4.40) means that the π-meson mass must be appreciably less than the quark mass. Further, neglecting all powers of μ higher than the first in (4.39), we can represent this formula as

$$\operatorname{ctg}\frac{amc}{\hbar}\sqrt{\frac{\mu}{2m}}=-\sqrt{\frac{2m}{\mu}}, \tag{4.41}$$

from which we obtain an approximate value of a (the "radius" of the π meson):

$$a\approx\frac{\pi\sqrt{2}\,\hbar}{c\sqrt{m\mu}}. \tag{4.42}$$

We now estimate the order of magnitude of the mean value of the quark momentum in the well. It follows from the uncertainty relation that

$$\chi a\approx\frac{\hbar}{mc}.$$

Replacing the radius a by its value from (4.42) in this relation and noting again that μ/m is small, we obtain

$$\frac{p^2}{m^2c^2}\sim\frac{\mu}{m}. \tag{4.43}$$

Therefore, quarks in a well can be regarded as nonrelativistic particles if the mass of the bound state is sufficiently small compared with the quark mass [the condition (4.40)].

Thus, the relativistic formulation of the two-body problem that we have developed leads to a justification of one of the principal hypotheses on which the quark model is based – the hypothesis that the motion of the quarks within the composite particle has a nonrelativistic nature.

13. Coulomb Potential

Let us consider an attractive Coulomb field [22, 24]:

$$V(r) = -\frac{e^2}{r}. \tag{4.44}$$

Equation (3.126) for the potential (4.44) takes the form

$$\left(H_0 - 2E_q - \frac{e^2}{r}\right)\Psi_{ql}(r) = 0. \tag{4.45}$$

Let us first find the solution of this equation that corresponds to the discrete spectrum. It is again convenient to take the parametrization (4.31). It is readily verified that Eq. (4.45) is satisfied by the function

$$\Psi_{ql}(r) = c_l(\chi_q, r)\, e^{-r\chi_q}(-r)^{(l+1)}\,\Phi\left[l+1-\frac{e^2}{2\sin\chi_q};\ 2l+2;\ 2\sin\chi_q\, e^{-i\chi_q};\ r+i(l+1)\right], \tag{4.46}$$

where $\Phi[\alpha;\ \gamma;\ c;\ z]$ is the generalized confluent hypergeometric function (3.96) and the explicit form of the "constant" $c_l(\chi_q, r)$ does not affect the formula for the energy levels. Obviously,

$$\Psi_{ql}(0) = 0. \tag{4.47}$$

The hypergeometric series in (4.46) increases not faster than a polynomial if

$$\frac{e^2}{2\sin\chi_q} = n, \quad n = 1, 2\ldots \tag{4.48}$$

With allowance for (4.48), this yields the rule for quantization of the energy levels:

$$E_q^n = \sqrt{1 - \frac{e^4}{4n^2}}. \tag{4.49}$$

In the continuous spectrum, $E_q = \cosh\chi_q \geq 1$, the solution of Eq. (4.45) has the form

$$\Psi_{ql}(r) = c_l'(\chi_q, r)\, e^{ir\chi_q}(-r)^{(l+1)}\,\Phi\left[l+1-\frac{i e^2}{2\,\mathrm{sh}\,\chi_q};\ 2l+2; -2i\,\mathrm{sh}\,\chi_q\, e^{-\chi_q};\ r+i(l+1)\right]. \tag{4.50}$$

As $r \to \infty$, Eq. (4.50) takes the asymptotic form

$$\Psi_{ql}^{ac}(r) = c_l'(\chi_q, r)\left\{\frac{\exp i\left[r\chi_q + \frac{e^2}{2\mathrm{sh}\,\chi_q}\ln r\,\mathrm{sh}\,\chi_q\right]}{\Gamma\left(l+1+\frac{ie^2}{2\mathrm{sh}\,\chi_q}\right)} + (-1)^{l+1}\frac{\exp i\left[-r\chi_q - \frac{e^2}{2\,\mathrm{sh}\,\chi_q}\ln r\,\mathrm{sh}\,\chi_q\right]}{\Gamma\left(l+1-\frac{i e^2}{2\,\mathrm{sh}\,\chi_q}\right)}\right\}. \tag{4.51}$$

As in the nonrelativistic case, the radial solution can be represented as the sum of ingoing and outgoing spherical waves distorted by logarithmic terms.

From the asymptotic expansion (4.51) we obtain an expression for the partial S matrix

$$s_l(q) = \frac{\Gamma\left(l+1-\frac{ie^2}{2\,\mathrm{sh}\,\chi_q}\right)}{\Gamma\left(l+1+\frac{ie^2}{2\,\mathrm{sh}\,\chi_q}\right)} = \frac{\Gamma\left(l+1-\frac{ie^2}{2\sqrt{E^2-1}}\right)}{\Gamma\left(l+1+\frac{ie^2}{2\sqrt{E^2-1}}\right)}. \tag{4.52}$$

In the nonrelativistic limit, the expressions (4.46), (4.49), (4.50), and (4.52) go over into the corresponding expressions for the quantum-mechanical Coulomb problem.

14. High-Energy Representation for the Relativistic Scattering Amplitude

We shall now obtain a closed expression for the relativistic quasipotential scattering amplitude that is valid at high energies [40]. This representation is the relativistic analog of the eikonal representation [41].

Consider Eq. (2.28). Assuming that the quasipotential $\tilde{V}(\mathbf{p}, \mathbf{k}; E_q)$ is local [19], we set in this equation

$$\left.\begin{aligned} &\mathbf{p}(-)\mathbf{q}=\boldsymbol{\Delta};\ \mathbf{k}(-)\mathbf{q}=\boldsymbol{\lambda};\\ &\Psi_q(\boldsymbol{\Delta}(+)\mathbf{q})\equiv\Phi_q(\boldsymbol{\Delta}).\end{aligned}\right\} \tag{4.53}$$

Since $(\mathbf{p}(-)\mathbf{k})$ is a relativistic invariant,

$$(\mathbf{p}(-)\mathbf{k})^2=(\boldsymbol{\Delta}(-)\boldsymbol{\lambda})^2. \tag{4.54}$$

Taking into account (4.54) and the invariance of the volume $d\Omega_k = d\Omega_\lambda$, we obtain the following equation for the function $\Phi_q(\boldsymbol{\Delta})$:

$$\Phi_q(\boldsymbol{\Delta})=(2\pi)^3\delta^3(\boldsymbol{\Delta})-\frac{1}{2(E_{\Delta(+)q}-E_q-i\varepsilon)}\frac{1}{(2\pi)^3}\int V((\boldsymbol{\Delta}(-)\boldsymbol{\lambda})^2)\Phi_q(\boldsymbol{\lambda})\,d\Omega_\lambda. \tag{4.55}$$

With allowance for (3.20), the Green's function in Eq.(4.55) can be rewritten as follows:

$$G=\frac{1}{2(E_q-E_{\Delta(+)q}+i\varepsilon)}=\frac{1}{2(E_q-E_\Delta E_q-\Delta q+i\varepsilon)}. \tag{4.56}$$

At high energies, $E_q \gg 1$, we have approximately

$$E_q\approx q+\frac{1}{2q}. \tag{4.57}$$

If the vector $\mathbf{q}$ is assumed parallel to the z axis [$\mathbf{q}$ = (0, 0, q)], then (4.56) is, with allowance for (4.57), replaced by

$$G=\frac{1}{2q\left(1-\Delta_0-\Delta_3-\dfrac{\Delta_0-1}{2q^2}+i\varepsilon\right)};\quad \Delta_0\equiv E_\Delta. \tag{4.58}$$

Further, we assume that at high energies the term $(\Delta_0-1)/2q^2$ in the denominator of (4.58) can be ignored. Then the Green's function (4.58) becomes

$$G\approx\frac{1}{2q(1-\Delta_0-\Delta_3+i\varepsilon)}. \tag{4.59}$$

If this approximate expression for the Green's function is used, we must ask for the conditions under which the neglect of the term $(\Delta_0-1)/2q^2$ is valid. To answer this question, we go over to the invariant Mandelstam variables s and t. Obviously,

$$\left.\begin{aligned} s&=4(\mathbf{q}^2+m^2)\approx 4\mathbf{q}^2;\\ t&=(E_p-E_q)^2-(\mathbf{p}-\mathbf{q})^2=\\ &=2(1-E_pE_q+\mathbf{pq})=2(1-\Delta_0),\end{aligned}\right\} \tag{4.60}$$

and therefore

$$\frac{\Delta_0-1}{2q^2}=\frac{|t|}{s}. \tag{4.61}$$

Therefore neglect of (4.61) means that the resulting expression is valid if the kinematic invariants s and t satisfy

$$\left|\frac{t}{s}\right|\ll 1.$$

Further, taking into account (3.58), we can interpret the integral term in Eq. (4.55) as a convolution on the group T(3) (see Sec. 8):

$$\frac{1}{(2\pi)^3}\int V\left((\Delta \oplus \lambda^{-1})^2\right) \Phi_q(\lambda)\, d\Omega_\lambda \equiv V(\Delta^2)\cdot \Phi_q(\Delta). \tag{4.62}$$

Finally, making the transition to the orispherical coordinates (3.49) and neglecting the term $(\Delta_0 - 1)/2q^2$, we transform Eq. (4.55) to

$$\Phi_q(a, \tilde{\gamma}) = \delta(a)\,\delta^2(\tilde{\gamma}) - \frac{1}{2q\,(e^a - 1 - i\varepsilon)} V(a, \tilde{\gamma}) * \Phi_q(a, \tilde{\gamma}). \tag{4.63}$$

We now apply the apparatus of Fourier transformation on T(3) to Eq. (4.63). First of all, we set by definition [cf. (3.64)]

$$\langle \tilde{\rho}_1 | \hat{\Phi}_q(z) | \tilde{\rho}_2 \rangle = \frac{1}{(2\pi)^3}\int \Phi_q(a, \tilde{\gamma}) \langle \tilde{\rho}_1 | \hat{U}_z(a, \tilde{\gamma}) | \tilde{\rho}_2 \rangle\, e^{2a}\, da\, d\tilde{\gamma}\,; \tag{4.64}$$

$$\langle \tilde{\rho}_1 | \hat{V}_q(z) | \tilde{\rho}_2 \rangle = \frac{1}{(2\pi)^3}\int V_q(a, \tilde{\gamma}) \langle \tilde{\rho}_1 | \hat{U}_z(a, \tilde{\gamma}) | \tilde{\rho}_2 \rangle\, e^{2a}\, da\, d\tilde{\gamma}. \tag{4.65}$$

Using formulas (3.62) and (3.65)-(3.68), we write Eq. (4.63) in the form

$$\langle \tilde{\rho}_1 | \hat{\Phi}_q(z) | \tilde{\rho}_2 \rangle = \langle \tilde{\rho}_1 | \tilde{\rho}_2 \rangle + \frac{1}{2qi}\int_{-\infty}^{\infty} dz'\, \hat{\vartheta}(z - z') \langle \tilde{\rho}_1 | \hat{V}_q(z') \hat{\Phi}_q(z') | \tilde{\rho}_2 \rangle, \tag{4.66}$$

where

$$\hat{\vartheta}(z - z') = \frac{1}{2\pi i}\int_{-\infty}^{\infty} \frac{e^{ia(z - z')}}{e^a - 1 - i\varepsilon}\, da \tag{4.67}$$

is the "step" function in the calculus of finite differences (see Sec. 9), which satisfies an inhomogeneous first-order differential equation:

$$\left.\begin{aligned} &\Delta_z \hat{\vartheta}(z - z') = \delta(z - z'); \\ &\Delta_z \equiv \frac{e^{-i\frac{d}{dz}} - 1}{-i}. \end{aligned}\right\} \tag{4.68}$$

Taking this equation into account, we obtain a difference equation that is equivalent to (4.66):

$$\Delta_z \hat{\Phi}_q(z) = \frac{1}{2qi} \hat{V}_q(z) \hat{\Phi}_q(z) \tag{4.69}$$

with the boundary condition for the operator $\hat{\Phi}_q(z)$

$$\hat{\Phi}_q(z)|_{z = -\infty} = 1. \tag{4.70}$$

Clearly, $\Phi_q(z)$ is the relativistic analog of the "slowly varying part" of the wave function (see for example, [40-42]).

A formal solution of Eqs. (4.66) and (4.69) is the "ordered exponential function"

$$\begin{aligned} \hat{\Phi}_q(z) &= \sum_{n=0}^{\infty} \frac{1}{(2qi)^n}\int \hat{\vartheta}(z - z_1)\, \hat{\vartheta}(z_1 - z_2) \ldots \hat{\vartheta}(z_{n-1} - z_n)\, \hat{V}_q(z_1)\, \hat{V}_q(z_2) \ldots \hat{V}_q(z_n)\, dz_1 \ldots dz_n \\ &\equiv P_z \exp\left[\frac{1}{2qi}\int_{-\infty}^{\infty} \hat{\vartheta}(z - z')\, \hat{V}_q(z')\, dz'\right]. \end{aligned} \tag{4.71}$$

For simple quasipotentials, this series can be calculated exactly.

We now turn to the quasipotential scattering amplitude A(p, q). In accordance with (3.27),

$$A(\mathbf{p}, \mathbf{q}) = -\frac{1}{4\pi} \cdot \frac{1}{(2\pi)^3} \int V_q((\mathbf{p}(-)\mathbf{k})^2)\Psi_q(\mathbf{k})\, d\Omega_k. \tag{4.72}$$

Hence, taking into account (4.53), (4.54), (4.58), and (4.62) and the fact that $\mathbf{q} = (0, 0, q)$, we find

$$A(\mathbf{p}, \mathbf{q}) = A(\Delta, \mathbf{q}) = -\frac{1}{4\pi} V_q(\Delta^2) * \Phi_q(\Delta). \tag{4.73}$$

Further, using the "convolution theorem" and (3.65), we conclude that

$$A(\Delta^2, q) = -\frac{1}{4\pi} \int dz\, d\tilde{\rho}_1\, d\tilde{\rho}_2 \langle \tilde{\rho}_1 | \hat{V}_q(z)\, \hat{\Phi}_q(z)\, \hat{U}_z(\Delta) | \tilde{\rho}_2 \rangle. \tag{4.74}$$

We can then show that in the high-energy limit [see (4.61)]

$$\left.\begin{array}{l} e^a \approx 1;\ a \approx 0; \\ \tilde{\gamma} \approx \tilde{\Delta};\ |t| \approx \tilde{\gamma}^2. \end{array}\right\} \tag{4.75}$$

In other words, in this approximation the momentum transfer vector Δ lies on the orisphere (3.50). This fact can be interpreted as a distinctive "transfersality condition" in the relativistic case. Finally, making simple calculations, in which (3.61), (4.68), (4.69), and (4.75) are used, we obtain the desired high-energy representation for the quasipotential scattering amplitude:

$$\begin{aligned} A(t, s) &= -\frac{iq}{2\pi} \int d\tilde{\rho}\, e^{-i\tilde{\Delta}\tilde{\rho}} \left[\int d\tilde{\rho}_1 \left\langle \tilde{\rho}_1 \middle| P_z \exp\left[\frac{1}{2qi} \int_{-\infty}^{\infty} \hat{V}(z)\, dz\right] \middle| \tilde{\rho} \right\rangle - 1 \right] \\ &= -iq \int_0^{\infty} \rho\, d\rho\, J_0(\sqrt{-t}\rho) \left[\int d\tilde{\rho}_1 \left\langle \tilde{\rho}_1 \middle| P_z \exp\left[\frac{1}{2qi} \int_{-\infty}^{\infty} \hat{V}_q(z)\, dz\right] \middle| \tilde{\rho} \right\rangle - 1 \right]. \end{aligned} \tag{4.76}$$

This formula can obviously be regarded as a direct generalization of the eikonal representation to the relativistic case [41].

Let us consider the case when the matrix $\langle \tilde{\rho}_1 | \hat{V}_q(z) | \tilde{\rho}_2 \rangle$ is diagonal:

$$\langle \tilde{\rho}_1 | \hat{V}_q(z) | \tilde{\rho}_2 \rangle = \delta^{(2)}(\tilde{\rho}_1 - \tilde{\rho}_2)\, V_q(z, \tilde{\rho}_1). \tag{4.77}$$

Then, proceeding from the expansion of the P_Z exponential function in powers of the quasipotential and applying the identity successively

$$\hat{\theta}(z - z')\hat{\theta}(z - z'') = \hat{\theta}(z - z')\hat{\theta}(z' - z'') + \hat{\theta}(z - z'')\hat{\theta}(z'' - z') + \frac{1}{i}\hat{\theta}(z - z')\,\delta(z' - z''), \tag{4.78}$$

we can show that

$$\left\langle \tilde{\rho}_1 \middle| P_z \exp\left[\frac{1}{2qi} \int \hat{V}_q(z)\, dz\right] \middle| \tilde{\rho} \right\rangle = \delta^{(2)}(\tilde{\rho}_1 - \tilde{\rho}) \exp\left\{ i \int_{-\infty}^{\infty} \ln\left(1 - \frac{V_q(z, \tilde{\rho})}{2q}\right) dz \right\}. \tag{4.79}$$

Finally, the representation (4.76) for the scattering amplitude can be written in the form

$$A(s, t) = -qi \int_0^{\infty} \rho\, d\rho\, J_0(\sqrt{-t}\rho) \left\{ e^{i \int_{-\infty}^{\infty} \ln\left(1 - \frac{V_q(z, \tilde{\rho})}{2q}\right) dz} - 1 \right\}. \tag{4.80}$$

In the special case when the quasipotential $V_q(z, \tilde{\rho})$ satisfies $V_q(z, \tilde{\rho})/2q \ll 1$, this expression is identical with the nonrelativistic eikonal representation of the scattering amplitude.

We are very grateful to Professor A. N. Tavkhelidze for his constant interest and valuable comments and to V. R. Garsevanishvili, A. D. Donkov, M. D. Mateev, V. A. Matveev, R. M. Muradyan, L. A. Slepchenko, I. T. Todorov, and R. N. Faustov for numerous helpful discussions.

LITERATURE CITED

1. J. Schwinger, Proc. Nat. Acad. Sci. USA, 37, 452 (1951).
2. E. Salpeter and H. Bethe, Phys. Rev., 54, 1232 (1951).
3. M. Gell-Mann and F. Low, Phys. Rev., 84, 350 (1951).
4. V. A. Fock, Phys. Assoc. Sowjetunion, 6, 425 (1934).
5. I. E. Tamm, Phys. USSR, 9, 449 (1945).
6. S. M. Dancoff, Phys. Rev., 78, 382 (1950).
7. A. A. Logunov and A. N. Tavkhelidze, Nuovo Cimento, 29, 380 (1963).
8. A. A. Logunov et al., Nuovo Cimento, 30, 134 (1963).
9. A. A. Logunov, A. N. Tavkhelidze, and O. A. Khrustalev, Phys. Lett., 4, 325 (1963).
10. B. A. Arbuzov et al., Zh. Éksp. Teor. Fiz., 44, 1409 (1963).
11. A. N. Tavkhelidze, Lectures on the Quasipotential Method in Field Theory, Bombay, Tata Institute of Fundamental Research (1964).
12. V. G. Kadyshevskii and A. N. Tavkhelidze, in: Problems of Theoretical Physics [in Russian] (Dedicated to N. N. Bogolyubov on the occasion of this 60th birthday), Nauka (1969) ([11, 12] contain detailed bibliographies).
13. V. A. Matveev, R. M. Muradyan, and A. N. Tavkhelidze, Preprint JINR E-3498 [in English], Dubna [see also: C. Fronsdal and L. E. Lundberg, Phys. Rev., 3, 3447 (1970)].
14. V. G. Kadyshevsky (Kadyshevskii), Nucl. Phys., B6, 125 (1968).
15. V. G. Kadyshevsky (Kadyshevskii) and M. D. Mateev, Nuovo Cimento, 55A, 275 (1967).
16. V. G. Kadyshevskii, Zh. Éksp. Teor. Fiz., 46, 654 (1964).
17. V. G. Kadyshevskii, Zh. Éksp. Teor. Fiz., 46, 872 (1964).
18. V. G. Kadyshevskii, Dokl. Akad. Nauk SSSR, 160, 573 (1965).
19. V. G. Kadyshevsky (Kadyshevskii), R. M. Mir-Kasimov, and N. B. Skachkov, Nuovo Cimento, 55A, 233 (1968).
20. V. G. Kadyshevskii, R. M. Mir-Kasimov, and N. B. Skachkov, Yad. Fiz., 9, 219 (1969).
21. V. G. Kadyshevskii, R. M. Mir-Kasimov, and N. B. Skachkov, Yad. Fiz., 9, 462 (1969).
22. V. G. Kadyshevskii, R. M. Mir-Kasimov, and M. Freeman, Yad. Fiz., 9, 646 (1969).
23. V. G. Kadyshevskii, M. D. Mateev, and R. M. Mir-Kasimov, Yad. Fiz., 11, 692 (1970).
24. M. Freeman, M. D. Mateev, and R. M. Mir-Kasimov, Nucl. Phys., B12, 197 (1969).
25. A. S. Wightman, Lectures on Invariance in Relativistic Quantum Mechanics (les Houches, 1960); in: Dispersion Relations and Elementary Particles, Paris (1960); I. M. Namislovskii, Phys. Rev., 160, 1522 (1967).
26. H. Shyder, Phys. Rev., 71, 38 (1947).
27. Yu. A. Gol'fand, Zh. Éksp. Teor. Fiz., 37, 504 (1954); 43, 256 (1962); 44, 1248 (1963).
28. V. G. Kadyshevskii, Zh. Éksp. Teor. Fiz., 41, 1885 (1961).
29. I. E. Tamm, in: Proc. 12th International Conference on High-Energy Physics [in Russian], Vol. II, Moscow, Atomizdat (1964).
30. R. M. Mir-Kasimov, Zh. Éksp. Teor. Fiz., 52, 533 (1967).
31. V. G. Kadyshevskii, Dokl. Akad. Nauk SSSR, 147, 588 (1962).
32. M. A. Naimark, Linear Representations of the Lorentz Group, Pergamon Press, London (1964).
33. I. S. Shapiro, Dokl. Akad. Nauk SSSR, 106, 647 (1956); Zh. Éksp. Teor. Fiz., 43, 1727 (1962).
34. N. Ya. Vilenkin, Special Functions and the Theory of Group Representations, AMS Translations of Math. Monogr., Vol. 22, Providence, R. I. (1968).
35. A. O. Gel'fond, Calculus of Finite Differences [in Russian], Fizmatgiz, Moscow (1967).
36. L. Brown et al., Ann. Phys., 23, 187 (1963).
37. E. Fermi and C. N. Yang, Phys. Rev., 76, 1739 (1949).
38. N. N. Bogolyubov, B. V. Struminskii, and A. N. Tavkhelidze, Preprint JINR D-1968 [in Russian], Dubna (1965).
39. A. N. Tavkhelidze, High-Energy Physics and Elementary Particles, IAEA, Vienna (1965), p. 753.

40. V. R. Garsevanishvili et al., Teor. Mat. Fiz., 7, 203 (1971).
41. R. J. Glauber, Lectures in Theoretical Physics, Vol. 1, Interscience, New York (1959).
42. V. R. Garsevanishvili et al., Phys. Lett., 29B, 191 (1969).
43. V. R. Garsevanishvili et al., Preprint IC/69/87, Trieste (1969).
44. V. R. Garsevanshvili, V. A. Matveev, and L. A. Slepchenko, in: Particles and Nuclei, Vol. 1, No. 1, Consultants Bureau, New York (1972).

CONSERVATION OF LEPTON CHARGE IN BETA DECAY

A. A. Borovoi,* Yu. A. Plis,
and V. A. Khodel'*

Recent data on the conservation of lepton charge in beta decay processes are reviewed. Experimental efforts to find neutrinoless double beta decay and theoretical estimates of the nuclear matrix elements for this process are discussed. The search for inverse beta processes which are forbidden by the law of lepton conservation is described. An estimate is given for the upper limit of a possible lepton nonconservation parameter which can be obtained from various experiments.

INTRODUCTION

Experimental data provide evidence that there are four different kinds of neutrino. Two of them – the electron neutrino and antineutrino (ν_e and $\tilde{\nu}_e$) – are associated with the electron, and two – the muon neutrino and antineutrino (ν_μ and $\tilde{\nu}_\mu$) – with the muon. No electromagnetic properties of these particles have been observed. It is well known that they participate only in weak interactions and behave completely differently. Thus, the electron antineutrino $\tilde{\nu}_e$, which is emitted by fission products, induces inverse beta decay (Reines – Cowan reaction); however, this process is not produced either by the muon neutrino, which appears in ($\pi \to \mu\nu_\mu$) decay, or by the electron neutrino ν_e, which appears in the decay of a proton in the nucleus.

The difference between neutrinos can be interpreted in terms of a fundamental quantum number – the lepton charge or lepton number – which is presumably conserved in all processes [1-3]. There are several methods for determining it.

Most customary is a scheme in which there are two independent additive lepton charges – the electron charge L_e and the muon charge L_μ:

$$L_e = +1 \text{ for } e^-, \nu_e; \quad L_\mu = +1 \text{ for } \mu^-, \nu_\mu;$$
$$L_e = -1 \text{ for } e^+, \tilde{\nu}_e; \quad L_\mu = -1 \text{ for } \mu^+, \tilde{\nu}_\mu;$$

$L_e = L_\mu = 0$ for all other particles.

As examples of allowed reactions, one can cite the following processes:

$$A(Z, N) \to A(Z+1, N-1) + e^- + \tilde{\nu}_e; \tag{B.1}$$
$$\nu_e + A(Z, N) \to A(Z+1, N-1) + e^- \tag{B.2}$$

*I. V. Kurchatov Institute of Atomic Energy.

Joint Institute for Nuclear Research, Dubna. Translated from Problemy Fiziki Élementarnykh Chastits i Atomnogo Yadra, Vol. 2, No. 3, pp. 691-716, 1972.

inverse beta decay induced by a neutrino:

$$A(Z,\ N) \rightarrow A(Z+2,\ N-2)+2e^- +2\tilde{\nu}_e \quad \text{(B.3)}$$

$2\beta^-$ (2ν) decay (two-neutrino double beta decay); and muon decay – $\mu^- \rightarrow e^- + \tilde{\nu}_e + \nu_\mu$.

Forbidden processes are:

$$A(Z,\ N) \rightarrow A(Z+2,\ N-2)+2e^- \quad \text{(B.4)}$$

$2\beta^-$ $(0\ \nu)$ decay (neutrinoless double beta decay);

$$\tilde{\nu}_e + A(Z,\ N) \rightarrow A(Z+1,\ N-1)+e^- \quad \text{(B.5)}$$

inverse beta decay induced by an antineutrino; and

$$\mu^\pm \rightarrow e^\pm + \gamma.$$

The question as to whether the lepton number is as universal a characteristic as electric charge or is only one of the approximately conserved quantities such as strangeness, for example, is very timely at the moment. Existing experimental data do not permit one to arrive at a definite conclusion, and this serves as one of the reasons for discussion of the consequences of a possible violation of the lepton charge-conservation law (LCCL). One of them consists of the creation of neutrino oscillations, i.e., transitions in vacuum between neutrino states $\nu_e \rightleftarrows \nu_\mu$, $\tilde{\nu}_e \rightleftarrows \tilde{\nu}_\mu$, $\nu_e \rightleftarrows \tilde{\nu}_e$, $\nu_\mu \rightleftarrows \tilde{\nu}_\mu$ similar to the oscillations in a beam of K^0 mesons. This idea was first put forward by B. M. Pontecorvo [3, 4] and elaborated by others [5-7]. The hypothesis of neutrino oscillations has a direct relation to experiments on the detection of solar neutrinos [8], since in the general case, oscillations lead to a reduction in the observed flux of solar neutrinos in comparison with the theoretical predictions based on the LCCL [9]. The consequences of violation of lepton conservation can also be considered applicable to problems in cosmology [10]. After the discovery of CP-parity nonconservation, it was proposed that CP-violating interactions failed to conserve lepton charge also [3, 11-14].

Thus a general consideration of the problem of lepton conservation encompasses an extremely broad class of questions. This review is limited to the presentation and analysis of recent data related to the conservation of lepton charge in beta decay. In the first part, experimental efforts to find neutrinoless double beta decay (B.4) are discussed and theoretical estimates of the nuclear matrix elements for this process are considered.

In the second part, a description is given of the experiments of the Davis group who looked for the type (B.5) reaction

$$\tilde{\nu}_e + {}^{37}\mathrm{Cl} \rightarrow {}^{37}\mathrm{Ar} + e^-, \quad \text{(B.6)}$$

and used a nuclear reactor as an antineutrino source. These experiments gave additional information about the upper limit of lepton charge nonconservation. The possibility of further increase in the sensitivity of experiments of this kind by looking for the process

$$\tilde{\nu}_e + {}^{19}\mathrm{F} \rightarrow {}^{19}\mathrm{Ne} + e^- \quad \text{(B.7)}$$

with a high-intensity pulsed reactor is discussed.

1. DOUBLE BETA DECAY

Theoretical Description

According to present ideas about the weak interaction, $2\beta^-$ decay* is a second-order process which occurs with simultaneous transformation of two neutrons in the nucleus into protons and the emission of two electrons and two antineutrinos (B.3). In the case of nonconservation of lepton charge, a version of double beta decay without antineutrino emission is possible. This process is considered as a combination

* All the most interesting experimental data for checking the LCCL are related to electron double beta decay. We shall therefore not concern ourselves with either $2\beta^+$ decay or double K capture in nuclei.

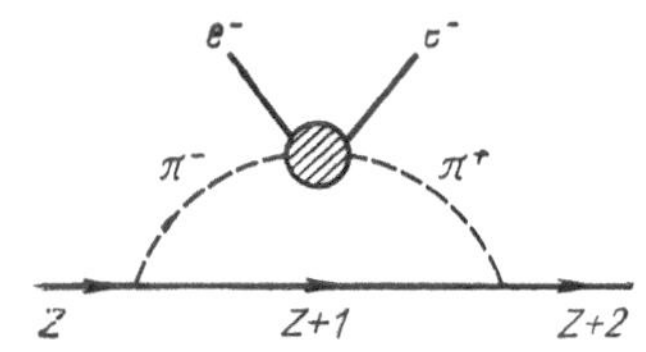

Fig. 1. Diagram of neutrinoless double beta decay produced in first order by a hypothetical ultraweak interaction.

of the two reactions (B.1) and (B.5) such that the virtual antineutrino emitted in the decay of one of the neutrons in the nucleus induces the decay of a second neutron. Furthermore, from the viewpoint of a search for nonconservation of lepton charge, the only fundamental difference between inverse beta decay (B.5) and $2\beta^-$ (0 ν) decay is that in the latter the creation and absorption of the neutrino occurs at a single point (inside the nucleus) while in process (B.5) they are separated by significant space and time intervals.

Another possibility, which is discussed in [11], is that neutrinoless double beta decay is induced in first order by some hypothetical ultraweak interaction which changes the lepton charge by two units at once.

One of the possible diagrams for this process is shown in Fig. 1.

In this case, experiments aimed at finding the process (B.5) and neutrinoless double beta decay give fundamentally different information. In particular, neutrinoless double beta decay may take place even with complete polarization of the neutrino while the process (B.5) would be forbidden.

We discuss the first possibility in greater detail. Modern theory of beta decay with the Hamiltonian

$$H_\beta^0 = \frac{G_v}{\sqrt{2}} [\overline{\Psi}_p \gamma_\mu (1 + x\gamma_5) \Psi_n] [\overline{\psi}_e \gamma_\mu (1 + \gamma_5) \Phi_{\nu_e}] + \text{complex conjugate}, \tag{1.1}$$

where x is the ratio between the axial vector and vector constants, assumes the existence of both types of completely polarized neutrinos with zero mass (neutrino and antineutrino). The neutrino has left helicity, and antineutrino right, and the transition from particle to antiparticle is simply accomplished by a change in the sign of the helicity. In this formalism, where the neutrino wave function has only two components, the assignment of lepton charge to the neutrino is equivalent to the assignment of helicity. It is obvious that both processes (B.4) and (B.5) are forbidden in this scheme since the particle created in one event cannot be absorbed in another because it has an "incorrect" helicity. These processes are also impossible in another scheme: the scheme of a massless Majorana neutrino ($\Phi_{\nu_e} = \Phi^C_{\nu_e} = C^{-1}\,\overline{\Phi}_{\nu_e}$, where C is the charge conjugation matrix) with complete nonconservation of parity because both theories are equivalent for $m_{\nu_e} = 0$ [14]. If processes (B.4) and (B.5) exist in nature, therefore, the general Hamiltonian for lepton–nucleon interaction must contain depolarization terms in addition to terms violating LCCL. In traditional form, the expression for H_β is [15]

$$H_\beta = \frac{G_v}{\sqrt{2}} \sum_i (\overline{\Psi}_p O_i \Psi_n) \{C_i [\overline{\psi}_e O_i (1 + \xi_i \gamma_5) \Phi_{\nu_e}] + D_i [\overline{\psi}_e O_i (1 + \eta_i \gamma_5) \Phi^C_{\nu_e}]\} + \text{c.c.} \tag{1.2}$$

where $\Phi^C_{\nu_e}$ is the neutrino charge-conjugation spinor, C_i and D_i are complex parameters, and i takes on five values corresponding to the various types of interaction (S, T, V, A, P).

The constants D_i characterize the amplitude of the admixture of the interaction not conserving lepton charge, and the constants ξ_i and η_i characterize the contribution of the right and left neutrinos.

In order to transform to the Hamiltonian (1.1), it is obviously necessary to set $\xi_i = 1$, $D_i = 0$ for all types of interaction, $C_V = 1$, $C_A = x$, and all remaining C_i equal to zero.

Analysis of existing experimental data on the basis of the Hamiltonian (1.2) is very complex and requires the introduction of simplifying (and often rather arbitrary) assumptions. We shall consider that all constants are real and, further, that V and A are the only types of interaction that take place. There remain eight constants in all (C_A, C_V, D_V, D_A, ξ_V, ξ_A, η_V, η_A). Some information about them can be obtained from experiments on simple beta decay (polarization, angular distribution of recoil electrons, Reines–Cowan experiment, etc.). Rather crude estimates for the upper limit of $\delta_i = \xi_i - 1$ (i = V, A) give a value $\delta \leq 0.2$. It is impossible to derive any estimates for D_i from experiments on single beta decay. They can be obtained only by studying a second-order process (double beta decay) or inverse beta processes (B.5).

For the existence of neutrinoless double beta decay, it is necessary to have nonzero values for the expressions

$$I_{ik}=I_{ki}=C_i D_k(1-\xi_i\eta_k)+C_k D_i(1-\xi_k\eta_i), \quad i,\ k=V,\ A. \tag{1.3}$$

The half-life for neutrinoless double beta decay is 10^{15}-10^{17} yr in the case of maximum violation of LCCL and a nonpolarized neutrino ($D_i = C_i$, $\xi_i = \eta_i = 0$). The experimental value for $T_{1/2}(2\beta)_{exp}$ is considerably greater than the second value. This means that the quantities I_{ik} must be considerably less than one.

There are two very simple possibilities. One of them is: a Majorana neutrino ($D_i = C_i$), and in addition, $\xi_A = \xi_V = \eta_A = \eta_V = 1 + \delta_M$, with $\delta_M \ll 1$. The Hamiltonian (1.2) then reduces to

$$H_\beta^M = \frac{G_v}{\sqrt{2}}\left[\overline{\Psi}_p\gamma_\mu(1+x\gamma_5)\Psi_n\right]\left\{\overline{\Psi}_e\gamma_\mu\left[1+(1+\delta_M)\gamma_5\right]\Phi_{\nu_e}^M\right\}+\text{c.c.};$$

$$\Phi_{\nu_e}^M=\frac{\Phi_{\nu_e}+\Phi_{\nu_e}^C}{\sqrt{2}}. \tag{1.4}$$

The probability of neutrinoless double beta decay is proportional to δ_M^2 in this case, i.e., to the depolarization term. Even now completed experiments on 2β (0 ν) decay give a value much less than 0.2 for the upper limit of δ_M.*

An alternative possibility is the assumption that the constants $D_A = D_V = D$, which violate the conservation of lepton charge, are small. In this case, it is convenient to consider $\eta_A = \eta_V = -1$ and $\xi_A = \xi_V = 1$. Then a neutrino created with a constant D_i has a helicity opposite to that of a neutrino created with a constant C_i. The Hamiltonian (1.2) transforms to

$$H_\beta^D=\frac{G_v}{\sqrt{2}}\left[\overline{\Psi}_p\gamma_\mu(1+x\gamma_5)\Psi_n\right]\left\{\overline{\Psi}_e\gamma_\mu\left[(1+\gamma_5)\Phi_{\nu_e}+\delta_D(1-\gamma_5)\Phi_{\nu_e}^C\right]\right\}+\text{c.c.};$$

$$\delta_D=\frac{D_i}{C_i}. \tag{1.5}$$

The probability of neutrinoless double beta decay is proportional to δ_D^2.

Finally, a more complex situation can be represented which corresponds to a simultaneous, but small, violation of both LCCL and two-component theory. Then the quantity analogous to δ which is responsible for neutrinoless double beta decay will be the product of two small parameters.

A measure of the quantity δ_D (or δ_M) is the ratio of the theoretical value for the half-life for $2\beta(0\ \nu)$ decay to the experimental value:

$$\delta^2=\frac{T_{1/2}(0\nu)_{\text{theor}}}{T_{1/2}(0\nu)_{\text{exp}}}. \tag{1.6}$$

At the present time, there are no data on the observation of neutrinoless double beta decay and there are only lower limits for $T_{1/2}(0\ \nu)_{exp}$.

Evaluation of the quantity δ^2 also requires a knowledge of $T_{1/2}(0\ \nu)_{theor}$. However, calculations of the half-life often contain significant errors because of uncertainties in the nuclear matrix elements for the transition A(Z, N) → A(Z + 2, N − 2). We shall therefore turn to this problem after describing the experiments in detail.

Experimental Situation

The work on double beta decay can be divided into two groups. In the first group are relatively few experiments in which this process is traced by the buildup of final products – the nuclei A(Z + 2, N − 2) – in samples containing the nucleus A(Z, N) (mass-spectrometric experiments).

*One of the possible sources of neutrino depolarization leading to $2\beta(0\ \nu)$ decay may be the existence of mass for this particle. The experimental mass value is $m_{\nu_e} \lesssim 100$ eV, and therefore the typical depolarization value associated with mass is very small: $m_{\nu_e}/E_{tran} < 10^{-4}$.

TABLE 1*

Transition	Transition energy, keV	Isotopic conc., %	$T_{1/2}(0\nu)_{theor}$, yr	$T_{1/2}(2\nu)_{theor}$, yr	$T_{1/2}(2\beta)_{exp}$, yr	δ
$^{82}Se \to {}^{82}Kr$ $(0^+ \to 0^+)$	2956 ± 73	9,19	$10^{16\pm2}$	$10^{22\pm2,5}$	$6 \cdot 10^{19\pm0,3}$[25]	$\leqslant 0,2$
$^{128}Te \to {}^{128}Xe$ $(0^+ \to 0^+)$	850 ± 19	31,79	$2 \cdot 10^{19\pm2}$	$2 \cdot 10^{27\pm2,5}$	$\geqslant 3 \cdot 10^{22}$ [23] $\geqslant 7,7 \cdot 10^{20}$ [24] $\geqslant 10^{23,3}$ [25]	$\leqslant 0,15$
$^{130}Te \to {}^{130}Xe$ $(0^+ \to 0^+)$	2509 ± 19	34,48	$2 \cdot 10^{16\pm2}$	$4 \cdot 10^{22\pm2,5}$	$1,4 \cdot 10^{21}$ [21] $3,3 \cdot 10^{21}$ [22] $(8,2 \pm 0,64) \times$ $\times 10^{20}$ [23] $(3,0 \pm 0,4) \times$ $\times 10^{20}$ [24] $10^{21,34\pm0,12}$ [26]	$\leqslant 0,05$

*Data on transition characteristics taken from [25].

In the majority of other experiments, the search for double beta decay was carried on by the detection of decay electrons. A very important feature of neutrinoless double beta decay for the experiment is the fact that the total energy of the two electrons is constant and equal to the transition energy while $2\beta(2\nu)$ decay is characterized by a continuous spectrum of electron energies.

There is a strong dependence of half-life on transition energy both for $2\beta(2\nu)$ and $2\beta(0\nu)$ decays. As a rule, therefore, the experiments were carried out with nuclei for which the transition energy is large and the decay probability relatively large.

Of most interest are the experiments from which a minimum upper limit of δ is obtained. More detailed information about experimental attempts to find double beta decays is contained in some other papers [16-18].

Mass Spectrometric Experiments. The rate of double beta decay is so low that a detectable amount of daughter products may be formed in a time comparable with the age of the earth. Therefore old minerals containing large concentrations of materials capable of double beta decay are selected to be used in mass-spectrometric experiments seeking to find that process.

The huge "observation" time is one of the main merits of this method. The basic deficiency is the lack of knowledge of the degree of predominance of the process over long periods of time. Because of ignorance of sample history, it is often difficult to determine what amount of material of interest was formed by double beta decay and what amount is associated with secondary processes (nuclear reactions, for example).

One uses as objects of investigation those elements that decay to the noble gases xenon and krypton. They are chemically inert and this makes it possible to separate even very small amounts of these gases from experimental samples comparatively easily and reliably. A decisive factor in the experimental detection of double beta decay is the circumstance that of all nuclei capable of double beta decay there are the nuclei yielding the transitions* $^{130}Te \to {}^{130}Xe$ and $^{82}Se \to {}^{82}Kr$ (Table 1) with an energy (about 3 MeV) close to the maximum possible for stable nuclei. A complete mass-spectrometric experiment includes the following operations: 1) determination of the age of the original rock and the tellurium (or selenium) content in it; 2) extraction of inert gas and mass-spectrometric analysis of its isotopic composition; measurement of the absolute amount of ^{130}Xe, ^{128}Xe (or ^{82}Kr) in the mineral; 3) estimate of the contribution from various background reactions which lead to the formation of inert gas isotopes, for example the spontaneous or

*The possibility of ^{130}Xe formation by the transition sequence $^{130}Te \xrightarrow{\beta^-} {}^{130}I \xrightarrow{\beta^-} {}^{130}Xe$ was discussed previously. Measurements that have been made up to this time [19, 20] show that the ^{130}I ground state is 477 ± 35 keV higher than ^{130}Te and $^{130}Te \xrightarrow{\beta^-} {}^{130}I$ decay is impossible.

neutron-induced fission of uranium present in the mineral; 4) estimate of the escape of gases from the mineral during the time since its formation ($\approx 10^9$ years); 5) calculation of the half-life from the expression

$$T_{1/2}(2\beta)_{\exp} = \frac{\ln 2 M_m t f}{M_t}, \tag{1.7}$$

where t is the age of the mineral, M_m is the amount of the parent double beta active element, M_t is the amount of the daughter (including correction as specified in 3 above), and $f \leq 1$ is a factor which defines the preservation of the daughter product in the crystal lattice of the mineral over a time t.

Even in the first studies [21, 22] in which the ^{130}Xe content in Bi_2Te_3 minerals was investigated, an excess of this isotope was discovered in comparison with the atmospheric xenon content. Assuming that this excess arose because of $^{130}Te \rightarrow {}^{130}Xe$ double beta decay, Inghram and Reynolds [21] estimated the $^{130}Te(2\beta)$ decay half-life to be $1.4 \cdot 10^{21}$ yr. Subsequently, a paper by Hayden and Inghram [22] gave the value

$$T_{1/2}\ (2\beta) = 3.3 \cdot 10^{21}\ \text{yr}. \tag{1.8}$$

In experiments by Takaoka and Ogata [23], the isotopic composition of xenon separated from a mineral also differed from the atmospheric composition. According to their estimates,

$$^{130}\text{Te}\ T_{1/2}\ (2\beta) = (8.2 \pm 0.64) \cdot 10^{20}\ \text{yr}; \tag{1.9}$$

$$^{128}\text{Te}\ T_{1/2}\ (2\beta) > 3 \cdot 10^{22}\ \text{yr}. \tag{1.10}$$

In all these papers, however, there was no analysis either of the possibility of ^{130}Xe formation by other process or of the loss of xenon from the mineral over geologic time because of diffusion. The anomalous content of other xenon isotopes indicated the need for such an analysis.

In the experiments of Gerling et al. [24], it was pointed out that the xenon separated from a telluro-bismuth mineral consisted of three components: atmospheric xenon, which produced an instrumental background; xenon from spontaneous fission of uranium, and xenon which was a product of nuclear reactions. Because of this, in addition to the small excess of ^{130}Xe, excesses of the other isotopes (^{128}Xe, ^{129}Xe, ^{131}Xe, ^{134}Xe, and ^{136}Xe) were also observed in comparison with atmospheric xenon (isotopic composition normalized to ^{132}Xe). The authors introduced a number of corrections in order to determine the amounts of ^{130}Xe and ^{128}Xe formed by double beta decay. As a result, if one neglects the loss of xenon over a time $t = 1.8 \cdot 10^9$ yr, the following half-lives are obtained:

$$^{130}\text{Te}\quad T_{1/2}(2\beta) = 4.4 \cdot 10^{21}\ \text{yr};$$

$$^{128}\text{Te}\quad T_{1/2}(2\beta) \geqslant 1.1 \cdot 10^{22}\ \text{yr}.$$

A basis for estimating the loss of ^{130}Xe and ^{128}Xe is the fact that their diffusion rate during thermal effects on the mineral is close to the diffusion rate of ^{136}Xe. The amount of ^{136}Xe formed in the mineral because of spontaneous fission of uranium (the main process) was found from the known uranium concentration in the sample. A value $f = 0.07$ was determined from the measured ^{136}Xe content. As a final result, including the loss of gas, we have

$$^{130}\text{Te}\ T_{1/2}\ (2\beta) = (3.0 \pm 0.4) \cdot 10^{20}\ \text{yr}; \tag{1.11}$$

$$^{128}\text{Te}\ T_{1/2}\ (2\beta) \geqslant 7.7 \cdot 10^{20}\ \text{yr}. \tag{1.12}$$

Kirsten et al. [25, 26] used a tellurium ore with a high tellurium content. A considerable excess of ^{130}Xe was observed which was about 70% of the total amount of xenon; this is more than 50 times the atmospheric isotopic concentration. Thus the initial ^{130}Xe excess is no small deviation and practically determines the xenon spectrum of a sample. Besides the increased content of ^4He, ^{40}Ar, and ^{130}Xe, no other isotopic anomalies were observed in the analysis of inert gases. Detailed investigations made by the authors conclusively indicated the impossibility of ^{130}Xe formation in this mineral by any process other than double beta decay. Determination of the age of the mineral by the potassium-argon method gave a value $(1.31 \pm 0.14) \cdot 10^9$ yr and $(2.05 \pm 0.55) \cdot 10^8$ yr by the uranium-helium method. The authors assumed that the "thermal" history of the mineral was such that the helium was gradually lost by diffusion while the amount of argon remained constant. On this basis, they neglected xenon loss and obtained a half-life

$$T_{1/2}\ (2\beta) = 10^{21.34 \pm 0.12}\ \text{yr.} \tag{1.13}$$

An upper limit was obtained for ^{128}Te (see [25]):

$$T_{1/2}\ (2\beta) \geqslant 10^{23.3}\ \text{yr.} \tag{1.14}$$

As is clear, the recent, most exact experiments give values of the ^{130}Te half-life which differ by approximately an order of magnitude [(1.9), (1.11), and (1.13)] although the accuracy given in the papers is 15-30%. We therefore use as the most reliable result

$$T_{1/2}\ (2\beta) = 10^{21}\ \text{yr.} \tag{1.15}$$

In addition to tellurium minerals, minerals containing selenium, $Cu_{4-x}Se_2$, [25] were also used in mass-spectrometric experiments. Krypton separated from a sample had the following isotopic composition: ^{82}Kr:^{83}Kr:^{84}Kr:^{86}Kr = 0.3:16.5:1:0.3 while for atmospheric krypton these ratios are 0.2:0.2:1:0.35 (normalized with respect to ^{84}Kr).

Thus a considerable excess of the isotope ^{83}Kr and a relatively small excess of ^{82}Kr are observed with respect to atmospheric composition. The authors made a careful analysis of possible reasons for the formation of those isotopes. ^{83}Kr may be formed in a mineral as the result of the following reactions: ^{80}Se(α, n)^{83}Kr and $^{82}\text{Se}(n,\gamma)^{83}\text{Se}\xrightarrow{\beta^-}{}^{83}\text{Br}\xrightarrow{\beta^-}{}^{83}\text{Kr}$. Evidently, these processes completely explain the excess of the isotope ^{83}Kr. Formation of ^{82}Kr is possible in the following ways:

from selenium isotopes by means of the reactions

$$^{78}\text{Se}\,(n,\ \gamma)\,^{79}\text{Se}\,(\alpha,\ n)\,^{82}\text{Kr};$$
$$^{80}\text{Se}\,(\alpha,\ 2n)\,^{82}\text{Kr};$$

in reactions occurring in low-level impurities

$$^{85}\text{Rb}\,(n,\ \alpha)^{82}\,\text{Br} \xrightarrow{\beta^-} {}^{82}\text{Kr};$$
$$^{81}\text{Br}\,(n,\ \gamma)\,^{82}\text{Br} \xrightarrow{\beta^-} {}^{82}\text{Kr};$$

and, finally, by double beta decay

$$^{82}\text{Se} \xrightarrow{2\beta^-} {}^{82}\text{Kr}.$$

Reaction 1 cannot lead to the formation of a noticeable amount of ^{82}Kr because of the relatively short lifetime of ^{79}Se ($T_{1/2} = 7 \cdot 10^4$ yr). Reaction 2 proceeds at α-particle energies above 11 MeV and can explain only an insignificant ^{82}Kr excess according to estimates based on ^{83}Kr formation in (α, n) reactions. It is impossible to make precise estimates with respect to reactions 3 and 4, which occur in impurity elements, but there are a number of well-known and convincing objections to the formation of marked amounts of the ^{82}Kr formed being associated with these processes. The age of the selenium mineral used in [25] was determined imprecisely and falls in a range from $60 \cdot 10^6$ to $230 \cdot 10^6$ yr. Because of this, the half-life of ^{82}Se may vary from $3 \cdot 10^{19}$ to 10^{20} yr. We take the average value

$$T_{1/2}\ (2\beta) = 6 \cdot 10^{19 \pm 0.3}\ \text{yr.} \tag{1.16}$$

An increase in accuracy of results obtained in experiments with selenium minerals is extremely desirable.

The half-life values obtained in mass-spectrometric experiments are not at variance with the theory of two-neutrino double beta decay. However, since these experiments do not distinguish the two types of decay, some contribution to the formation of daughter products may be made by neutrinoless double beta decay which proceeds by way of lepton charge nonconservation.

An upper limit for the quantity δ [see Eq. (1.6)] can be obtained by setting $T_{1/2}(0\nu)_{\text{exp}}$ equal to the half-life measured experimentally

$$\delta^2 \leqslant \frac{T_{1/2}(0\nu)_{\text{theor}}}{T_{1/2}(2\beta)_{\text{exp}}}. \tag{1.17}$$

Calculated values of δ derived from the experiments mentioned are given in the last column of Table 1. Uncertainties in the theoretical values for the half-lives leads to a variation of two orders of magnitude in the estimates [thus for ^{130}Te, $\delta \leq 5 \cdot 10^{-2}$ for the maximum value of $T_{1/2}(0\nu)_{\text{theor}}$ and $\delta \leq 5 \cdot 10^{-4}$ for the minimum value].

The most important conclusion to be drawn from analysis of the mass-spectrometric experiments is that the very process of double beta decay was first observed in them. Particularly reliable evidence for this is given in [24-26].

Experiments on Double Beta Decay Using Electron Detection. Direct observation in double beta decay of two electrons with a total energy equal to the decay energy would be irrefutable proof of neutrinoless double beta decay. The most intense experimental effects were made in just that field. The most varied experimental techniques were used for electron detection (photographic emulsion, Gieger-Mueller counters, Wilson chamber, scintillation counters, etc.). This type of experiment is extremely difficult. With a half-life $T_{1/2} \approx 10^{21}$ yr, there would occur a total of 1 to 2 decays per year in a gram of material. It is therefore no wonder that neutrinoless decay was "discovered" many times; however, the results of earlier experiments were not confirmed by subsequent ones.

We consider the latest experiments, from which one can obtain the best limit for the quantity δ (1.6). Data for the transitions are given in Table 2.

The double beta decay $^{76}Ge \rightarrow {}^{76}Se$ was studied in the experiments of Fiorini et al. [27]. A Ge(Li) crystal was used as source and detector of radiation. This detector has a low intrinsic background because of the low impurity level in the original material and has high energy resolution. The experimental equipment was set up at a depth of 70 m water equivalent. The detector was additionally shielded by layers of mercury (3 cm), copper (4 cm), lead (20 cm), cadmium (2 cm), and paraffin (20 cm). In addition, to reduce the cosmic-ray background, they used an anticoincidence technique with plastic scintillators located above the detector. All these measures made it possible to reduce the background in the operating range to a value of $(5.3 \pm 0.5) \cdot 10^{-3}$ counts/h-keV. An analysis of the energy spectrum leads to the conclusion that a half-life for ^{76}Ge less than $1.2 \cdot 10^{21}$ yr is excluded with 68% confidence.

The most intense recent search for double beta decay was made with the nucleus ^{48}Ca, which has a maximum decay energy.

In the experiments of der Mateosian and Goldhaber [28], the electrons were detected by means of scintillations in a specially grown crystal of calcium fluoride containing 11 g of ^{48}Ca. A second crystal, enriched in ^{40}Ca, was used to evaluate the background. Effective shielding against external background was also provided. The experiments yielded a lower limit of the half-life for neutrinoless double beta decay:

$$T_{1/2}(0\nu) \geqslant 2 \cdot 10^{20} \text{ yr}. \tag{1.18}$$

These experiments, like those of Fiorini et al. (which were completed later), used only two properties of neutrinoless double beta decay: simultaneity of electron emission and selection of events according to the sum of particle energies.

The experiments set up by the Wu group [29] have the advantage that the methods used are based on all the characteristic properties of neutrinoless double beta decay. A disc of CaF_2 containing 10.6 g of ^{48}Ca enriched to 97% was the source in this experiment (diameter of the disc was 46 cm and the thickness was about 20 mg/cm^2). The source was located between two aluminum foils and was the central electrode of a cylindrical spark chamber, the discharge gap of which was located on both sides of the disc. For background measurements, the source was replaced by a disc made of natural calcium. The experiment was performed at a depth of approximately 600 m. Track photography was carried out from both sides through spaces shielded by water and discs of scintillating plastic each of which was divided into 16 segments independently scanned by photoelectron multipliers. The scintillator was covered by a transparent nickel mesh which acted as the external electrode of the chamber.

TABLE 2*

Transition	Transition energy, keV	Isotopic conc., %	$T_{1/2}(0\nu)_{theor}$, yr	$T_{1/2}(2\nu)_{theor}$, yr	$T_{1/2}(0\nu)_{exp}$, yr	δ
$^{76}Ge \to {}^{76}Se$ ($0^+ \to 0^+$)	2045±4	7,67	$1\cdot10^{17\pm2}$	$8\cdot10^{22\pm2}$	$\geqslant 1,2\cdot10^{21}$ [27]	$\leqslant 0,1$
$^{48}Ca \to {}^{48}Ti$ ($0^+ \to 0^+$)	4240±10	0,185	$5\cdot10^{15\pm2}$	$8\cdot10^{19\pm2}$	$\geqslant 2\cdot10^{20}$ [28] $\geqslant 1,6\cdot10^{21}$ [29]	$< 5\cdot10^{-2}$ $\leqslant 1,7\cdot10^{-2}$

* Data on transition characteristics taken from [18].

The plastic scintillator gave information about the simultaneous emission of two electrons (from coincidence of pulses in two segments) and about their energies. The spark chamber, which was in a magnetic field, made it possible to distinguish events in which the electrons were emitted from a single point in the source from pulses produced by random "external" particles.

The chamber was triggered when the following conditions were satisfied:

1) two coincident events were recorded in the scintillator (2τ = 100 nsec);

2) the energy deposition in each event was greater than the discrimination level;

3) the total energy of the particles was within the limits set by the discriminator.

Under these circumstances, the chamber fired approximately 120 times per hour. A detailed analysis of the photographs showed that of all the events recorded in 1103 h of operation, only one could be assigned to neutrinoless double beta decay. Based on this, one can assert with 80% confidence that the lower limit of the half-life for neutrinoless double beta decay in ^{48}Ca is

$$T_{1/2}(0\nu) \geqslant 1.6 \cdot 10^{21} \text{ yr}. \qquad (1.19)$$

A search for two-neutrino double beta decay was also carried out on this same equipment and a limit established for that half-life:

$$T_{1/2}(2\nu) \geqslant 1.9 \cdot 10^{19} \text{ yr}. \qquad (1.20)$$

The experimental results of the search for neutrinoless double beta decay in ^{48}Ca and ^{76}Ge are given in Table 2.

Theoretical Evaluation of Nuclear Matrix Elements for Neutrinoless Double Beta Decay

In the traditional approach, the matrix element $M^{0\nu}$ for neutrinoless double beta decay is written in the form

$$M^{0\nu} = \sum_n \langle f | H_\beta | n \rangle \langle n | H_\beta | i \rangle (E_n - E_i)^{-1}, \qquad (1.21)$$

where H_β is the general Hamiltonian in beta decay (Eq. 1.2); $E_n = \varepsilon_s + \varepsilon_p + \omega_k$ is the energy of the intermediate state $|n\rangle$, which is equal to the sum of the nuclear energy (ε_s) for the s state, the energy of the emitted electron (ε_p) with momentum p, and the energy (ω_k) of the virtual neutrino with integration (summation) being performed over the momentum k of the neutrino (as well as over these states). The maximum value of k in the integral (1.21) is of the order of the inverse distance between neutrons ($\omega_{k_{max}}$ = $k_{max} \sim 1/r_0 \sim$ 150 MeV). The region of integration over momenta greater than k_{max} makes a negligible contribution to (1.21) because the nuclear matrix moments are negligibly small for $k > k_{max}$. For the same reasons, the summation over s states is cut off at energies $(\varepsilon_s - \varepsilon_i) \sim \varepsilon_F$ which are significantly

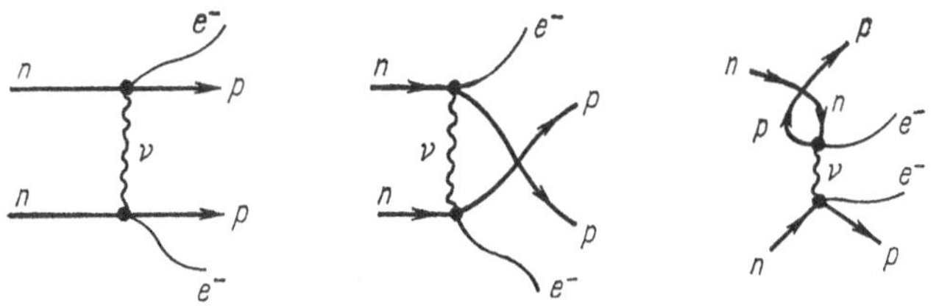

Fig. 2. Diagrams of neutrinoless double beta decay with a virtual neutrino in the intermediate state.

less than ω_{max}. Therefore the difference ($\varepsilon_s - \varepsilon_i$) in the denominator ($E_n - E_i$) can be neglected and one can then sum over all intermediate states of the nucleus using the completeness condition [15, 30]. In this approximation, the matrix element $M^{0\nu}$ for the decay breaks up into two factors, one of which depends on the kinematic characteristics of the problem and the other (the nuclear matrix element M) only on the initial and final states of the nucleus.

A large number of papers [15, 17, 30, 32], beginning with the work of Furry [31], have been devoted to the calculation of the kinematic factors. The most exact is the paper of Greuling and Whytten [30]. There, for the first time, sufficiently rigorous consideration was given to the effect of the Coulomb field of the nucleus on the motion of electrons emitted during neutrinoless double beta decay. Although there are three particles in the final state in this process as in ordinary beta decay, a product-nucleus and a pair of leptons, allowed emission of both members of the pair is impossible (because of identity). This leads to the appearance in the formula for transition probability of additional factors depending on particle energy and on the magnitude of the nuclear Coulomb field. Since the decay energy is usually small, the Coulomb terms are dominant and the matrix element for the decay is proportional to the square of the nuclear charge. Even in ^{48}Ca decay, where the electron energy is maximum, consideration of the Coulomb terms increases the transition probability by an order of magnitude.

We now turn to the consideration of the nuclear matrix elements which are the main source of uncertainty in the calculations. In the nonrelativistic limit with respect to nucleon momenta, there are three types in all:

$$M_n = \left\langle \Psi_f \left| \sum_{l,\, l'} \frac{\tau_l^+ \tau_{l'}^-}{r_{ll'}} m_{ll'}^{(n)} \right| \Psi_i \right\rangle, \tag{1.22}$$

where

$$m_{ll'}^{(1)} = 1; \quad m_{ll'}^{(2)} = (\sigma_l \sigma_{l'}); \quad m_{ll'}^{(3)} = \frac{(\sigma_l r_{ll'})(\sigma_{l'} r_{ll'})}{r_{ll'}^2}; \tag{1.23}$$

and $\Psi_f(\Psi_i)$ is the exact wave function for the initial (final) nuclear state. Unfortunately, there are as yet no actual calculations of M_n for most nuclei. A generally accepted technique for estimating it is to take the dimensional factor R^{-1} ($R = r_0 A^{1/3}$ is the nuclear radius) outside the integral and replace the remaining integral by some universal number for all nuclei, 0.1-0.15 (with an error of ±1 order of magnitude) [15, 30].

Diagrammatic methods were used to calculate nuclear matrix elements in [33]. Some diagrams from perturbation theory for the matrix element for neutrinoless double beta decay are shown in Fig. 2.

As is clear from the diagrams, one can draw a parallel between the calculation of the nuclear matrix elements M_n and the calculation of the matrix elements Γ^c for the Coulomb interaction between nucleons:

$$\Gamma^c = \left\langle \Psi_f \left| \sum_{l,\, l'} \frac{(1+\tau_l^{(3)})(1+\tau_{l'}^{(3)})}{r_{ll'}} \right| \Psi_i \right\rangle. \tag{1.24}$$

A quantitative analogy only exists between the calculations of Γ^c and of the matrix elements M_1 since M_2 and M_3 contain the additional operators σ. If Γ^c is broken up into isoscalar, isovector, and isotensor

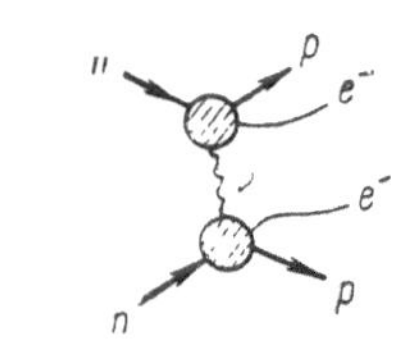

Fig. 3. Set of "long-range" diagrams for neutrinoless double beta decay.

components, the isotensor component $\Gamma_T^c = \sum_{ll'} \frac{\tau_l^{(3)} \tau_{l'}^{(3)}}{r_{ll'}}$ and the vector matrix element M_1 turn out to be different components of the same isotensor. Therefore, knowing some matrix elements of the amplitude Γ_T^c from experiment, let us say, one can calculate the corresponding values of M_1 also. To illustrate this, we consider the double beta decay $^{42}Ti \rightarrow {}^{42}Ca$. The nuclei ^{42}Ti and ^{42}Ca together with ^{42}Sc are members of an isotriplet. The spectra of these nuclei have been carefully studied experimentally, and from that data we find $M_1 = 1.6/R$. Similar values of M_1 are also obtained from analysis of other isotriplets. As is clear, the M_1 value obtained is approximately an order of magnitude greater than that usually assumed in half-life evaluations. Calculations show that so large a value for nuclear matrix elements is characteristic of those transitions where either no change occurs in the spatial portion of the nucleon wave functions during decay or where the change is minimal (for example, the transition of a nucleon pair within a single spin-orbital doublet). The equations obtained in [33, 34] for the matrix elements M_n were based on the principles of the theory of finite Fermi systems [35]. As in the solution of other problems by these methods, the main computational problems are: 1) separation of the amplitude for neutrinoless double beta decay into two parts; one is universal for all nuclei, and the other is nonuniversal, varying from nucleus to nucleus; 2) characterization of the main nonuniversal components of this amplitude; 3) their expression through the same parameter for all nuclei.

The universal component of the decay amplitude corresponds to neutrino exchange at small distances between nucleons. The evaluation of this component is made in the same way as the evaluation of the amplitude of any local interaction between nucleons. It is of the order of $A^{-2/3}$. The principal nonuniversalities arise in long-range diagrams, slowly decreasing at large distances $r_{12} \sim R$. The set of these diagrams for decay amplitude is shown in Fig. 3. The shaded circles correspond to the set of diagrams of the vertex portion of ordinary beta decay, which can be calculated by the methods of finite Fermi-system theory.

The contribution of this portion of the decay amplitude depends on the single-particle quantum numbers ν_1 and ν_2 (ν = n, l, j), which annihilate neutrons and create protons. When the quantum numbers of the initial and final states are close to one another, the contribution of the long-range component dominates the matrix element. It is proportional to $A^{-1/3}$. If ν_1 and ν_2 are markedly different, this component has negligible value and the quantity M_n is significantly reduced. It is determined by the contribution from other components of the decay amplitude (in particular, by its local component) and it is proportional to $A^{-2/3}$.

On the other hand, even for identical quantum numbers $\nu_1 = \nu_2 = \nu$, the value of M_n (Table 3) changes significantly with changes in the number of particles in the level ν. In this table, which is taken from [33], the values of the nuclear matrix elements M_1 and M_2 calculated for double beta transitions between the various Ca and Ti isotopes are given (in units of R^{-1}).

In these isotopes, the neutrons and protons fill the same level, $1f_{7/2}$, i.e., the double beta transition occurs without a change in single-particle quantum numbers. The main contribution to the nuclear matrix elements is made by the long-range component of the decay amplitude and the contribution from the remaining terms is small. Calculations show that the matrix elements M_3 are negligibly small in the case of double beta transitions within the $1f_{7/2}$ shell. That is why there are no values for M_3 in Table 3. It is also clear from the table that the value of the vector matrix element M_1 is considerably less than the value of the axial matrix element M_2 everywhere (except for the $^{42}Ti \rightarrow {}^{42}Ca$ transition). This occurs for the simple reason [15] that the contribution from the main component of M_1, which is proportional to $\sum_{l,l'} \tau_l^+ \tau_{l'}^+$, vanishes for transitions between states of different isotopic spin. Therefore the decay probability is determined mainly by the single matrix element M_2.

Very much more reliable information about the properties of weak interactions can be obtained by calculating theoretical nuclear matrix elements for the decay and comparing the theoretical data to the experimental results obtained by Wu and associates [29]. Using the Hamiltonian in the form (1.5) and substituting in the formula for $T_{1/2}$ values of the nuclear matrix elements derived in [30], for example, we find $\delta_D \leq 3 \cdot 10^{-4}$ [33]. As is evident, refinement of the nuclear matrix elements leads to significantly smaller values of the upper limit δ_D than follows from the approximate evaluations [29]. In the model

TABLE 3

A	42	44	46	48	50
M_1	+1,20	+0,07	+0,07	−0,05	+0,005
M_2	−1,25	−0,60	−0,44	+0,18	−1,70

with the Hamiltonian (1.4) the value of δ_M for depolarization of a massless Majorana neutrino also is estimated to be $\delta_M \leq 3 \cdot 10^{-4}$. If the neutrino does indeed have mass, neutrinoless double beta decay can also take place with $\delta_M = 0$. The upper limit of the mass m_{ν_e}, calculated on the basis of equations in [30], is comparatively small: $m_{\nu_e} \leq 200$ eV.

As is clear, it is necessary to make a rather large number of assumptions (about V and A types of decay, about the form of the violation of lepton conservation, etc.) in a detailed analysis of ^{48}Ca decay. A detailed study of the problem is of course impossible at the present time because of the lack of a sufficient amount of experimental information. Therefore a precise study of as large a number as possible of the nuclei capable of double beta decay is extremely desirable. The great popularity of ^{48}Ca results from the fact that $^{48}Ca \rightarrow {}^{48}Ti$ decay energy is maximum. At the same time, the nuclear matrix elements M_n for this decay are rather small as can be seen in Table 3. It is therefore of prime interest to study those decays for which the M_n are sufficiently large despite a reduction in emitted energy. Obviously, these are decays of nuclei with a large Z ($M_n \sim Z^2$) in which the single-particle quantum numbers for neutron annihilation and proton creation are not very different. Especially favorable decays, from this viewpoint, are those in which the nucleon transitions take place within a single spin-orbital doublet. Examples one can give are $^{100}Mo \rightarrow {}^{100}Ru$ ($E_0 \approx$ 3 MeV, abundance of ^{100}Mo is 9.6%) and $^{136}Xe \rightarrow {}^{136}Ba$ ($E_0 \approx$ 2.7 MeV, abundance of ^{136}Xe is 8.9%).

As already stated, the class of mass-spectrometric experiments described above gives incomplete information about double beta decay because it does not allow one to distinguish neutrinoless decay from two-neutrino decay. The experimental half-lives for ^{130}Te and ^{82}Se are of the order of 10^{20}-10^{21} years, which is characteristic of two-neutrino decay. However, quantitative analysis of results on the basis of this assertion is still impossible because the theoeretical values for $T_{1/2}$ (2ν) contain large uncertaintities (even greater than those in neutrinoless decay of the elements). It is usually considered that these matrix elements are proportional to the nuclear matrix elements for neutrinoless decay with the coefficients of proportionality varying over a rather wide range. This arbitrary assumption does little because there are the same uncertainties in the evaluations of the matrix elements for neutrinoless decay. Although the calculations of the matrix elements for the double beta decays $^{130}Te \rightarrow {}^{130}Xe$ and $^{82}Se \rightarrow {}^{82}Kr$ are lacking and it is difficult to arrive at any definite conclusions about their nature, it nevertheless seems odd that the half-life of ^{82}Se is an order of magnitude less than the half-life of ^{130}Te although the matrix elements for the ^{82}Se decay must be considerably less. Thus a detailed study of the $^{82}Se \rightarrow {}^{82}Kr$ decay and refinement of $T_{1/2}$ values is a paramount problem.

2. EXPERIMENTS WITH ANTINEUTRINOS FROM REACTORS

Experiments of the Davis Group

Powerful neutrino fluxes are required for the observation of inverse beta processes. However, the neutrino radiation incident on the earth from space is of insufficient intensity, and it is only recently that attempts have been made to detect neutrinos from the most intense natural source – the sun. As yet experiments [8] have given only an upper limit for the neutrino flux: $2 \cdot 10^6$ ν/cm²-sec. Thus direct observation of reactions caused by neutrinos only became possible after the construction of powerful nuclear reactors with a neutrino flux of approximately 10^{13} $\tilde{\nu}$/cm²-sec. A nuclear reactor is a source of antineutrinos which are produced in the beta decay of fission products containing an excess of neutrons (the average number of antineutrinos is 6.06 per fission [36]). This makes it possible to study reactions like (B.5) which are forbidden by LCCL.

B. M. Pontecorvo pointed out in 1946 [37] that if ν_e and $\tilde{\nu}_e$ are identical, a reactor antineutrino can give rise to the reaction

$$\tilde{\nu}_e + {}^{37}\mathrm{Cl} \rightarrow {}^{37}\mathrm{Ar} + e^-, \tag{2.1}$$

which is the inverse of electron capture in ^{37}Ar ($T_{1/2}$ = 35 days):

$$^{37}\mathrm{Ar} + e^- \rightarrow {}^{37}\mathrm{Cl} + \nu_e. \tag{2.2}$$

To observe the process (2.1) one can physically separate argon from a volume of material containing chlorine and detect the decay products.

Assuming the identity of ν_e and $\tilde{\nu}_e$, the cross section for the reaction (2.1) calculated by Alvarez [38] is $\sigma_{\mathrm{theor}} = 2.6 \cdot 10^{-45}$ cm^2. Experiments were set up by the Davis group first at the Brookhaven reactor [39] and then at the Savannah River reactor [40-42].

Reaction (2.1) was chosen for the following reasons: 1) argon is an inert gas and does not enter into chemical reactions after its formation. This makes it possible to separate a small amount of radioactive ^{37}Ar atoms from a large target volume; 2) the antineutrino spectrum emitted by fission fragments has a maximum at approximately 1.5 MeV and falls exponentially at higher energies. Therefore the threshold for inverse beta decay is of great importance. As is well known, the lowest threshold can be obtained in inverse electron capture. The threshold energy E^{th} = 0.814 MeV for the $^{37}\mathrm{Cl} \rightarrow {}^{37}\mathrm{Ar}$ transition; 3) the ^{37}Ar decay is an allowed transition (log ft = 5.058) and the value of the cross section for the inverse beta process is therefore comparatively large.

Despite the large neutrino flux and the favorable reaction conditions, performance of this experiment was extremely difficult and required many years of effort. We give here only a rough outline of the experimental method.

After irradiation of a large amount of CCl_4 (from 4 to 10 m^3) outside the reactor shield for approximately two half-lives of ^{37}Ar, the radioactive atoms produced were separated from the target together with a carrier (0.1-1 cm^3 of inactive ^{36}Ar) by flushing the CCl_4 volume with helium. The argon was separated by adsorption in an activated charcoal trap cooled to liquid nitrogen temperatures, impurities were removed, and the argon was introduced into a miniature Geiger-Muller counter. In it, 2.8-keV Auger electrons, which are emitted in 80% of the decays, were obtained. The detector was surrounded by a ring of other counters connected in anticoincidence for protection against cosmic-ray background. In addition, the entire system was surrounded by a steel and mercury shield. These measures resulted in a counter background of a few pulses per day. The efficiency of the argon separation was better than 90%. Thus the total efficiency for separation and detection was sufficiently great.

An ever decreasing value was given for the upper limit of the cross section for the reaction (2.1) in a series of successive papers: $\sigma_{\exp} \leq 2 \cdot 10^{-42}$ cm^2 (1955); $\leq 0.9 \cdot 10^{-45}$ cm^2 (1956); $\leq 1.1 \cdot 10^{-45}$ cm^2 (1957); $\leq 0.25 \cdot 10^{-45}$ cm^2 (1959). The lowest value, which was obtained when using 11,700 liters of CCl_4, was reported in the Bulletin of the American Physical Society, but a detailed publication has not appeared. We use as the final result $\sigma_{\exp} = (0.1 \pm 0.6) \cdot 10^{-45}$ cm^2, which is usually given in review articles (see [43], for example).

As shown by experiments on the detection of solar neutrinos [44], a residual ^{37}Ar activity is induced by protons produced in nuclear reactions with cosmic-ray muons. Any significant increase in accuracy of the ^{37}Ar experiments is attended with considerable difficulties (increase in reactor power, burial underground for shielding against muons, etc.).

Planned Experiments at a Pulsed Nuclear Reactor

At the present time, consideration is being given to the possibility of studying inverse beta decay by using a high-power pulsed nuclear reactor as a source of antineutrinos [45]. The effective length of a neutino pulse is determined by the lifetime of fission fragments with sufficiently high maximum beta-spectrum energy and is a few seconds. The expected $\tilde{\nu}_e$ flux intensity at the location of the equipment amounts to approximately $2 \cdot 10^{16}\ \tilde{\nu}_e$/cm^2 per reactor burst with the interval between bursts being about 2.5 h while it is about $10^{13}\ \tilde{\nu}_e$/cm^2-sec in an ordinary reactor.

The use of a pulsed reactor makes it possible to markedly improve the ratio of signal to the background created by the formation of radioactive atoms by cosmic rays (the possibility is realized if the separation of the reaction products occurs in a time much less than the interval between bursts); for rapid separation, it also makes possible selection of reaction leading to the formation of short-lived isotopes, which shortens counting time and reduces the contribution from the intrinsic counter background.

This requirement is satisfied by the reaction

$$\tilde{\nu}_e + {}^{19}\mathrm{F} \rightarrow {}^{19}\mathrm{Ne} + e^-, \tag{2.3}$$

which can be detected through the positron decay of ^{19}Ne in the reaction

$$^{19}\mathrm{Ne} \xrightarrow{T_{1/2}=18\ \mathrm{sec}} {}^{19}\mathrm{F} + e^+ + \nu_e. \tag{2.4}$$

The maximum energy of the positron spectrum $E^{max}_{\beta+} = 2.2$ MeV. The $\tilde{\nu}_e$ threshold energy for the reaction (2.3) is 3.2 MeV. Since ^{19}F and ^{19}Ne are mirror nuclei, the transition (2.4) is superallowed: $\log ft = 3.29$ and $\sigma_{theor} \approx 1.0 \cdot 10^{-44}$ cm^2 (assuming the identity of ν_e and $\tilde{\nu}_e$).

It is proposed to use a solution of KF as the target material. The chief experimental difficulty is the necessity of effective separation of ^{19}Ne from a large volume of target material in a time less than 20 sec. Experiments performed with models showed that the problem of rapid separation can be solved. In an experiment at a pulsed reactor, this makes it possible to achieve a sensitivity better than 10^{-46} cm^2 in searching for the forbidden reaction (2.3). If one considers that the theoretical value of the cross section is approximately 10^{-44} cm^2, this means an increase by an order of magnitude with respect to the Davis experiments.

Estimate of Upper Limit of Possible Lepton Nonconservation

We consider the conclusions which might be arrived at from experiments seeking to find inverse beta processes which take place with violation of the law of lepton conservation. In the known calculations of the cross section for reaction (2.1), transitions to excited levels of ^{37}Ar were not taken into account. Bahcall [46] first drew attention to this possibility in connection with the problem of detection of solar neutrinos:

$$\nu_e + {}^{37}\mathrm{Cl} \rightarrow {}^{37}\mathrm{Ar} + e^-. \tag{2.5}$$

The greatest contribution to this reaction is made by four states in ^{37}Ar (see level diagram):

^{37}Ar	5.15	MeV	—3/2+, 3/2
	1.61	MeV	—5/2+, 1/2
	1.42	MeV	—1/2+, 1/2
	0	MeV	—3/2+, 1/2
^{37}Cl			3/2+, 3/2
			$J^{(\pi)}$, T

where J is the moment, T is the isotopic spin, and π is the parity.

The level with an excitation energy of 5.15 MeV is an analog with respect to the ground state of ^{37}Cl, and therefore the transition to it is superallowed. Using the data of Bahcall [46] for transition matrix elements in the ^{37}Ar excited state and the spectrum of reactor antineutrinos calculated by Avignon [36], one can find the transition cross section for this state. The results are given in Table 4.

The total cross section for the reaction (2.1) is

$$\sigma_{theor} = 9{,}2 \cdot 10^{-45}\ \mathrm{cm}^2, \tag{2.6}$$

which is a markedly larger value than assumed previously ($\sigma_{theor} = 2.6 \cdot 10^{-45}$ cm^2). The accuracy of the result (2.6) is determined by the uncertainty in the estimates of the matrix elements and the antineutrino spectrum. In any case, the error is no more than 25%.

TABLE 4*

Level parameters, ^{37}Ar			Type of transition	Cross section 10^{-45} cm²/fission $\tilde{\nu}$
E, MeV	J^{π}	T		
0	3/2⁺	1/2	Allowed Gamow–Geller	2,68
1,42	1/2⁺	1/2	The same	2,75
1,61	5/2⁺	1/2	»	3,2
5,15	3/2⁺	3/2	Superallowed Fermi and Gamow–Teller	0,53

*Since ν_e and $\tilde{\nu}_e$ are considered unpolarized in the calculations, the factor two introduced for the two-component neutrino [47] does not enter into the value of the cross sections.

TABLE 5

Experimental method	Process studied	δ (values from known experiments)
Mass spectrometry	$^{130}Te \rightarrow {}^{130}Xe$	$\leqslant 0{,}05$
	$^{128}Te \rightarrow {}^{128}Xe$	$\leqslant 0{,}15$
	$^{82}Se \rightarrow {}^{82}Kr$	$\leqslant 0{,}2$
Detection of decay electrons	$^{48}Ca \rightarrow {}^{48}Ti$	$\leqslant 3 \cdot 10^{-4}$
	$^{76}Ge \rightarrow {}^{76}Se$	$\leqslant 0{,}1$
Radiochemistry	$\tilde{\nu}_e + {}^{37}Cl \rightarrow {}^{37}Ar + e^-$	$\leqslant 0{,}3$

One can obtain from (2.3) and (2.6) an estimate of the upper limit for violation of the lepton charge conservation law

$$\delta^2 = \frac{\sigma_{\exp}}{\sigma_{\text{theor}}}, \quad \delta \leqslant 0{,}3. \tag{2.7}$$

As is evident, the results of this experiment are of poorer accuracy than those obtained from double beta decay. However, the total information obtained in these two types of experiment may not be equivalent. In the first place, this is associated with the possibilities discussed in the first section (consideration of double beta decay as a first-order process without neutrino participation, etc.). In the second place, this may be the result of the properties of the particular nuclei. As an illustration, we consider the following speculative possibility. Let the terms which do not conserve lepton charge make a contribution only to the vector portion of the interaction. For double beta decay of a nucleus under investigation, the leading components of the Fermi matrix elements tend to zero because of the difference in isotopic spins of the final and initial nuclear states. Then the experiments on neutrinoless double beta decay and the Davis experiment with ^{37}Cl are only slightly sensitive to violation of the LCCL. At the same time, the Fermi matrix element makes a large contribution on experiments on $\tilde{\nu}_e$ capture by ^{19}F because the transition is superallowed. A summary of the basic results is given in Table 5.

In conclusion, the authors are grateful to B. M. Pontecorvo and L. A. Mikaélyan for fruitful discussions of the problems considered here, and to L. M. Soroko, T. A. Strizh, G. A. Nezhdanov, Yu. P. Chertov, and G. V. Shavel'zon for help in preparation of the review.

LITERATURE CITED

1. Ya. B. Zel'dovich, Dokl. Akad. Nauk SSSR, 86, 505 (1952).
2. E. J. Konopinski and H. Mahmoud, Phys. Rev., 92, 1045 (1953).
3. B. M. Pontecorvo, Zh. Éksp. Teor. Fiz., 53, 1717 (1967).
4. B. M. Pontecorvo, Zh. Éksp. Teor. Fiz., 34, 247 (1958).
5. B. Pontecorvo and V. Gribov, Phys. Letters, 28B, 493 (1969).
6. B. M. Pontecorvo, Izv. Akad. Nauk SSSR, Ser. Fiz., 33, 1787 (1969).
7. J. N. Bahcall and S. C. Frautschi, Phys. Letters, 29B, 623 (1969).
8. R. Davis, Jr., D. S. Harmer, and K. C. Hoffman, Phys. Rev. Letters, 20, 1205 (1968).
9. J. N. Bahcall et al., Phys. Rev. Letters, 20, 1209 (1968).
10. L. Oster, Phys. Rev. Letters, 23, 987 (1970).
11. B. Pontecorvo, Phys. Letters, 26B, 630 (1968).
12. H. Primakoff and S. Rosen, Phys. Rev., 184, 1925 (1969).
13. H. Primakoff and D. H. Sharp, Phys. Rev. Letters, 23, 501 (1969).
14. C. Ryan and S. Okubo, Suppl. Nuovo Cimento, 2, 3, 234 (1964).
15. H. Primakoff and S. P. Rosen, Repts. Progr. Phys., 22, 121 (1959).
16. J. Allen, The Neutrino [Russian translation], Izd-vo Inostr. Lit., Moscow (1960).
17. Dell Antonio and E. Fiorini, Suppl. Nuovo Cimento, 17, 1, 132 (1960).
18. V. Lazarenko, Usp. Fiz. Nauk, 90, 601 (1966).
19. H. Daniel et al., Nucl. Phys., 63, 145 (1965).
20. N. Zeldes et al., Nucl. Phys., 63, 1 (1965).
21. M. G. Inghram and J. H. Reynolds, Phys. Rev., 76, 1265 (1949); Phys. Rev., 78, 822 (1950).
22. R. J. Hayden and M. G. Inghram, Nat. Bur. Stand. Circ., 522, 189 (1953).
23. N. Takaoka and K. Ogata, Z. Naturforsch., 21a, 84 (1966).
24. E. K. Gerling et al., Yad. Fiz., 6, 311 (1967).
25. T. Kirsten et al., Z. Physik, 202, 273 (1967).
26. T. Kirsten et al., Phys. Rev. Letters, 20, 1300 (1968).
27. E. Fiorini et al., Lett. Nuovo Cimento, 3, 149 (1970).
28. E. der Mateosian and M. Goldhaber, Phys. Rev., 146, 810 (1966).
29. R. K. Bardin et al., Phys. Letters, 26B, 112 (1967).
30. E. Greuling and R. C. Whytten, Ann. Phys., 11, 510 (1960).
31. W. H. Furry, Phys. Rev., 56, 1184 (1939).
32. L. A. Sliv, Zh. Éksp. Teor. Fiz., 20, 1035 (1950).
33. V. A. Khodel', Phys. Letters, 32B, 583 (1970).
34. V. A. Khodel', Yad. Fiz., 12, 11 (1970).
35. A. B. Migdal, Theory of Finite Fermi Systems and Properties of Atomic Nuclei [in Russian], Nauka, Moscow (1965).
36. F. T. Avignon et al., Phys. Rev., 170, 931 (1968).
37. B. M. Pontecorvo, Chalk River Lab. Rep. PD-205 (1946).
38. L. W. Alvarez, Rep. UCRL-328, Univ. of Calif. Rad. Lab. (1949).
39. R. Davis, Jr., Phys. Rev., 97, 766 (1955).
40. R. Davis, Jr., Bull. Amer. Phys. Soc., 1, 219 (1956).
41. R. Davis, Jr., Proc. Int. Conf. on Radioisotopes, 1, 728 (1958).
42. R. Davis, Jr. and D. S. Harmer, Bull. Amer. Phys. Soc., 4, 217 (1959).
43. C. S. Wu and S. Moszkowsky, Beta Decay, Interscience Publishers (1966).
44. R. Davis, Jr., Phys. Rev. Letters, 12, 303 (1964).
45. A. A. Borovoi et al., Yad. Fiz., 11, 790 (1970).
46. J. N. Bahcall, Phys. Rev., 135B, 137 (1964).
47. T. D. Lee and C. N. Yang, Phys. Rev., 105, 1671 (1957).

HYDROGEN TARGETS IN THE PHYSICS OF HIGH-ENERGY PARTICLES

L. B. Golovanov

Liquid-hydrogen targets for experiments based on electronic methods in high-energy physics are discussed. Information relating to the targets is systematized, methods of determining the principal parameters are presented, and references intended to help in design calculations are given. Hydrogen targets developed in the Joint Institute for Nuclear Research for physical experiments in Dubna and Serpukhov are described. A detailed description of a precision apparatus with a liquid-hydrogen target is presented, and a jet target is also described.

INTRODUCTION

In studying the interaction of elementary particles with protons, the latter are usually obtained from hydrogen in the densest possible solid or liquid state. The targets used in the physics of high-energy particles may be divided into three categories as regards their functional relationship with the detectors: 1) autonomous targets, i.e., targets entering into the experimental installation as an independent, separate and distinct element (targets of this kind are used in experiments based on the electronic method); 2) targets for film detectors (streamer and bubble chambers); 3) detector targets, i.e., targets which simultaneously constitute detectors, or have detectors inside them. The two latter categories differ from the first in that they serve as an integral part of the actual detectors. A number of papers have already been devoted to targets for film chambers and detectors targets [1-3]; in this review we shall therefore only consider autonomous hydrogen targets intended for experiments based on the electronic method.

The length of the target is mainly determined by the energy of the particles. Thus, for energies of several hundred megaelectron-volts the target length required is no greater than 10 cm; for an energy of ten gigaelectron-volts or over it extends to several meters. The determination of the exact dimensions of the target, the choice of the ratio between the amount of hydrogen and wall material in the path of the particles, and the calculation of the solid angle within which the target walls are as thin as possible depend on the problems underlying the physical experiments and on the scientific foresight of the research worker. We shall therefore not touch on these questions, but shall simply consider ways in which to satisfy the demands made on the target as an instrument of physical investigation. The targets also present considerable engineering problems, e.g., the construction of cryogenic vessels, which are designed to contain a liquid having a low boiling point (20.4°K) and a low heat of vaporization (107.1 kcal/kg) as well as being extremely liable to explode in contact with air.

1. TARGETS AS INSTRUMENTS FOR PHYSICAL INVESTIGATIONS

1.1. Construction of the Targets

The targets consist of an internal vessel containing liquid hydrogen and around this a vacuum jacket isolating the inner vessel from the surrounding medium. Between the inner vessel and the jacket is a

Joint Institute for Nuclear Research, Dubna. Translated from Problemy Fiziki Élementarnykh Chastits i Atomnogo Yadra, Vol. 2, No. 3, pp. 717-762, 1972.

TABLE 1

Material	Wall thickness, mm	
	inner vessel	vacuum jacket
Stainless steel 1Kh18N9T	0,07	0,7
Copper M-3	0,13	0,8
Aluminum AMts	0,20	1,1
Dacron	0,15	3,5
Titanium VT3-1	0,05	0,9

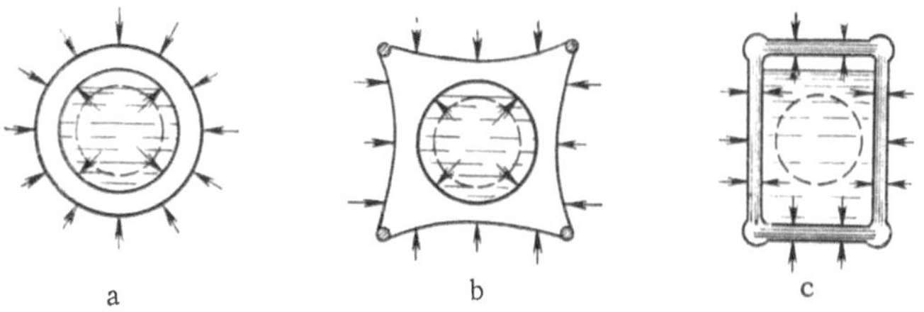

Fig. 1. Types of target constructions. a) Rigid (without framework); b) rigid framework-type; c) soft. The arrows show the forces acting on the inner vessel and the vacuum jacket of the target. The broken lines indicate the working zone of the target.

vacuum space. Exceptions are certain targets made of a thermally insulating material, such as foam plastic [4]. In order to reduce the possibility of interaction between the elementary particles and the walls of the target and thus to reduce the background, the inner vessel and vacuum jacket have to be made as thin as possible. The wall thickness is determined by the strength characteristics of the materials composing the target and also by the shape, size, and structure of the target vessels. The inner vessel operates at 20.4°K and is designed for a pressure of one atmosphere, since the inside of the vessel contains hydrogen at approximately atmospheric pressure and the outside is in vacuum. The vacuum jacket has to withstand atmospheric pressure on the outside and provides stability. These conditions of operation of the vessels determine the wall thickness. Targets of this construction may be classed as rigid (without framework), because when the insulating space is evacuated all the material of the vacuum jacket acts as a stress-bearing unit and suffers no visible deformations (Fig. 1a). Calculations for such vessels in relation to their strength and stability have been set out in full detail in the special literature [5, 6]. Table 1 gives the wall thicknesses of the inner vessel and vacuum jacket. For calculation purposes we assume that both the inner vessel and vacuum jacket have the shape of a cylinder 100 mm in diameter.

The wall thickness of the vacuum jacket may be reduced by 30% if it is strengthened with bracing ribs. We see from Table 1 that the wall thickness of the vacuum jacket is 5-10 times smaller than that of the inner vessel. In order to reduce the wall thickness of the vacuum jacket, framework-type structures of the target must be employed, the vacuum jacket being made in the form of a rigid, stress-bearing framework with thin walls fixed to it (Fig. 1b); alternatively, the inner vessel and vacuum jacket may be made of thin-walled shells with insulating material between them [7, 8] (Fig. 1c). In soft target constructions, the jackets themselves do not provide strength and rigidity, but only separate the vacuum space from the hydrogen and the surrounding medium. Individual elements of the target (or the target in its entirety) may have rigid, soft, or framework construction. For physical problems in which the minimum amount of wall material is required in an angle of 4π, framework-type and soft constructions of the targets are most frequently employed. In cases in which a minimum amount of material is simply required along the axis of the beam, rigid, frameless targets, with thin-walled windows at the ends, are used.

1.2. Choice of Material for the Target Walls

The choice of material for the target walls through which the particles pass depends mainly on the strength of the material, the background obtained as a result of interaction with the target walls, and the relation between the mean-square angles of multiple scattering in different materials.

TABLE 2

Name of substance	Atomic weight [9]	Density ρ, g/cm³ [9, 10]	Permissible stress $\sigma_{per} = \sigma_B^* / 2.6$, kg/cm² [11]	Radiation length t_0, g/cm² [12]
Stainless steel 1Kh18N10T	55,85	7,9	1470	13,9
Copper M-3	63,54	8,96	770	13,0
Aluminum AMts (annealed)	26,98	2,7	500	24,3
Aluminum AMg (annealed)	26,98	2,7	730	24,3
Dacron (polyethyleneterephthalate)	12,01	1,38	675	43,3
Glass	28,09	2,4	115	27,4
Titanium VT3-1	47,90	4,5	1920	—

*σ_B – tensile strength.

The relation between the number of interactions with different materials is determined by the equation

$$\varphi = \frac{X_I}{X_{II}} \sqrt[3]{\frac{A_I}{A_{II}}}, \tag{1.1}$$

where X_I and X_{II} are the amounts of materials I and II in the path of the particles in g/cm², A_I and A_{II} are the corresponding atomic weights.

From (1.1) we find that the relation between the background and the effect for small-angle scattering will be

$$\varphi = \frac{\Sigma l_i \rho_i}{L_p \rho_p} A^{1/3}, \tag{1.2}$$

where l_i is the thickness of the i-th wall in cm, ρ_i is the density of the material of the i-th wall in g/cm³, L_p is the length of the inner vessel of the target filled with hydrogen in cm, ρ_p is the density of hydrogen in g/cm³, and A is the atomic weight of the wall material. From Eq. (1.2) we readily find that, for a target 50 cm long with four windows made from copper foil 150 μ thick in the path of the particles, the relation between the background and the effect is $\varphi = 0.62$. On replacing the foil by Dacron of the same thickness $\varphi = 0.054$.

By using Eq. (1.1) we may obtain a relation for the background effects from target walls of identical thickness but made of different materials:

$$\varphi = \frac{\sigma_{II} \rho_I}{\sigma_I \rho_{II}} \left(\frac{A_I}{A_{II}}\right)^{1/3}, \tag{1.3}$$

where σ_I and σ_{II} are the permissible stresses in walls made of materials I and II in kg/cm². We see from Eq. (1.3) that on replacing stainless steel windows by a Dacron film of the same strength the background falls by a factor of 4.4.

For certain work, an important criterion in the choice of wall material is the ratio between the mean-square multiple-scattering angles. This is determined by the equation

$$\frac{\langle\theta_I\rangle}{\langle\theta_{II}\rangle} = \sqrt{\frac{l_I}{l_{II}} \cdot \frac{t_{0\,II}}{t_{0\,I}}} \quad \text{or} \quad \frac{\langle\theta_I\rangle}{\langle\theta_{II}\rangle} = \sqrt{\frac{\sigma_{II}}{\sigma_I} \cdot \frac{t_{0\,II}}{t_{0\,I}}}, \tag{1.4}$$

where $\langle\theta_I\rangle$ and $\langle\theta_{II}\rangle$ are the mean-square multiple-scattering angles in materials I and II, l_I and l_{II} are the corresponding wall thicknesses in cm, t_{0I} and t_{0II} are the radiation lengths in the corresponding materials in g/cm². From Eq. (1.4) we find that, on replacing stainless steel windows by a Dacron film of equal strength, the mean-square multiple-scattering angle will be only a factor of 1.2 smaller. Tables 2

TABLE 3

Substance	Atomic weight [9]	Radiation length* g/cm² [12]	State	Density ρ, g/cm³ [13, 14, 22, 49, 51]	Temp., °K	Pressure, mm Hg
Normal hydrogen (75% o-H_2)	1,008	62,8	Solid	0,08902	4,2	—
			Liquid	0,0709735	20,380	760
			Gaseous	0,0013316	20,380	760
			The same	0,00008988	273	760
Parahydrogen (0.21% o-H_2)	—	—	Solid (crystalline)	0,0873	11,49	0 atm
			Liquid	0,707749	20,268	759,92
			Gaseous	0,0013376	20,268	759,92
Normal deuterium (66,67% o-D_2)	2,0147	—	Solid	0,2060	4,2	
			Liquid	0,1690	21 (on the saturation line)	
			Gaseous	0,00018	273	760
Helium	4,03	93,1	Liquid	0,1249	4,2	760
			Gaseous	0,0155	4,2	760
			The same	0,00017846	273	760
Nitrogen	14,008	38,6	Solid	1,026	20,66	760
			Liquid	0,804	77,4	760
			Gaseous	0,00498	77,4	760
			The same	0,001225046	273	760
Air	—	37,1		0,0012928	273	760

* The radiation length is numerically equal to the target thickness in which the electron energy falls by roughly e times owing to bremsstrahlung losses. The radiation length of a multicomponent substance is given by the expression

$$\left(\frac{1}{t_{0i}}\right)_{\text{mix}} = \sum_i \left(\frac{P_i}{t_{0i}}\right),$$

where P_i is the weight proportion of the i-th component; t_{0i} is the radiation length of the i-th component.

and 3 give the constants for several materials; these are required in choosing wall materials for the target and determining the ratio between the effect and the background.

1.3. Determination of the Amount of Hydrogen in the Target on the Path of the Particles

The amount of hydrogen on the path of the particles is determined by the product of the length of the target and the density of the liquid hydrogen. For ordinary measurements not requiring very high accuracy the distance between the two end points of the inner vessel may be taken as the length, and the density of the hydrogen at 760 mm Hg may be taken (depending on its ortho/para composition) as between 0.07077 g/cm³ (for parahydrogen) and 0.07097 g/cm³ (for normal hydrogen). In order to determine the amount of hydrogen in the path of the particles to a higher accuracy, in measuring the length we must allow for the shape of the target windows and also make a correction for the nonuniformity arising from the formation of a gaseous phase by the thermal flux.

In bending, the thin-walled target windows assume an almost spherical shape (Fig. 2) except in the case of windows shaped in accordance with a prespecified profile. Allowing for the bending of the windows, the length of the target at various distances from the axis lies within the range

$$L = (L_c + 2L_0) \pm w, \tag{1.5}$$

where L_c is the length of the cylindrical part, L_0 is the mean displacement of the windows in the working zone of the target, and w is the (relative) displacement of the windows over the working zone of the target.

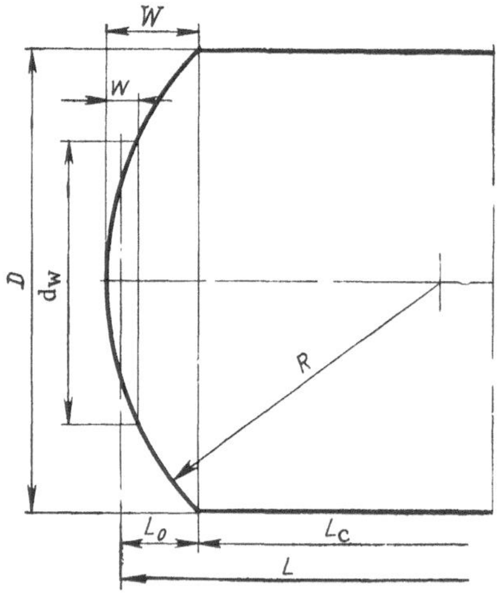

Fig. 2. Determination of the length of the target, allowing for the bending of the windows.

The quantities L_0 and w are given by the equations

$$w = R - \sqrt{R^2 - \frac{d_W^2}{4}}; \quad R = \frac{W}{2} + \frac{D^2}{8W}; \quad L_0 = W - \frac{w}{2}; \qquad (1.6)$$

here R is the radius of curvature, D is the diameter of the target window, W is the total bending of the target window, and d_W is the diameter of the working zone of the target.

For example, for a 50-cm target with a cylinder length of L_C = 500 mm, a window diameter of D = 100 mm (with a sag of W = 10 mm), and a diameter of the working zone d_W = 70 mm (defined by the diameter of the beam "thrown" onto the target), the length of the target (on allowing for the bending of the windows at different distances from the axis) will be L = 515 ± 5 mm.

In working with liquid-hydrogen targets, we often encounter the following problem: By how much does the amount of hydrogen on the path of the particles diminish if it contains a gaseous phase? The total thickness of the gas phase in the working zone of the target due to the inflow of heat from the surrounding medium may be calculated from the equation

$$l = \frac{Q}{\pi v r \rho D \arcsin \frac{d}{D}}, \qquad (1.7)$$

where Q is the flow of heat to the target in cal/sec, v is the velocity with which the bubbles rise in the hydrogen (20 cm/sec), r is the heat of vaporization of the hydrogen (107.1 cal/g), ρ is the density of the gaseous hydrogen at T = 20.4°K (0.0013 g/cm^3), D is the target diameter in cm, and d is the diameter of the working zone in cm.

This equation applies to the case in which the heat inflow only passes to the cylindrical part of the target. For example, from Eq. (1.7) we easily find that the thickness of the gas phase for a cylindrical target of 100 mm and a heat inflow of Q = 10 kcal/h equals 0.2 mm.

For precision measurements (Section 3.6), however, in determining the amount of hydrogen we must allow for the change in target dimensions on cooling and the manner in which the density varies with the pressure and the ortho/para composition of the hydrogen. Apart from this we should take steps to reduce the effect of the bending of the windows, to eliminate the gas phase from the working zone of the target, and to keep the pressure over the liquid in the target constant, independent of any changes in the barometric pressure or changes in the column of liquid hydrogen over the working zone.

1.4. Background Measurements

In order to eliminate the influence of the walls and to determine the effect in hydrogen in pure form, measurements are made alternately with hydrogen-filled and empty targets. The actual methods of carrying out the background measurements differ for different installations. For example, one method is that of pumping the hydrogen from the inner vessel into a special container and then evacuating the gas from the target. This gives an exact measurement of the background but is complicated in execution.

Recently background measurements have often been carried out by replacing the target in the beam with plates simulating the window, or simulators representing the internal vessel of the target, or by using a special empty target made for the sole purpose of background measurements.

The replacement of the target in the beam by a simulator also takes different forms; this may be done by motion in the vertical place with the help of pulleys, motion in the horizontal plane by pneumatic pressure, or by suspending the targets on long rods and imparting a pendulum motion.

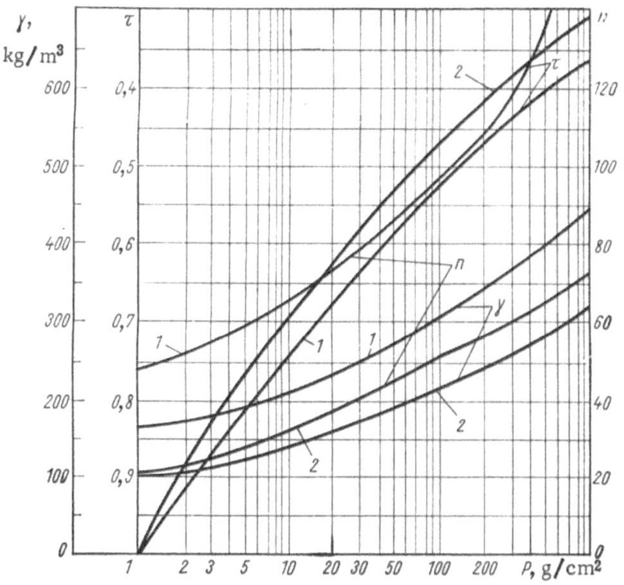

Fig. 3

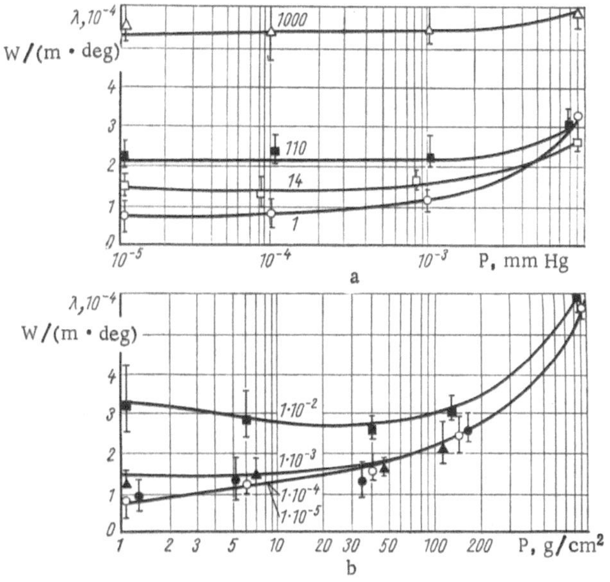

Fig. 4

Fig. 3. Dependence of the bulk weight γ of the insulation, the number of screens n, and the compression coefficient of the insulation τ on the pressure p applied to it: 1) Glass paper SBR 50 μ thick and aluminized Dacron film 12 μ thick; 2) glass cloth ÉVTI 100 μ thick and aluminized Dacron film 12 μ thick.

Fig. 4. Dependence of the thermal conductivity λ of insulation consisting of glass paper SBR 50 μ and aluminized Dacron film 12 μ thick on the pressure in the insulated space P for various pressures on the insulation p (a), and on the pressure applied to the insulation p for various pressures in the insulated space P (b); boundary temperatures 293 and 20.4°K; residual gas hydrogen.

2. THE TARGET AS A CRYOGENIC VESSEL

2.1. Thermal Insulation of the Targets

It is well known that heat passes to cold surfaces by three mechanisms of heat transfer: radiation, heat conduction by the residual gases, and heat conduction by solids (bridges). In order to reduce the inflow of heat by way of the residual gases, a vacuum of 10^{-5}-10^{-6} mm Hg is required between the heat-transmitting surfaces. In order to reduce the inflow of heat by radiation, these surfaces must be carefully polished, and one of more heat screens must be placed between them. The most effective type is a screen cooled by liquid nitrogen. Frequently one or more uncooled (floating) screens or screens cooled by the emerging vapor are employed. The methods of calculating the heat inflow to the cryogenic surfaces are set out in detail in [13, 14]; they will not be repeated here.

Recently a new type of insulation has been widely used: a multilayered vacuum screen. This consists of alternating heat screens with good reflecting surfaces, for example, aluminized Dacron, and a heat-insulating material such as glass paper. This insulation has a low thermal conductivity and substantially reduces the heat inflow, without greatly complicating the target construction. Figures 3-5 show the thermophysical properties of the best multilayered insulations [15]. Table 4 gives the heat inflows for various forms of thermal insulation used in the construction of targets. The heat flows are calculated for parallel surfaces at 20 mm from each other. The area of the surface is 1 m². The boundary temperatures are 293-20.4°K.

2.2. Method of Producing the Vacuum Insulation in the Targets

In order to ensure the smallest possible inflow of heat to the cryogenic surfaces, it is essential to keep the insulating vacuum in the targets down to 10^{-2}-10^{-6} mm Hg, depending on the type of insulation employed. The vacuum may be obtained in two ways: 1) external evacuation with vacuum or adsorption pumps; 2) the use of the cold target surface as a cryogenic and adsorption pump. In evacuating the insula-

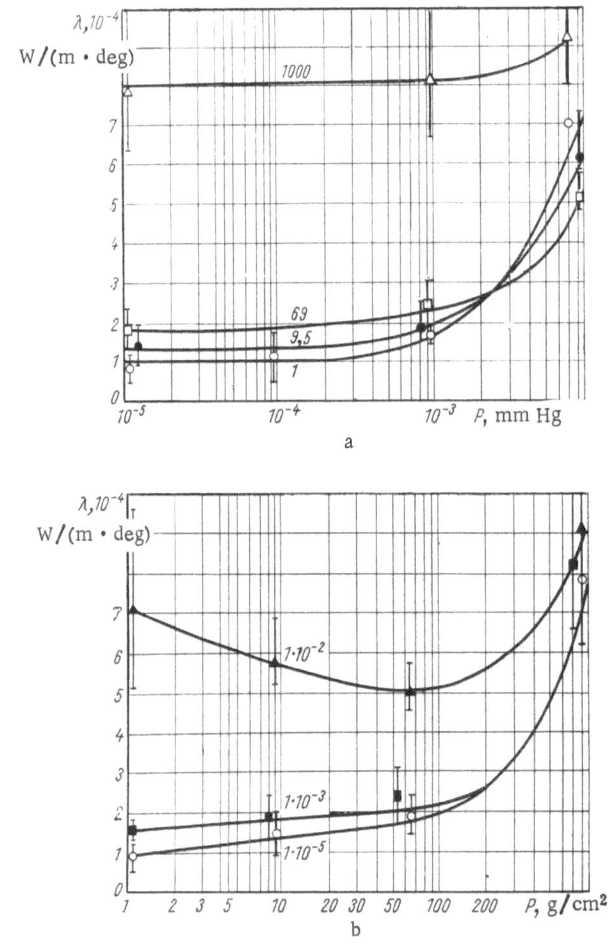

Fig. 5. Dependence of the thermal conductivity λ of insulation comprising glass cloth (ÉVTI) 100 μ thick and aluminized Dacron foil 12 μ thick on the pressure in the insulated space P for various pressures on the insulation p (a), and on the pressure applied to the insulation p for various pressures P in the insulated space (b); boundary temperatures 293 and 20.4°K; residual gas hydrogen.

tion space of liquid-hydrogen targets with vacuum pumps, it is essential that the electrical equipment of these pumps should be made safe from explosions. If the thin windows should rupture, it is important that a trip closing the pumping line should be operated. All this greatly complicates the construction and use of the target. The only advantage of such a system is that it is able to operate continuously for an almost unlimited time.

The insulating vacuum in liquid-hydrogen targets is maintained most easily if the surfaces of the inner vessels are used as adsorption pumps. For this purpose an adsorbent for absorbing the residual gases is placed on the cold surface of the target. Materials widely used for this function include activated charcoal, silicagel, and others. The adsorption capacities of the charcoal (BAU) at 20.4 and 88°K are illustrated in Fig. 6 [16]. The period of continuous operation of this kind of adsorption pump is determined by the properties and volume of the adsorbent, the size of the leak into the vacuum space, and the degree of excellence with which the apparatus has been prepared for operation. Before pouring the hydrogen into the target, the inner surfaces, thermal insulation, and adsorbent should be properly conditioned, i.e., evacuated with a backing pump while heating to 380°K. This preliminary operation occupies initially no

TABLE 4

Form of insulation	Heat inflow, W/m^2	Note
Gaseous hydrogen (p = 760 mm Hg)	1460	For λ = 0.092 W/(m • deg) without allowing for convection
High-vacuum insulation (p = 10^{-6} mm Hg)	8.7	For a reduced emissivity of ε_{red} = 0.02, thermal radiation equals 98.5%
High-vacuum insulation (p = 10^{-6} mm Hg) with nitrogen screen	0.07	Thermal radiation equals 50%
Multilayer insulated (glass paper 50 μ thick and aluminized Dacron film at p = 10^{-4} mm Hg, with compression of the insulation by a force of 1 g/cm^2)	1.1	λ = 0.8 • 10^{-4} W/(m • deg) n = 50 screens/cm
Multilayer insulation (the same but with a compression of 1000 g/cm^2)	7.9	λ = 5.8 • 10^{-4} W/(m • deg) n = 135 screens/cm
Foam plastic (foam polystyrene, density ρ = 40 kg/m^3)	400	λ = 0.03 W/(m • deg)

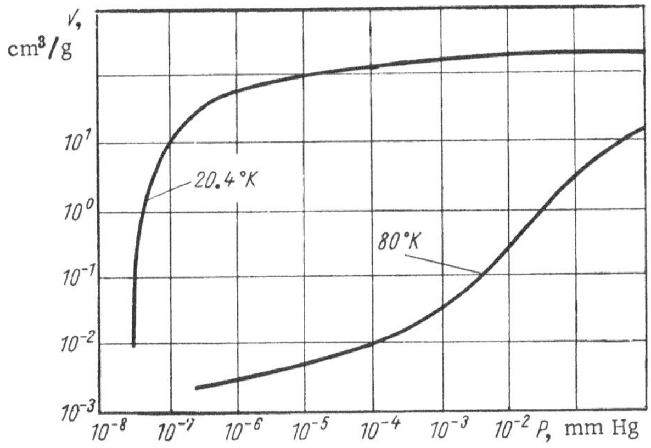

Fig. 6. Hydrogen and deuterium adsorption isotherms on BAU charcoal (volume under normal conditions).

more than two days. Working experience shows that, in conditioned targets, adsorption pumps maintain the required insulating vacuum for 10-30 days. The service life of the adsorbent may be greatly reduced by diffusion through thin-walled organic windows. The adsorbent is usually fixed to the walls by means of a grid. The layer of adsorbent is made no thicker than 5-7 mm. A detailed calculation of adsorption pumps is given in the specialized literature [17].

2.3. Filling the Target with Hydrogen and Monitoring the Level of Liquid

The amount of liquid in the targets diminishes on account of the flow of heat from the surrounding medium. A constant level of hydrogen may be maintained in a variety of ways. One of these lies in making the inner vessel of the target somewhat larger than its working zone. In other constructions an additional vessel is placed over the inner vessel of the target, the cryogenic flowing in automatically from the former. This additional vessel may be made an integral part of the target, i.e., it may have a common vacuum insulation with the latter, or it may be joined to the target and have an independent vacuum. In the latter case the same additional (intermediate) vessel may work with different targets. The hydrogen passes along pipelines to the targets from the liquefiers or is conveyed in Dewars. The inner vessel of the target may be filled directly from a Dewar, in which case a reliable automatic system is required. It is simpler initially to fill an intermediate vessel from a portable source and then let the liquid flow automatically into the target. At the present time, liquid-helium cooling is employed to condense the hydrogen in the target and maintain it there in the solid or liquid state. This has one particular advantage: The apparatus contains less explosive hydrogen. The disadvantages are the great expenditure of helium and the fact that, if the specified conditions are not exactly obeyed, the overflow line may be blocked by solid hydrogen and the target may then be destroyed. The greatest disadvantage is the fact that the density of the solid hydrogen in the target cannot be determined exactly, and errors are therefore committed in determining the amount of hydrogen in the target. The simplest procedure is to fill the target with hydrogen and keep it at the required level by means of compact, reliable cooling systems of the Philips type.

The level of hydrogen in the target may be checked visually, if the target has transparent windows (e.g., Dacron), or else by some kind of level meter. Widely used for this purpose are level meters of the condensation [18] and capacitive [19, 20] types, as well as devices based on the change in electrical resistance with temperature [21]. The simplest and most reliable of these are the condensation type. The scale of the device may be taken a long way away from the sensor. No electricity is used, and this makes for greater safety when dealing with hydrogen.

2.4. Precautionary Measures when Working with Hydrogen

A mixture of air and hydrogen burns for concentrations of 18-59 vol.% hydrogen and explodes for 4-75%. The minimum ignition energy of the mixture is 0.02 mJ. For comparison, the electric charge arising when the operator takes off a coat made of synthetic material is several times greater than the minimum ignition energy. The damage which occurs when 1 kg of hydrogen mixed with air explodes is equivalent to the explosion of 10-15 kg of TNT. The heating capacity of hydrogen is 68 kcal/mole, the temperature of the flame 2045°C, and the rate of flame propagation 2.7 m/sec [22]. These properties of hydrogen clearly demonstrate why it is essential to take special protective measures when working with it: The main points are that hydrogen/air mixtures should not be allowed to form or accumulate even in the smallest quantities, and that certainly no sparks should be tolerated in places where the formation of such a mixture is conceivable.

The formation of an explosive mixture is only too possible if leaks occur in the vessels, and it is therefore vital to adhere strictly to the design prescribed for these [5, 6], and before starting work to make a careful examination of any possibly dangerous units in order to determine the way in which they will behave in any emergency situation.

All the rooms in which hydrogen is used should be supplied with a ventilation (blower) system; in rooms in which emergency situations may lead to the formation of an explosive concentration throughout the room, emergency extractor fans exchanging the air at least twelve times an hour should be provided. The space around the hydrogen-containing apparatus is to be treated as a danger zone. In the Joint Institute for Nuclear Research, the radius of the danger zone is taken approximately as the radius of the hemisphere within which an explosive concentration is formed when the whole of the hydrogen present is mixed with air. The boundary of the danger zone is formed by the protective wall of the blower system, provided that the flow rate of the hydrogen-evacuation system exceeds the rate of outflow of the hydrogen in an emergency situation. The hydrogen-ejection pipe, ending in a flame-interceptor, passes into the atmosphere outside the room.

All the working parts used in places in which explosive concentrations are liable to form are made explosion-resistant in accordance with the existing rules [24], or else taken outside the danger zone.

Rooms in which hydrogen is used are equipped with gas analyzers which indicate the development of hydrogen concentrations of no more than half the dangerous level.

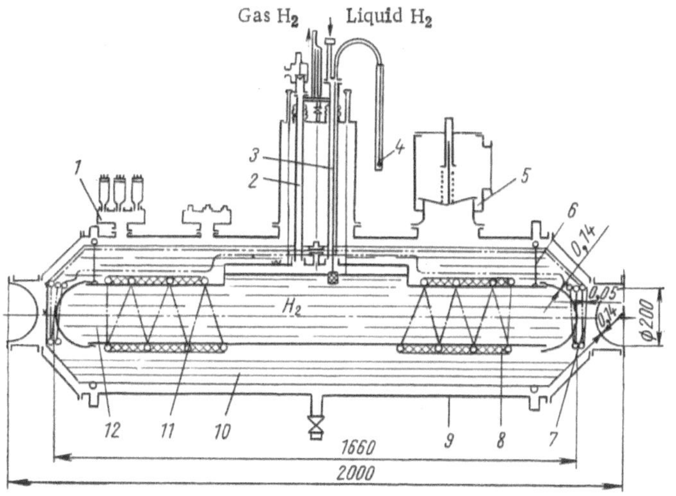

Fig. 7. Construction of the 150-cm liquid-hydrogen target. 1) Vacuum unit; 2) emergency outlet tube; 3) tube for filling with liquid hydrogen; 4) level indicator; 5) safety valve; 6) suspension; 7) screen; 8) adsorbent; 9) vacuum jacket; 10) multilayer insulation; 11) coil for cooling inner vessel; 12) inner vessel; the broken lines show the path of the evaporating hydrogen passing from the inner vessel through the screen to the ejection tube.

3. TARGETS AND INSTALLATIONS WORKING IN THE HIGH-ENERGY LABORATORY OF THE JOINT INSTITUTE FOR NUCLEAR RESEARCH

3.1. The 150-cm Liquid-Hydrogen Target [25]

In the High-Energy Laboratory of the Joint Institute for Nuclear Research, the 150-cm liquid-hydrogen target is used to measure the total cross sections of π^- mesons colliding with protons at momenta of 3.4-9.2 GeV/c and also the scattering cross sections of π^+ mesons scattered through an angle of 180° [26-30].

The principal parts of the target are the inner vessel 12 and the vacuum jacket 9 (Fig. 7). The inner vessel of the target has the shape of a cylinder 200 mm in diameter and 1500 mm long. In the middle of the upper part of the inner vessel is an additional container; the level of liquid hydrogen in this varies during the experiment. Thin-walled end pieces of stainless steel 0.14 mm thick, spherical in shape, are sealed to the ends of the cylinder. A coil 11 is wound on the inner vessel of the target. Liquid nitrogen flows through this in order to cool the target before filling with hydrogen. The vacuum jacket has a cylindrical shape with conical extension pices; end pieces similar to those of the inner vessel are also fitted to these. In the target jacket is a safety valve 5 which opens when the pressure in the jacket exceeds 0.03 atm. The inner vessel of the target is suspended on two wire suspensions 6, 1.5 mm in diameter, and these in turn are fixed to the walls of the vacuum jacket by tension members.

The cylindrical part of the inner vessel is insulated by the multilayer insulation 10 consisting of glass fiber and aluminum foil. The mean thickness of the insulation is 90 mm. The ends of the inner vessel of the target are protected from heat inflow by polished copper screens 7, 0.05 mm thick, cooled with hydrogen vapor. The screen temperature is under 100°K. Preliminary evacuation of the vacuum jacket is effected with a backing pump; this is switched off before target cooling begins. In the hydrogen-filled target, the insulating vacuum is maintained by means of the adsorbent 8 placed between the turns of the coil on the inner vessel. The adsorbent employed is activated charcoal of the BAU type. A vacuum of 10^{-5} mm Hg is obtained with the aid of the charcoal and remains intact in the target for 10 days.

The inner vessel of the target is connected to the atmosphere by two stainless steel tubes 30 mm in diameter. Tube 2 is designed to carry the gaseous hydrogen away from the target when there is a breakdown of the vacuum in the jacket; tube 3 is provided for filling the target with hydrogen. In order to reduce the heat flow to the target, 70% of the evaporating hydrogen is passed through the labyrinth of screens and 30% through the tubes 2 and 3. For this flow ratio the heat inflow to the target is minimal and equals 6.2 W. A target of similar size was constructed earlier [31]. Our own target differs from this in being simpler, and also (fundamentally) in the absence of a nitrogen screen.

Technical data

Length of working volume of target, mm	1660 ± 10
Diameter of target window, mm	200
Amount of hydrogen on target axis, g/cm^2	11.7
Amount of material in the path of the particles on entering and leaving the target, g/cm^2:	
stainless steel	0.22
copper	0.04
Volume of inner vessel of target, liters	50
Volume of additional space, liters	5
Heat flow (experimental) to target, W	6.2 (0.7 liter liquid H_2/h)
Period of continuous operation of target without refilling with hydrogen, h	12
Overall size, mm	1160 × 2000 × 538
Weight, kg	200

3.2. Three-Meter Liquid-Hydrogen Target [32]

The installation with the 3-m liquid-hydrogen target constitutes the cryogenic component of certain physical devices designed for studying the regeneration of high-energy K^0 mesons in hydrogen [33, 34]; it is installed in one of the channels of the Serpukhov accelerator. Apart from the target, the installation incorporates (Fig. 8): a simulator for background measurements, a backing pump, a gas control board, and siphons and Dewars [35] needed for preparing the target for operation and filling it with liquid hydrogen and nitrogen. A special table with the target and simulator mounted upon it enables these to be set in the path of the particles. The target and simulator are remote-controlled by a pneumatic system. All the elements of the apparatus are placed in a metal framework with two flat stages, upper and lower. On the lower stage are the target, the simulator, and the equipment required to prepare the target for operation; on the upper stage are the gas board and Dewars. The target is filled with hydrogen from the upper stage. In the framework the sides are tightly covered with incombustible cloth; the top and bottom are "lined" with stainless steel sheet. The extraction fan system is attached to the upper part of the framework. The frame stands on rails and may be moved 10-15 m along the beam. The length of the frame is 5, the width 2, and the height 7 m.

The target (Fig. 9) consists of a cylindrical inner vessel 1 for liquid hydrogen 250 mm in diameter, 3 m long, with an annular vessel (the nitrogen screen 2) placed around it in order to reduce the heat flow. The vessels for the hydrogen and nitrogen lie inside a demountable vacuum jacket consisting of a cylindrical corrugated shell 3 2 mm thick, two end flanges 15, and two rings 6 and 14. Through ring 6 pass all the tubes connecting the inner vessel to the outer services of the target; through ring 14 pass the tubes connecting the nitrogen vessel to the atmosphere. At the ends of the inner vessel and the vacuum jacket are windows 7, 120 mm in diameter; these are included to reduce the amount of wall material in the path of the particles. The windows are made of Dacron film 190 μ thick. The Dacron is fixed to the vessels with pressure flanges (Sec. 3.4).

The fixing of the vessels containing liquid hydrogen and nitrogen inside the vacuum jacket was one of the most difficult problems in constructing the 3-m target. On cooling the inner vessel and nitrogen screen, their length diminishes considerably (by approximately 10 mm); the cooling takes place at different rates and not everywhere at the same time. For this reason the vessels are suspended in the following manner: The inner vessel is connected on one side to tube 9 of the gaseous hydrogen outlet, which is welded to the end of

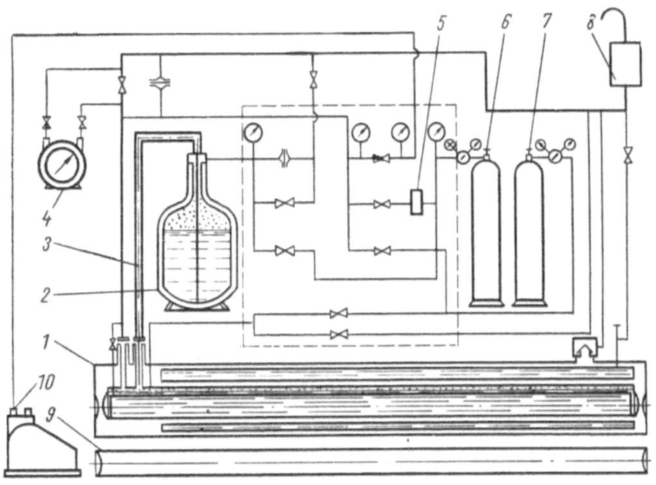

Fig. 8. Arrangement of the installation with the three-meter hydrogen target: 1) Target; 2) Dewar; 3) siphon; 4) gas counter; 5) automatic device for filling the system with gaseous hydrogen; 6) hydrogen cylinder; 7) nitrogen cylinder; 8) flame-arrester; 9) simulator; 10) pump.

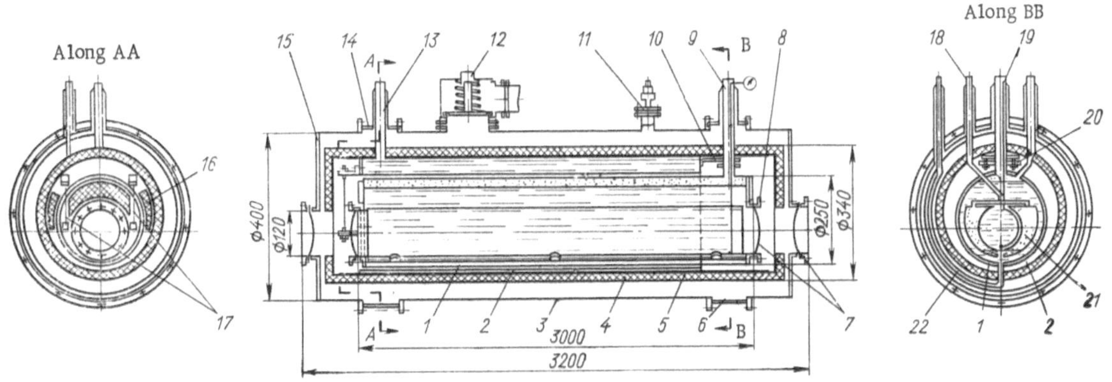

Fig. 9. Construction of the target: 1) Inner vessel; 2) nitrogen vessel; 3) shell of vacuum jacket; 4) multilayer insulation; 5) shell for fixing the insulation; 6) ring of vacuum jacket; 7) Dacron window; 8) pressure flange; 9) gaseous hydrogen exit tube; 10) bracket; 11) vacuum meter unit; 12) safety valve; 13) nitrogen exit tube; 14) ring of vacuum jacket; 15) flange of vacuum jacket; 16) adsorbent; 17) stays of inner vessel; 18) gas outlet tube for supplementary vessels; 19) tube for filling the target with hydrogen; 20) support plate; 21) supplementary vessels; 22) hydrogen decanting tube.

the vacuum jacket 6, and on the other to the nitrogen screen by way of two stainless steel pull rods 17. The nitrogen screen, like the inner vessel, hangs, on one side, on the nitrogen ejection tube 13, which is welded to the ring of the vacuum jacket 14; on the other side (by way of the bracket 10) it rests freely on the plate 20, which is fixed to the hydrogen ejection tube. Thus, while cooling, one end of the vessels remains stationary, while the other moves freely.

In order to reduce the heat flow to the hydrogen, the outer surface of the inner vessel and the surface of the nitrogen screen turned toward it are carefully polished. The nitrogen vessel is insulated from the surrounding medium with multilayered vacuum-screen insulation comprising 80 layers of foiled glass paper of the FBS type. The vacuum in the insulation space is maintained by means of the adsorbent is fixed to the ends of the inner vessel and nitrogen screen. In the hydrogen-filled target the vacuum equals $\sim 10^{-5}$ mm Hg. For measuring the vacuum a vacuum-meter unit 11 is attached to the jacket. The rate of evaporation of the hydrogen from the target is ~1 liter/h and that of the nitrogen no more than 0.5 liter/h.

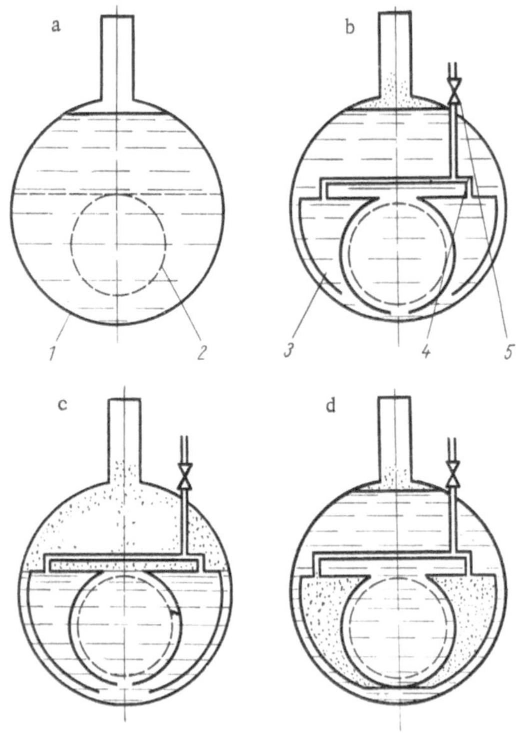

Fig. 10. Internal vessel of the target. a) Without additional vessels; b, c, d) with additional vessels; 1) inner vessel; 2) working zone; 3) additional vessels; 4) connecting tube; 5) stop valve.

In order to keep a store of hydrogen to facilitate prolonged operation of the target (Fig. 10), the diameter of the inner vessel 1 is made a little greater than the diameter of the working zone. The diameter of the working zone is determined by the dimensions of the Dacron windows. The windows of the inner vessel are moved down as far as possible. In order to ensure the efficient use of the hydrogen, to the left and right of the working zone of the target two additional vessels 3 are placed inside the main vessel. The additional vessels are connected by a tube 4, which is closed outside the target by a valve 5. The inner vessel and the additional vessels are filled with hydrogen with the valve 5 open. During the operation of the target, when the hydrogen level falls to the upper part of the additional vessels (Fig. 10c), the valve 5 is closed, and the hydrogen is forced out of the additional vessels into the inner vessel (Fig. 10d). After this, work continues until the level again falls to the working zone. The additional vessels have a volume of 50 liters, i.e., the same as the volume over the working zone of the target. The total volume of the inner vessel of the target is 140 liters. The additional vessels placed within the target enable the target working time to be doubled. The pressure in the additional vessels is raised as a result of ortho/para conversion and the heat flow along the contacts between the additional and inner vessels.

The 3-m liquid-hydrogen target may be used for physical experiments requiring the maintenance of a constant amount of hydrogen in the path of the particles over the whole cross section to an accuracy of 0.05%. For this purpose the installation should incorporate a hydrogen-pressure stabilizer, and the windows of the inner vessel should contain double Dacron films as in the UMVP-1 apparatus (Sec. 3.6).

During the filling of the target with hydrogen (Fig. 11) the level is checked visually through the transparent Dacron windows or by means of a condensation-type level indicator 3, the scale 5 of which is taken out to the upper section of the framework. During the beam throw, the operation of the target is checked from the experimental room, which lies at a distance of 50 m from the target 1. The target 1 is in the working state if the level of hydrogen in it lies above the working zone 2. Remote-control monitoring of the rate of evaporation is based on measuring the temperature (by means of a thermocouple) at a fixed point A in the tube 4 through which the evaporating hydrogen emerges. As the evaporation rate rises, the temperature of the tube at the thermocouple fixing point falls, as may be seen from the readings of the

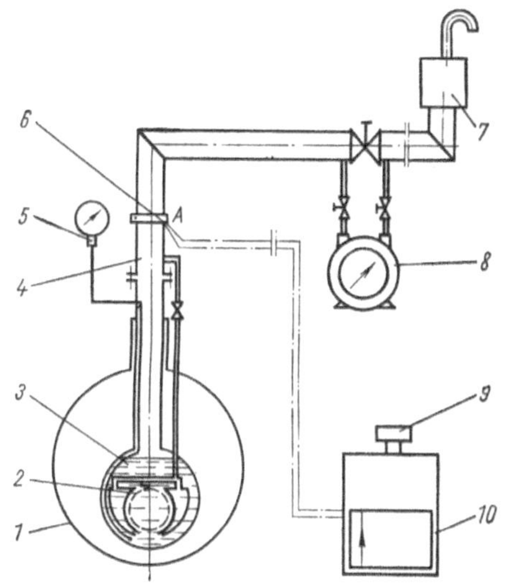

Fig. 11. Arrangement for monitoring the rate of evaporation of hydrogen from the target: 1) Target; 2) working zone; 3) condensation-type level indicator; 4) hydrogen outlet tube; 5) scale of level indicator; 6) thermocouple fixing point; 7) flame arrester; 8) gas counter; 9) signal panel; 10) potentiometer.

potentiometer 10. When the rate of evaporation doubles, the signal panel 9, previously cut off by a suitable potential, lights up. The monitoring and signalling system is calibrated by means of the gas counter 8. The system gives a fairly accurate indication of the rate of evaporation, and hence the instant at which the target has been filled with hydrogen. This method of checking the state of the target is simple and reliable; it may be carried out from any distance and is safe in operation.

In setting up the installation with the 3-m liquid-hydrogen target, great attention was paid to its safety in use. Possible emergency situations were considered, and arrangements for evacuating the hydrogen from the installations were provided in each case. Let us consider these cases:

1. Hydrogen pours into the vacuum jacket of the target when the Dacron window of the inner vessel breaks. In this case, at a pressure of 0.05 atm the safety valve 12 opens (Fig. 9) and the gaseous hydrogen passes into the exhaust line along a tube 100 mm in diameter.

2. The insulating vacuum is damaged when the Dacron window of the vacuum jacket breaks and the thermal flux to the hydrogen rises sharply. The diameter of the exhaust line from the inner vessel is designed in such a manner as to ensure that in such a case the pressure in the vessel should not exceed 0.5 atm.

3. Hydrogen is spilled onto the open surface around the target when the Dacron windows of the vacuum jacket and inner vessel break at the same time. In this case the complete evacuation of the hydrogen from the closed space surrounding the installation must be provided for by extraction fans.

In order to prevent explosions in emergency situations, all the elements of the installation are made from nonsparking materials. During work with hydrogen, the electrical supply of the installation is completely disconnected except for the lighting, which is designed in an explosion-resistant manner.

Technical data of the target

Length of working volume of target, mm	3000 ± 5
Diameter of Dacron windows, mm	120
Amount of hydrogen on the target axis, g/cm^2......	21.3
Amount of material in the path of the particles at the entrance and exit from the target (Dacron), g/cm^2	0.0525 each
Capacity of the inner vessel, liters.............	140
This includes:	
capacity of the working zone	40
capacity of the upper part of the inner vessel	50
volume of the additional capacities (vessels)	50
Thermal flux (experimental), W:	
to the inner vessel	8.4 (1 liter liquid H_2/h)
to the nitrogen screen	22.2 (0.5 liter liquid N_2/h)
Period of continuous operation of the installation without recharging with H_2 and N_2, h	100

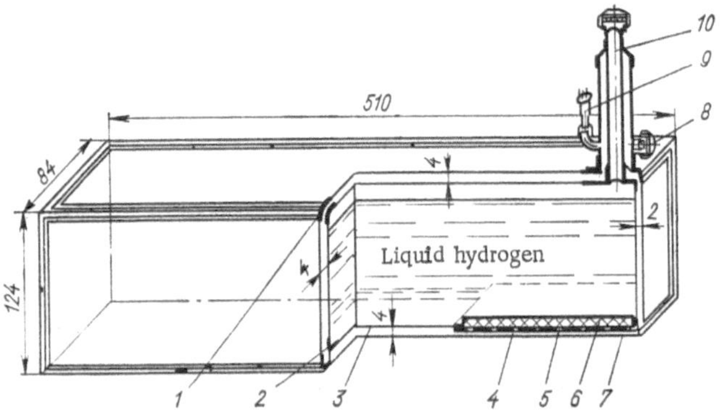

Fig. 12. Schematic construction of the thin-walled liquid-hydrogen target: 1) Upper frame; 2) inner framework; 3) foil; 4) multilayered insulation; 5) grid; 6) adsorbent; 7) lower frame; 8) vacuum pump valve; 9) vacuum-meter tube; 10) throat.

Amount of liquid required for cooling the target, liters:	
hydrogen	60
nitrogen.	100
Time for the initial filling of the target with hydrogen, h	1
Overall target dimensions, mm	3200 × 648 × 480
Weight of target, kg.	400

3.3. Plane Thin-Walled Liquid-Hydrogen Target [8]

A plane thin-walled liquid-hydrogen target was used to measure π^-–p scattering close to 180° in a beam of π^- mesons with a momentum of 4-7 GeV/c [36]. Special features of the target include flat walls for siting the particle detectors in the immediate vicinity of the liquid hydrogen and a minimum amount of wall material over an angle of 4π.

The inner vessel of the target (Fig. 12) consists of a framework 2 in the shape of a parallelepiped made of angle iron 7 × 7 mm with copper foil walls 3, 40-80 μ thick, welded to it in a vacuum-tight manner on all sides. The vacuum jacket of the target has the same construction as the inner vessel.

Between the inner vessel and the vacuum jacket is some multilayered thermal insulation 4 consisting of glass paper and aluminized Dacron film. The initial evacuation of the vacuum space is carried out through a valve with a detachable head 8. The insulating vacuum is maintained by means of an adsorbent 6. On evacuating the insulation space of the target, the flexible inner and outer walls distort under atmospheric pressure and compress the insulation between them. This yields a vessel in which the framework and the insulation between the two flexible walls carry the load arising from the pressure inside the target. The mean thermal conductivity of the multilayered insulation on compression with a stress of 1 kg/cm^2 equals $5.8 \cdot 10^{-4}$ W/(m · deg). The thermal flux to the target is 18.6 W.

The target has a throat 10 for filling with liquid hydrogen and extracting the evaporating gas. The presence of hydrogen in the target is monitored by a condensation-type level indicator. The target is constantly refilled with hydrogen by means of an additional vessel of 30 liters capacity from which the hydrogen flows automatically into the target (Fig. 13).

The target withstands an internal pressure of 0.4 atm. In emergency situations when the vacuum is contaminated with air the maximum pressure inside the target is 0.18 atm; or on contamination with hydrogen 0.2 atm. The maximum thermal load when air enters the vacuum is 4400 W/cm^2, or when hydrogen enters 2560 W/m^2.

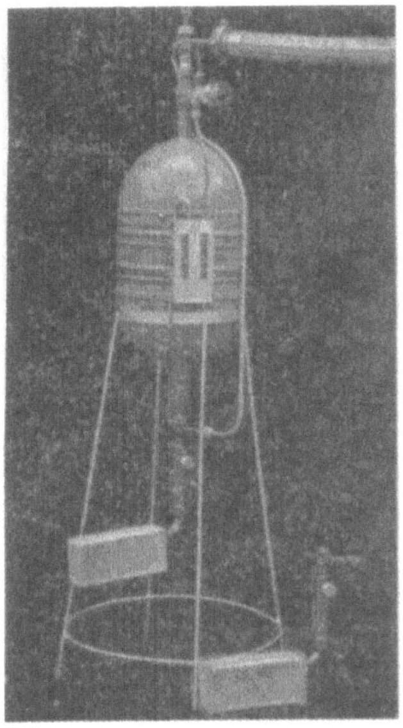

Fig. 13

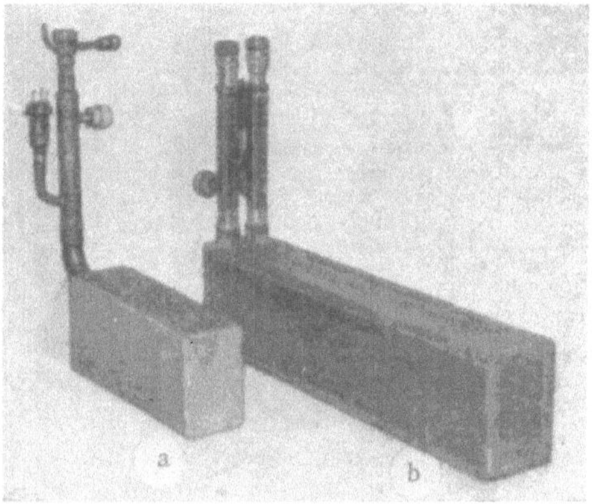

Fig. 14

Fig. 13. General view of the target with a reservoir of 30 liters capacity for continuous filling with liquid hydrogen.

Fig. 14. General view of the 25-cm (a) and 50-cm (b) targets.

Two targets have been developed: one 50 and the other 25 cm long (Fig. 14).

Technical Data Relating to the Targets

	50-cm	25-cm
Length of inner vessel, mm	503 ± 1	253 ± 1
Cross section of inner vessel, mm	75 × 115	70 × 110
Amount of hydrogen on target axis, g/cm^2	3.54	1.77
Amount of material in the path of the particles, g/cm^2, through the side walls:		
aluminum	0.143	0.143
copper	0.142	0.142
glass	0.112	0.112
Through the end walls:		
aluminum	0.071	0.071
copper	0.071	0.071
glass	0.056	0.056
Volume of inner vessel, liters	4	2
Thermal flux to target (corresponding to the evaporation rate), W (liters/h)	18.5 (2.1)	10.9 (1.2)
Time of continuous operation of the target before refilling with hydrogen (together with the intermediate vessel of 30 liters capacity), h	12	13
Overall size, mm	512 × 395 × 125	398 × 310 × 115
Weight, kg	3.4	1.9

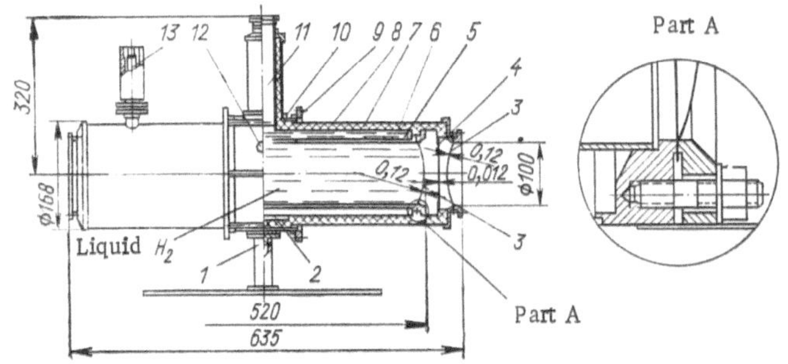

Fig. 15. Construction of the 50-cm liquid-hydrogen target: 1) Support; 2) adsorbent; 3) Dacron windows; 4) screen; 5) false shell; 6) inner vessel; 7) multilayered insulation; 8) cylindrical can; 9) cover flange; 10) ring; 11) throat; 12) vacuum valve with removable head; 13) vacuum-meter tube.

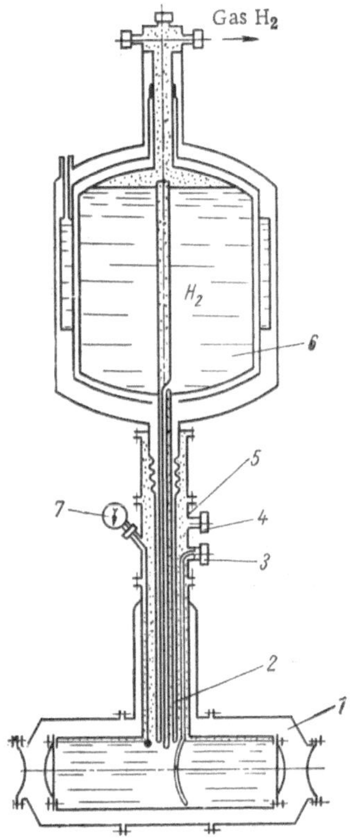

Fig. 16. Arrangement for attaching the target to the intermediate vessel: 1) target; 2) tube for pouring the hydrogen into the target; 3) tube for removing liquid hydrogen from the target; 4) pulse tube for pressure measurement; 5) sleeve; 6) intermediate vessel; 7) condensation-type level indicator.

3.4. Fifty-Centimeter Liquid-Hydrogen Target with Dacron Windows [37]

The inner vessel of the 50-cm target (Fig. 15) is made in the form of a cylinder, 144 mm in diameter and 1 mm thick, to the ends of which flanges for fixing the Dacron windows 3 are welded. The vacuum jacket of the target is demountable and consists of three parts: a ring 10 for fixing the inner vessel and two cylinders 8, 168 mm in diameter and 1 mm thick. The construction of the Dacron windows at the ends of the vacuum jacket is the same as in the inner vessel. The inner vessel of the target is suspended in the throat 11, which is soldered at the upper end to the tube of the vacuum jacket.

The cylindrical part of the inner vessel of the target is insulated with a vacuum-screen multilayered insulation 7 comprising 40 layers of glass paper 12 μ thick. The vacuum in the insulation space is maintained by means of an adsorbent 2 fixed to the inner vessel of the target. Activated charcoal BAU is employed as adsorbent. The jacket is initially evacuated by means of a backing pump through a valve with a detachable head 12. The vacuum is monitored by means of a vacuum-meter tube of the LT-4M type 13.

In order to obtain a uniform mass of liquid hydrogen in the working space, without any bubbles, a false shell 5 is placed inside the target. The bubbles formed on boiling are lifted up through the annular gap. In order to reduce boiling from the ends, the Dacron windows are protected by screens 4, which are in contact with the cylindrical part of the target. The target is filled with hydrogen and the evaporated hydrogen is allowed to escape through the target throat 11. The level of hydrogen in the target is monitored visually through the transparent Dacron windows and a special slot in the screen, and also by means of a condensation-type level indicator.

In making the target, the fixing and sealing of the Dacron windows cause considerable difficulty. In other target constructions the Dacron windows are sealed in the vacuum jacket by means of rubber gaskets, and in the inner vessel, which works at a temperature of 20.4°K, with epoxy resin. In the present target construction the fixing and sealing of the Dacron windows to the vacuum jacket and inner vessel are effected by pressing the Dacron between two flanges: one welded to the vessel and the other removable. The sealing surfaces of the flanges are made in the form of teeth 1 mm high and 2 mm wide. Before sealing, these are lapped on a flat plate and the sharp edges are blunted. The number of pins for fixing the film between the flanges is chosen in accordance with the diameter of the windows and the rigidity of the flanges. For the present target with a window diameter of 100 mm, a Dacron thickness of 120 μ, and a flange thickness of 15 mm, the number of M8 pins equals twenty. Many tests on strength and density as well as experience in using the targets showed that this form of sealing was simple and reliable (Fig. 15, part A).

The target is filled with hydrogen by automatic flow from the intermediate vessel 6, 50 liters in capacity, placed above the target 1 (Fig. 16). The intermediate vessel and target have an independent insulating vacuum and are connected to each other by the sleeve 5, which is also used for the leads of the condensation-type level indicator 7, the tube for decanting the hydrogen from the target 3, the tube for ejecting the gas when cooling the target, and the pulse tube for measuring the pressure in the target 4.

The intermediate vessel is made in the form of a Dewar flask (50 liters), differing from the latter in having a tube 2 at the bottom, along which the hydrogen pours into the target. In order to increase the rate of flow, the liquid hydrogen and evaporating gas leaving the target travel along different channels. The level of hydrogen in the target depends on the depth to which the tube taking the gaseous hydrogen from the target is immersed in it. This system for filling the target with hydrogen is simple and reliable and requires no additional automatic equipment.

Technical Data Relating to the Target

Length of the inner vessel of the target, mm	515 ± 5
Diameter of target window, mm	100.0
Amount of hydrogen along target axis, g/cm^2	3.7
Amount of material in the path of the particles at the entrance and exit of the target, g/cm^2:	
Dacron	0.033
aluminum............................	0.33
Volume of inner vessel of the target, liters	7.7
Thermal flux to the target, W	4.2 (0.5 liquid H_2/h)
Time of continuous operation of the target together with the intermediate 50-liter vessel before refilling with hydrogen, h	50
Overall dimensions, mm	635 × 650 × 210
Weight, kg	24

Two other targets have been constructed on the basis of the fundamental one just described (Fig. 17). The first of these differs from the original in that it has an elongated jacket. This, firstly, enables the Dacron window of the vacuum jacket to be taken further away from the hydrogen, and secondly eliminates some of the air in the path of the particles before and after the target. The second target is distinguished by the fact that the throat is made, not in the middle, but toward the edge of the target. This enables a greater part of the target to be surrounded by a single circular counter.

The existence of the autonomous intermediate vessel for filling the target with hydrogen and the fact that all the targets have couplings of the same sizes make it an easy matter to interchange targets while leaving the intermediate vessel and all the services unaltered.

Work associated with determining the real part of the amplitude of elastic forward π^-–p scattering [38-41] has been carried out in cylindrical targets with Dacron windows.

3.5. Liquid-Hydrogen Conical Target [42]

For measuring the partial probabilities of the decay of vector mesons into e^+e^- pairs in the reaction $\pi^- + p \to v^0 + n$, $v^0 \to e^+e^-$ for incident particle momenta of P = 4.0 GeV/c [43-45], conical liquid-hydrogen targets are employed (Fig. 18). The target has the shape of a truncated cone with its axis lying horizontally.

Fig. 17. General view of the 50-cm liquid-hydrogen target.

Fig. 18. General view of the conical hydrogen target.

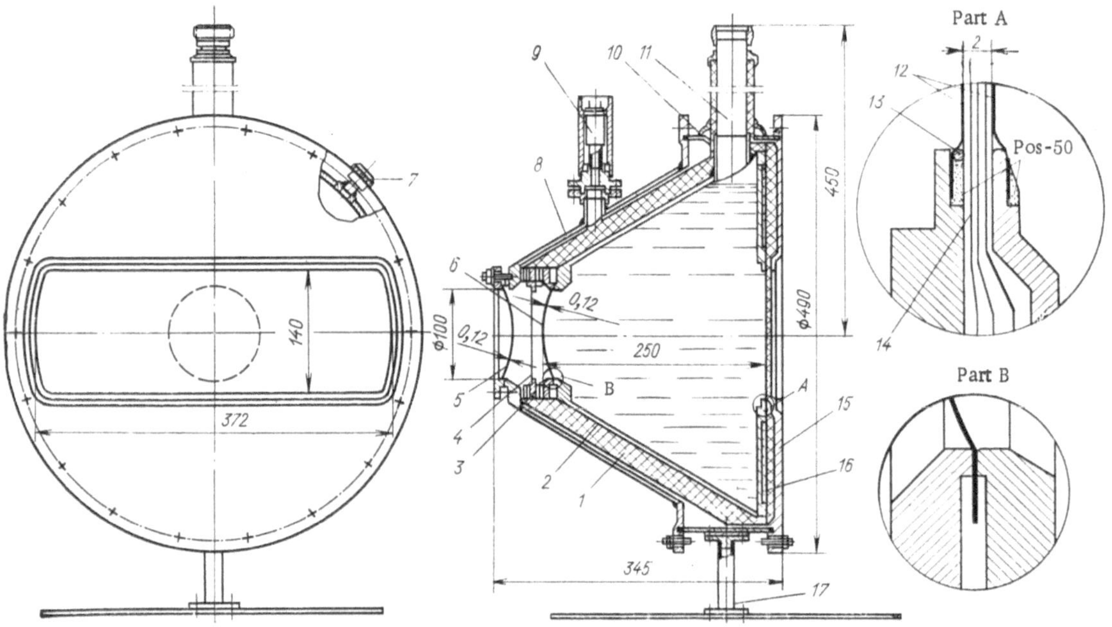

Fig. 19. Construction of target: 1) multilayered insulation; 2) inner vessel; 3) support unit; 4) screen; 5, 6) Dacron film; 7) vacuum valve with a removable head; 8) vacuum jacket; 9) thermocouple manometer; 10) ring; 11) throat; 12) foil; 13) frame; 14) multilayered insulation (compressed by a pressure of 1 kg/cm^2); 15) flange; 16) adsorbent; 17) support.

In order to reduce the interaction of the elementary particles with the target walls, thin windows are made in the ends (bases): an entrance window 100 mm in diameter and an exit window in the shape of a rectangle 372 × 140 mm in size. The length of the target is 250 mm.

The construction of the target is illustrated in Fig. 19. The target consists of an inner vessel 2, a vacuum jacket 8, and multilayered insulation 1 between them. The conical shell of the inner vessel is made of stainless steel 2 mm thick. Welded to the large base of the shell is a flange containing a milled window 372 × 140 mm in size. The window is sealed with copper foil 50 μ thick. At the other end a flange with an aperture 100 mm in diameter is welded to the conical shell. This flange carries a Dacron film 6,

120 μ thick, through which the particle beam enters the target. The inner vessel of the target hangs in the throat 11, which is welded at the top to the tube of the vacuum jacket. In order to fix the position of the inner vessel relative to the vacuum jacket, a support unit 3 consisting of multilayered insulation is placed between them on the side of the smaller base.

The vacuum jacket consists of a ring 10 and fixed to one side of this a flange 15 with a window exactly the same as that in the inner vessel; fixed to the other side of the ring is a conical jacket of stainless steel 2 mm thick terminated by a Dacron film 5. The Dacron film is fixed in exactly the same way to the inner vessel and vacuum jacket by means of a pressure flange (part B in Fig. 19; also see Sec. 3.4).

Between the inner vessel and the vacuum jacket is some multilayered insulation comprising 40 layers of glass paper alternating with 40 layers of metallized Dacron. The working vacuum in the insulation space is obtained by means of the adsorbent 16 fixed to the large flange of the inner vessel. The vacuum is measured with a thermocouple manometer 9. In order to reduce the heat flow to the hydrogen, an aluminum foil screen 4 is placed between the Dacron windows. The level of hydrogen in the target is checked visually through transparent Dacron windows and a slot in the screen specially made for this purpose.

On evacuating the insulation space between the inner vessel and the vacuum jacket, the copper foil sheets 12 are fixed to the target flanges; these bend under atmospheric pressure and compress the insulation between them. The thickness of the multiple-layer insulation between the foil sheets after evacuation is 2 mm. So as to ensure a smooth transition at the junction point when the foil bends, a special frame 13 is welded to the flange of the inner vessel, while the edges of the outer window are rounded (part A in Fig. 19). This improves the technology of soldering and increases reliability. The hydrogen is poured in and the evaporating gas allowed to escape through the throat 11. The target is filled with hydrogen by automatic feed from the additional 50-liter vessel; the time of continuous operation of the target without replenishing the liquid hydrogen is 40 h. The calculated values of the thermal flux through the various target elements are:

Thermal flux through the insulation, W:	
to the conical part of the inner vessel	0.4
to the large flange of the inner vessel	3.6
of the support unit	0.3
Thermal flux by radiation to the entrance window, W	0.9
Thermal flux by residual gases to the entrance window, W	0.1
Thermal flux through the slot in the screen, W	0.2
Thermal flux through the throat, W	1.5

Technical data

Volume of the inner vessel of the target, liters	13.6
Diameter of the entrance window of the target, mm	100
Size of the exit window of the target, mm	372 × 140
Length of the inner vessel of the target, mm	250
Amount of hydrogen in the path of the particles, g/cm^2	1.76
Amount of wall material in the path of the particles on entering the target (two Dacron film windows 120 μ thick, aluminum screen 10 μ thick), g/cm^2	0.034
Amount of wall material in the path of the particles on leaving the target (two copper foils 50 μ thick, 40 layers of metallized Dacron 12 μ thick with aluminum coating 0.01 μ thick, 40 layers of glass paper 50 μ thick with a bulk weight of 0.3 g/cm^3) g/cm^2	0.2
Rate of evaporation in terms of liquid hydrogen, liters/h	1.2
Target weight, kg	40

3.6. Precision Apparatus with a Liquid-Hydrogen Target [46, 47]

Description of the Apparatus. The precision apparatus was used to measure the total cross sections of the π^-p interaction with a systematic error of less than 30 μb for energies of between 4 and 6 GeV [48].

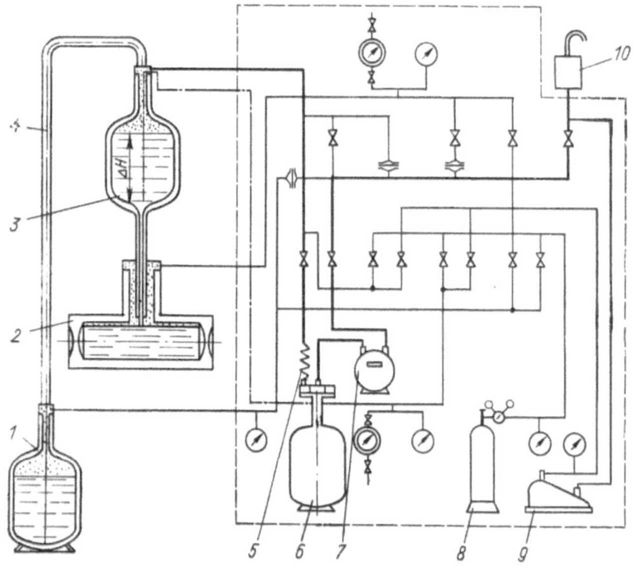

Fig. 20. Arrangement of the precision apparatus with a liquid-hydrogen target: 1) Dewar vessel; 2) target; 3) intermediate vessel; 4) siphon; 5) heater coil; 6) pressure stabilizer; 7) gas counter; 8) gas cylinder with reducing valve; 9) vacuum pump; 10) flame-arrester; the parts of the apparatus (control and measuring instruments and valves) shown within the broken line are placed in the control desk and the rest behind the desk. The thick line indicates the course of the evaporating hydrogen out of the apparatus, with the target in the working state. The dotted and dashed line indicates the tube of the pressure stabilizer for compensating the fall in level in the intermediate vessel.

This apparatus enables the amount of hydrogen in the path of the particles to be kept constant to a high accuracy (0.05%) over long periods.

The apparatus (Fig. 20) incorporates two targets, a working one 2, filled with liquid hydrogen, and an empty one for background measurements (the background target is not shown in Fig. 20). The working target is filled with hydrogen by automatic flow from an intermediate vessel of 50 liters capacity, 3, placed above it. This amount of liquid hydrogen is sufficient for 50 h continuous operation of the apparatus. The targets and intermediate vessel are placed in a movable framework allowing the targets to be placed alternately along the particle beam. The intermediate vessel and target are filled with liquid hydrogen from portable 100- and 50-liter Dewar vessels 1 by means of the siphon 4. The valves, controlling and measuring instruments, and auxiliary equipment required in preparing the apparatus for operation and maintaining the specified conditions are placed in the control desk. The desk is linked to the target and intermediate vessels by flexible bellows hoses, so that the desk may be placed in a position convenient for working purposes.

The principal element in the precision apparatus is the target (Fig. 21). This consists of a cylindrical inner vessel 1, 500 mm long, a vacuum jacket 2, and multilayered insulation 3 between them. In order to reduce the amount of material in the path of the particles, Dacron film windows 5 (125 μ thick) are placed at the end of the target. The windows are fixed mechanically to the inner vessel and vacuum jacket by means of the pressure flanges 6 (Sec. 3.4). The insulating vacuum, of the order of $1 \cdot 10^{-5}$ mm Hg, is maintained with the aid of activated charcoal 4, fixed to the cold surface of the target. Initial evacuation is carried out through a valve 7 situated in the jacket. The hydrogen is poured into the target and the evaporating hydrogen allowed to escape through the target neck 8.

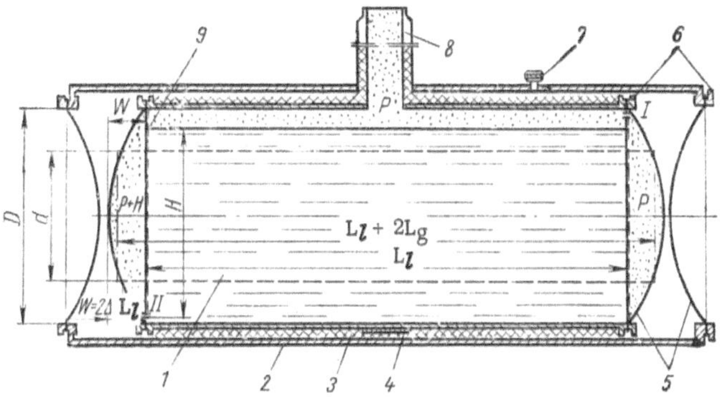

Fig. 21. Liquid-hydrogen cylindrical target: 1) Inner vessel; 2) vacuum jacket; 3) multilayer insulation; 4) adsorbent; 5) Dacron windows (bearing); 6) pressure flanges; 7) valve; 8) throat; 9) Dacron window (limiting).

The principal requirement laid upon the apparatus is that of maintaining the amount of hydrogen in the path of the particles constant to a high accuracy (around 0.05%).

The amount of hydrogen is determined by the equation

$$q = L\rho, \tag{3.1}$$

where L is the length of the target and ρ is the density of the hydrogen. We see from this that for an exact determination of the amount of hydrogen it is essential that: 1) We should know the mean length of the inner vessel of the target precisely; 2) the density in the working part of the target, which mainly depends on the pressure over the boiling liquid, should be kept constant.

Reducing the Effect of the Bending of the Dacron Windows of the Inner Vessel of the Target and Determining the Amount of Hydrogen in the Path of the Particles. On one side of the windows of the inner vessel of the target is hydrogen at atmospheric pressure and on the other side the insulating vacuum. The Dacron window takes a pressure drop of 1 atm and the film sags accordingly. The sag (Fig. 21) may amount to a considerable distance – from 10 to 15 mm, depending on the film thickness, the diameter of the target windows, and the pressure inside the target. As a result of the bending, the amount of hydrogen in the path of the elementary particles passing through the center of the target is greater than in the path of those passing through the edge of the window. In order to reduce the error associated with the bending of the film when determining the length of the inner vessel, the windows are made of two films: a bearing film 5 and a limiting film 9. The bearing film, some 125 μ thick, withstands the 1 atm pressure drop; it has a spherical shape and separates the hydrogen from the vacuum. The limiting film, 15 μ thick, separates the liquid hydrogen from the gaseous; it is almost plane and simply withstands the pressure drop equal to the height of the column of liquid hydrogen. In order to prevent the limiting film from bending, apertures I or II, 0.3-0.5 mm in diameter, are made in it to equalize the pressure between the films and the inner vessel. The diameter of the apertures is chosen in order to ensure that the liquid falling between the films should be able to evaporate as a result of the heat flow to the window. Since the density of gaseous hydrogen in equilibrium with liquid is 50 times smaller than the density of the latter, errors arising from the different lengths of the target are reduced by a factor of 50 with this construction of the windows.

Dacron has a greater coefficient of linear expansion than the stainless steel used for the inner vessel; hence on cooling it is tightened and the film limiting the liquid hydrogen becomes plane. In order to eliminate the effects of bending due to the pressure of the column of liquid, the apertures in the limiting films are made in the upper part of the film I on one side of the target and in the lower part II on the other. If the apertures are placed at the bottom, the pressure between the films will be greater than the average pressure in the target by an amount equal to the half height of the column of liquid in the target. If the apertures are placed at the top, the pressure between the films will be smaller than the mean pressure in the target by the same amount. Thus if the apertures are placed at different heights the limiting films will bend in the same direction (as indicated by the broken lines in Fig. 21), maintaining a constant amount of hydrogen over the whole cross section of the target.

Protection of the Working Zone of the Target from Bubbles. Owing to the flow of heat from the surrounding medium to the inner vessel, the hydrogen in the target boils. The bubbles of gaseous hydrogen rise up in the liquid, creating a nonuniform density in the working zone of the target (the working zone is limited by the diameter of the particle beam). In order to avoid nonuniformity in the boiling of the hydrogen, a shell is placed inside the target to protect the working zone from bubbles falling into it. The working volume is protected from bubbles forming at the edges of the Dacron windows by a cylindrical shield fixed to the limiting film.

Maintenance of Constant Pressure in the Target. In order to maintain a constant density of the hydrogen, a constant pressure must be held above the boiling liquid. The pressure in the target may fluctuate when the barometric pressure changes, or when the height of the column of liquid hydrogen in the intermediate vessel varies as the hydrogen is used up (Fig. 20). The pressure in the target is determined by the pressure over the boiling liquid in the intermediate vessel and the height of the column of liquid in the latter.

A constant pressure may be maintained over the boiling liquid in the intermediate vessel by means of a pressure stabilizer 6 (Fig. 20). This consists of two chambers separated by a membrane. The lower chamber is connected to a closed, thermally-insulated volume at a specified (master) pressure, while evaporating hydrogen from the intermediate vessel passes through the upper chamber. If the master pressure in the stabilizer is greater than that in the intermediate vessel, the membrane will bend under the impetus of the former and create resistance to the passage of gas through the working chamber of the stabilizer. The pressure in the intermediate vessel will accordingly rise. The pressure over the boiling liquid in the intermediate vessel will rise until it balances the master pressure. This construction of the stabilizer keeps a constant pressure over the boiling liquid in the intermediate vessel to an accuracy of ±0.5 mm Hg. The pressure in the target, however, is determined by the pressure in the intermediate vessel together with the height of the column of liquid hydrogen, so that when the amount of liquid in the intermediate vessel diminishes the pressure in the target will fall.

In order to compensate the fall of level in the intermediate vessel and thus keep a constant pressure in the target, a tube is passed down to the bottom of the intermediate vessel, the aperture at the end of this tube being sealed, while the other end is connected to the stabilizer space with the master pressure (in Fig. 20 the tube is shown by a dotted and dashed line). The stabilizer and tube are filled with gaseous hydrogen to the pressure which it is desired to maintain in the target. On filling the intermediate vessel with liquid hydrogen, hydrogen condenses in the stabilizer tube. The level of the liquid in the tube will be the same as the level in the intermediate vessel 3. Since some of the gas from the stabilizer space condenses in the tube, the master pressure in the stabilizer will fall. When the level of liquid in the intermediate vessel falls, the liquid in the tube evaporates and the pressure in the stabilizer increases.

The internal diameter of the tube is chosen in such a way as to ensure that, when the level of liquid in the intermediate vessel falls by an amount ΔH, the master pressure in the stabilizer rises by an amount Δp equal to the fall in the column of liquid; the pressure in the target will then be constant.

The internal diameter of the stabilizer tube is given by the equation

$$d=\sqrt{\frac{4gV_0}{\pi RT_0}},$$

where V_0 is the volume of the stabilizer vessel with the master pressure, m^3; T_0 is the temperature of the gas in the vessel with the master pressure, °K; R is the gas constant; for hydrogen, $R = 420.56 \cdot 9.8\ m^2 \cdot (sec^2 \cdot {}^\circ K)^{-1}$; g is the gravitational acceleration, m/sec^2.

The gas in the closed volume of the stabilizer is thermally insulated from the surrounding medium. For a daily room-temperature fluctuation of ±5° the temperature of the gas in the stabilizer varied by ±0.2°. The pressure stabilizer, together with the auxiliary compensation for the change in the hydrogen level in the intermediate vessel, keeps the pressure in the target constant to within ±1 mm Hg.

Determination of the Amount of Hydrogen in the Target in the Path of the Particles. The amount of hydrogen in the path of the particles is determined from Eq. (3.1). When the construction incorporates windows with double walls, the particles pass through both liquid and gaseous hydrogen, and Eq. (3.1) therefore takes the form

$$G = L_l \rho_l + 2L_g \rho_g. \quad (3.2)$$

Allowing for the accuracy in the determination of the linear dimensions of the target and the density of the hydrogen, Eq. (3.2) may be written

$$G = (L_l \pm \Delta L_l)(\rho_l \pm \Delta\rho_l) + 2(L_g \pm \Delta L_g)(\rho_g \pm \Delta\rho_g)$$

or

$$G = (L_l\rho_l + 2L_g\rho_g) \pm (L_l\Delta\rho_l + \Delta L_l\rho_l + 2L_g\rho_g + 2\Delta L_g\rho_g), \quad (3.3)$$

where L_l is the mean length of the cylindrical path of the target filled with liquid hydrogen, cm; ΔL_l is the deviation from the mean length of the cylindrical part of the target; L_g is the mean sag in the Dacron window of the inner vessel of the target in the working zone of diameter d, cm (Fig. 21); ΔL_g is the deviation from the mean sag in the Dacron windows, cm; ρ_l is the density of the liquid hydrogen in the target in g/cm^3; ρ_g is the density of the gaseous hydrogen in equilibrium with liquid, g/cm^3; $\Delta\rho_l$, $\Delta\rho_g$ are the absolute errors in determining the density of the liquid and gaseous hydrogen, associated with fluctuations in the pressure over the boiling liquid, the accuracy of the temperature measurement, and the accuracy in determining the para-hydrogen concentration, g/cm^3.

Determination of the Length of the Cylindrical Part of the Target. The average length of the cylindrical part of the target is found from the equation

$$L_0 = \frac{L_{max} + L_{min}}{2},$$

where L_{max} and L_{min} are the maximum and minimum lengths of the target at two diametrically situated points on the sealing surfaces of the inner vessel.

The deviation from the average length in the working part of the target is

$$\Delta L_0 = \frac{L_{max} - L_{min}}{2} \cdot \frac{d}{D},$$

where d is the diameter of the working zone of the target (defined by the beam size), cm; D is the diameter of the windows of the inner vessel of the target, cm.

For example, in the case of one of the targets with D = 10 and d = 7 cm, we measured L_{max} = 50.004 cm, L_{min} = 49.998 cm, and deduced L_0 = 50.001 cm and ΔL_0 = 0.002 cm. The measurements were made with a micrometer at a room temperature of 18°C.

The length of the target at liquid-hydrogen temperature (20.4°K) is given by the equation

$$L_l = L_0 - L_0 \cdot 10^{-5}\left[\left(\frac{\Delta L}{L_0} 10^5\right)_{290°K} - \left(\frac{\Delta L}{L} 10^5\right)_{20°K}\right],$$

where $[(\Delta L/L_0) \cdot 10^5]_{290°K}$ = 288 and $[(\Delta L/L_0) \cdot 10^5]_{20°K}$ = −1.1 are the relative elongations of heating the stainless steel from 0 to 290 and 0 to 20°K, respectively [49]. On substituting the numerical values, we find that the length of the target falls by 0.149 cm on cooling from room temperature to the temperature of liquid hydrogen, so that we have L_l = 49.852 cm.

The determination of the average bending of a Dacron window in the inner vessel of a target was set out in detail in Sec. 1.3. For one of the targets, the bending of the Dacron film measured for a pressure drop of 1 atm and a temperature of 77°K in the film equalled 1 cm. Hence by Eqs. (1.5) and (1.6) the sag in the working zone of the target w = 0.5 cm while the maximum deviation from the average sag of the Dacron windows over the diameter of the working zone will be ΔL_g = 0.25 cm.

The densities of the liquid and gaseous hydrogen were taken from data supplied by the American National Bureau of Standards [22, 50] (Figs. 22-24). The densities of the liquid and gaseous parahydrogen at

TABLE 5

Variable parameter and density change measuring unit	Change in density		Notes
	liquid hydrogen	gaseous hydrogen	
Pressure, 10^{-6} (g/cm^3) per mm Hg	4.92	1.43	For parahydrogen
Temperature, 10^{-4} (g/cm^3)/deg	11.6	3.7	For parahydrogen
Parahydrogen concentration, 10^{-6} (g/cm^3)/%(p-H_2)	2.65	0.08	At a pressure of 760 mm Hg

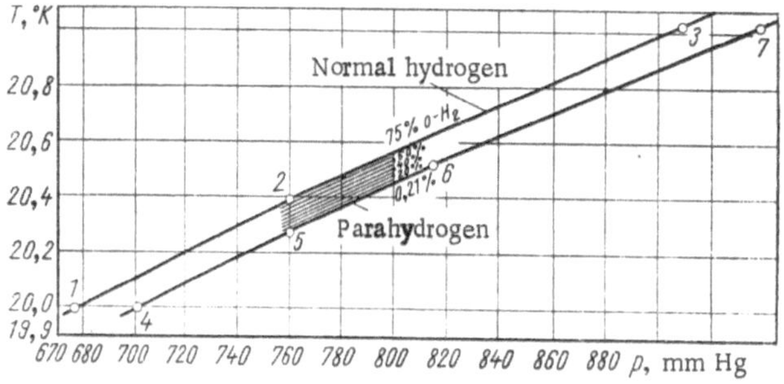

Fig. 22. Dependence of temperature on the pressure of liquid normal hydrogen and parahydrogen.

Point	Temperature (for $T_{f.p.}$ H_2O = 273.16°K)	Pressure	
		atm	mm Hg
1	20,000†	0,891†	677,16
2	20,380†	1,000†	760,00
3	21,000†	1,196†	908,96
4	20,000*	0,9228*	701,33
5	20,268*	0,9999*	759,92
6	20,507*	1,0723*	814,95
7	21,000*	1,2334*	937,38

* Cryogenics, 1963, 3, 1, 17.

† Technology and Uses of Liquid Hydrogen (edited by R. B. Scott, p. 381).

a pressure of 780 mm Hg (the pressure stabilized in the target) equal ρ_l = 0.07068 g/cm^2, ρ_g = 0.00137 g/cm^3.

The dependence of the densities of liquid and gaseous hydrogen on changes in pressure, temperature, and proportion of parahydrogen (Table 5) may be determined from the NBS data [22, 50] (Figs. 22-24). The results are averaged for the temperature range 20-21°K.

The pressure in the target is maintained to an accuracy of ±1 mm Hg. The temperature of the hydrogen in the target is measured to an accuracy of 0.01°, and the concentration of parahydrogen is determined to an accuracy of 1%.

Allowing for the accuracy of the measurement of the hydrogen parameters and the dependence of the hydrogen density on changes in these parameters (Table 5), we find that the error in measuring the density of liquid hydrogen is $\Delta\rho_l = 19.6 \cdot 10^{-6}$ g/cm^3 and the error in measuring the density of gaseous hydrogen $\Delta\rho_g = 4.2 \cdot 10^{-6}$ g/cm^3.

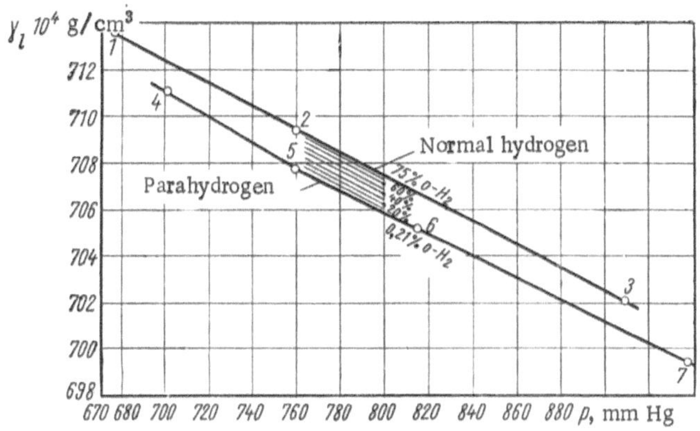

Fig. 23. Dependence of density on pressure for liquid normal and parahydrogen.

Point	Pressure		Density	
	atm	mm Hg	g-mole/cm^3	g/cm^3
1	0,891*	677,16	0,03540*	0,0713655
2	1,000*	760,00	0,03519*	0,0709735
3	1,196*	908,96	0,03483*	0,0702000
4	0,9228†	701,33	0,0352753†	0,0711051
5	0,8999†	759,93	0,0351115†	0,0707749
6	1,0723†	814,95	0,0349837†	0,0705173
7	1,2334†	937,38	0,0346992†	0,0699438

* Technology and Uses of Liquid Hydrogen (edited by R. B. Scott, p. 381).

† Cryogenics, 1963, 3, 1, 17.

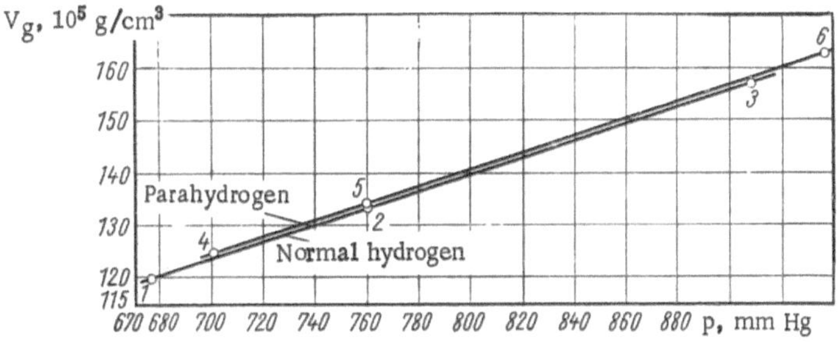

Fig. 24. Dependence of density on pressure for gaseous normal and parahydrogen.

Point	Pressure		Density	
	atm	mm Hg	g-mole/cm^3	g/cm^3
1	0,891*	677,16	0,000595*	0,0011994
2	1,000*	760,00	0,000660*	0,0013316
3	1,196*	908,96	0,000776*	0,0015641
4	0,922*	701,33	0,0006176*	0,0012449
5	1,000*	759,92	0,0006636*	0,0013376
6	1,233*	937,38	0,0008030*	0,0016186

* Technology and Uses of Liquid Hydrogen (edited by R. B. Scott, p. 381).

Substituting the numerical values into Eq. (3.3), we obtain $G = 3.525 \pm 0.00171$ g/cm^2. The relative error in measuring the amount of material in the path of the particles is 0.05%.

Amount of Material of the Dacron Windows in the Path of the Particles. In the path of the particles we have four Dacron windows 125 μ thick and two limiting Dacron films 15 μ thick. The total Dacron thickness is 530 μ and the amount of Dacron in the path of the particles 0.07314 g, accounting for about 2% of the amount of hydrogen in the particles path. Thus the ratio between the background and the effect for the 50-cm target will be $\varphi = 0.048$.

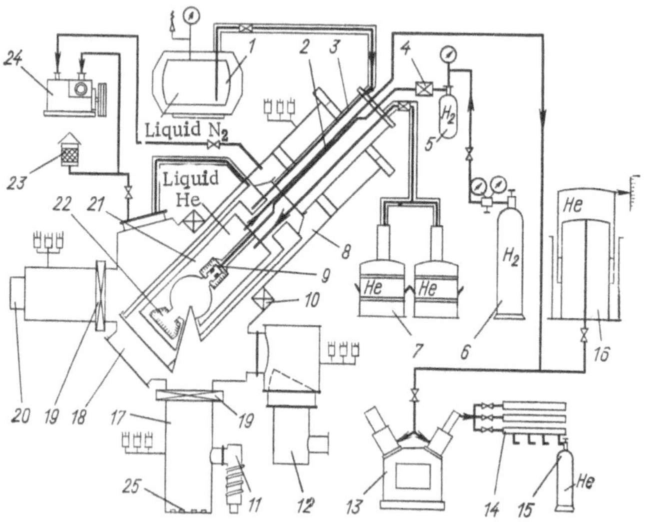

Fig. 25. Apparatus for forming a gas target in the acceleration chamber: 1) "Tank" containing liquid nitrogen; 2) heat exchanger; 3) system for forming the gas jet; 4) electromagnetic valve; 5) buffer space; 6) compressed hydrogen cylinder; 7) liquid helium Dewars; 8) vacuum lock; 9) device for forming and trapping the gas jet; 10) vacuum gate; 11) diffusion pump; 12) diffusion pump; 13) helium compressor; 14) filling ramp; 15) helium cylinders; 16) gas-holder; 17) ion guide; 18) accelerator chamber; 19) vacuum gate; 20) film target; 21) vessel for liquid helium; 22) copper trap; 23) flame-arrester; 24) backing pump; 25) semiconducting silicon detectors.

3.7. Hydrogen Jet Target

Some interesting work on p–p and p–d scattering has been carried out by a number of research workers in the Joint Institute for Nuclear Research, using the Serpukhov accelerator [57]. The experiments were carried out in the inner beam of the accelerator, using a hydrogen jet target. This target has the following advantages: a high probability of the nuclear interaction of the primary protons (0.9), owing to the repeated passage of the beam through the target; a complete absence of secondary interactions in the target owing to the small dimensions of the latter; a prolonged and uniform extension of the primary beam [52]. The chief problems which arise in making the gas target are the formation and capture of the jet in vacuum. In the vacuum chamber of the accelerator the hydrogen is trapped by means of a helium condensation pump.

The apparatus for forming and trapping the jet (Fig. 25) works in the following way. Compressed hydrogen passes from a buffer space through an electromagnetic valve and a heat-exchanger coil to a jet system and flows into the accelerator chamber. The formation of the jet is achieved by means of three coaxial nozzles (Fig. 26) with intermediate pumping chambers between them. Evacuation is achieved by the condensation method. The hydrogen jet intersects the zone of the accelerator beam and falls into a helium condensation pump, which consists of a vessel for liquid helium and a copper trap made in the form of a chamber 1800 cm^3 in capacity with a neck 50 mm in diameter. The inner surface of the trap is covered with ribs (thickness 2.5, height 7.5, pitch 5 mm) and has an area of 1000 cm^2. The whole of this system is situated in a vacuum lock before work begins. After the liquid helium has been poured in and the necessary preparatory work carried out for all the systems, the trap is opened and the target is introduced into the accelerator chamber, occupying the working position illustrated in Fig. 25. When the trap in the chamber of the jet apparatus has collected the limiting amount of solid hydrogen, the target is again returned to the lock, cut off from the accelerator chamber by the vacuum gate, and heated. The evaporating hydrogen is evacuated by means of backing pumps.

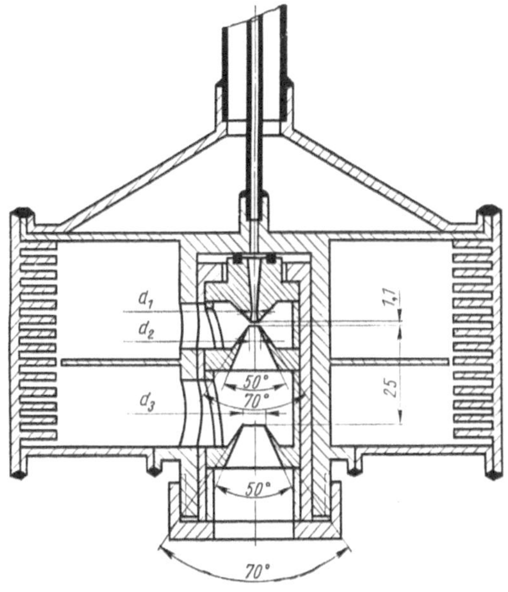

Fig. 26. Arrangement of jet apparatus.

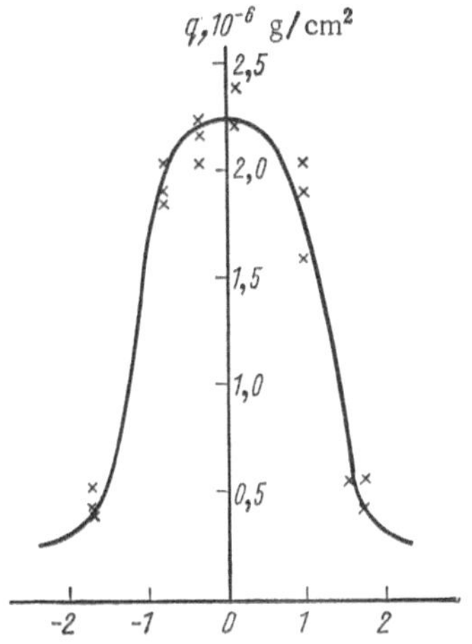

Fig. 27. Distribution of the amount of hydrogen in the gas jet.

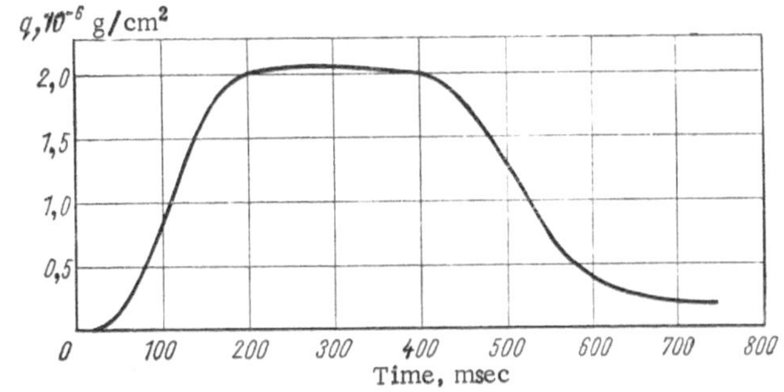

Fig. 28. Time dependence of the amount of hydrogen in the jet.

In order to measure the shape of the jet as well as its density, we use an electron gun with a well-focused electron beam. The electron energy is 3 keV and the current 1 mA. Figure 27 shows the distribution of the amount of hydrogen in the gas jet; Fig. 28 shows the time dependence of the amount of hydrogen in the jet. Methods of measuring the various parameters of the apparatus were set out in detail in [53-56], together with the corresponding results.

Principal Parameters of the Jet Apparatus and the Jet

Number of nozzles (jets) in series	3
Diameter of the nozzles	d_1 = 0.4 mm
	d_2 = 1.5 mm
	d_3 = 6 mm
Diameter of the diaphragm at a distance of 40 mm from the third nozzle	d_4 = 12 mm
Gas pressure in front of valve	7.5 atm
Period of opening of the valve	70 μsec
Hydrogen temperature at nozzle inlet	30-40°K

Flow of gaseous hydrogen.	200 cm^3/cycle
Jet width 70 mm from the diaphragm.	40-50 mm
Amount of hydrogen in particle path	$2.2 \cdot 10^{-6}$ g/cm^2

In conclusion, the authors are very happy to thank Professor A. G. Zel'dovich for valuable advice and useful discussions in setting up the complex of cryogenic targets, laboratory colleagues Yu. T. Borzunov, V. I. Vinogradov, A. A. Demin, V. S. Il'in, A. I. Kalmykova, É. V. Komogorov, A. A. Kulikov, B. K. Kuryatnikov, V. L. Mazarskii, N. I. Nikonov, N. D. Rylov, I. A. Sychkov, G. G. Khorev, and especially V. P. Mokrinov and A. P. Tsvinev for constant participation and creative contributions in the construction, preparation, testing and practical use of the targets, and also V. S. Stavinskii for useful advice in preparing the manuscript for the press.

LITERATURE CITED

1. V. Eckardt and A. Ladage, International Symposium on Nuclear Electronics, Paris, Versailles (1968).
2. E. Bertolucci et al., Liquid-hydrogen target interaction point localization by Cerenkov effect, Nucl. Insts. Methods, 69, 21-24 (1969).
3. A. Ladage et al., A Streamer Chamber with Liquid-Hydrogen Target, DESJ, Hamburg (1969).
4. V. I. Petrukhin, Yu. D. Prokoshkin, and V. M. Soroko, Pribory i Tekh. Éksperim., No. 2, 22 (1964).
5. A. A. Lashchinskii and A. R. Tolchinskii, Fundaments of the Construction and Design of Chemical Apparatus, Handbook [in Russian], Mashgiz, Moscow (1963).
6. RTM 42-62, Vessels and Apparatus, Norms and Methods of Strength Calculation for Parts and Components, No. 1 [in Russian], Izd. Standartov, Moscow (1964).
7. L. B. Golovanov, Transactions of the Fourth International Conference on Low-Temperature Physics and Technology [in Russian], Dresden (1965).
8. N. I. Balandiko et al., Cryogenics, 6, No. 3 (June), 158 (1966).
9. J. Kaye and T. Laby, Tables of Physical and Chemical Constants [Russian translation], Gostekhizdat, Moscow (1962).
10. V. V. Korshak and S. V. Vinogradova, Heterocyclic Polyesters [in Russian], Izd. AN SSSR, Moscow (1958).
11. Short Handbook on Engineering Materials [in Russian], Mashgiz, Moscow (1963).
12. O. I. Dovzhenko and A. A. Pomanskii, Zh. Éksp. Teor. Fiz., 45, 268 (1963).
13. N. P. Malkov et al., Handbook on the Physicotechnical Fundaments of Deep Cooling [in Russian], Gosénergoizdat, Moscow (1963).
14. H. W. Wooley, R. B. Scott, and F. G. Brickwedd, Res. Nat. Bur. Standards, 41, No. 5, 379-475 (1948).
15. L. B. Golovanov, Preprint JINR, R8-3237 [in Russian] (1967).
16. M. F. Fedorova, Zh. Tekh. Fiz., 33, No. 5, 587 (1963).
17. M. G. Kaganer, Thermal Insulation in Low-Temperature Technology [in Russian], Mashinostroenie, Moscow (1966).
18. L. L. Bewilogua and F. K. Lange, Pribory i Tekh. Éksperim., No. 5, 144 (1960).
19. Yu. I. Nechaev, Pribory i Tekh. Éksperim., No. 4, 174 (1961).
20. Yu. M. Kholikov, Pribory i Tekh. Éksperim., No. 1, 192 (1963).
21. G. K. White, Experimental Techniques in Low-Temperature Physics [Russian translation], Gostekhizdat, Moscow (1961).
22. R. B. Scott, Technology and Uses of Liquid Hydrogen, Pergamon Press (1964).
23. I. I. Strizhevskii and V. F. Zakaznov, Industrial Fire Arresters [in Russian], Khimiya, Moscow (1966).
24. Rules for the Construction of Electrical Installations [in Russian], Kemerovsk. Knizh. Izd. (1965).
25. L. B. Golovanov, Proc. Third Regional Conference on Physics and Techniques of Low Temperatures, Prague (1963).
26. A. S. Vovenko, V. A. Kulakov, et al., Preprint JINR D-72 [in Russian] (1961).
27. A. S. Vovenko, L. B. Golovanov, et al., Zh. Éksp. Teor. Fiz., 42, 715 (1962).
28. B. A. Kulakov et al., International Conference on High-Energy Physics at CERN, Geneva (1962), p. 584.
29. A. S. Vovenko et al., International Conference on High-Energy Physics at CERN, Geneva (1962), p. 385.
30. I. A. Savin et al., Phys. Lett., 17, June, 1 (1965).
31. Bigz Cork, Phys. Rev., 107, 248 (1957).

32. L. B. Golovanov, V. L. Mazarskii, and A. P. Tsvinev, Preprint JINR R-8-5416 [in Russian](1970).
33. Z. V. Borisovskaya, A. S. Vovenko, et al., "Preliminary results of a study of K°–K° regeneration at high energies," Trans. of the International Conference on High-Energy Physics [in Russian], Kiev (1970).
34. S. Basiladze et al., Preprint JINR, R1-5361 [in Russian], Dubna (1970).
35. Yu. T. Borzunov and L. B. Golovanov, Kriogen. Kislorod. i Avtogen. Mashinostroenie, No. 1 (1969).
36. B. A. Kulakov et al., Yadernaya Fiz., 6, 1010 (1967).
37. Borzunov et al., Preprint JINR R8-5418 [in Russian] (1970).
38. I. V. Chuvilo, Nucl. Instrum. and Meth., 54, 217 (1967).
39. I. A. Colutvin et al., Report No. 95, Fourteenth Internat. Conf. on High-Energy Physics, Vienna (1968).
40. S. G. Vorob'ev et al., Preprint JINR R1-4445 [in Russian], Dubna (1968).
41. I. M. Ivanchenko et al., Report No. 411 at the Lund Internat. Conf. on Elementary Particles (1969).
42. Yu. T. Borzunov, L. B. Golovanov, and V. L. Mazarskii, Preprint JINR, R8-5417 [in Russian](1970).
43. M. A. Azimov et al., Preprint E1-3148, Dubna (1967).
44. M. A. Azimov et al., Yadernaya Fiz. 6, 515 (1967).
45. R. G. Astvakaturov et al., Preprint E1-3770, Dubna (1968).
46. Yu. T. Borzunov et al., Trans. of the Internat. Conf. on High-Energy Physics Apparatus [in Russian], Dubna (1970).
47. Yu. T. Borzunov et al., Preprint JINR R8-5212 [in Russian], Dubna (1970).
48. N. Giordenesku et al., Trans. of the Internat. Conf. on High-Energy Physics [in Russian], Kiev (1970).
49. R. B. Scott, Cryogenics Engineering, 333 (1959).
50. H. M. Rodez et al., Cryogenics, 3, No. 1, 16 (1963).
51. Properties of Liquid and Solid Hydrogen, Reference Reviews, No. 1 [in Russian], Izd. Standartov, Moscow (1969).
52. K. D. Tolstov, Preprint JINR, 1689 [in Russian] (1964).
53. L. S. Zolin, V. A. Nikitin, and Yu. K. Pilipenko, Preprint JINR R13-3225 [in Russian] (1967).
54. L. S. Zolin et al., Cryogenics, 8, No. 3 (1968).
55. V. D. Bartenev et al., in: All-Union Conference on Charged-Particle Accelerators 1968, Vol. 1 [in Russian], Moscow (1970), p. 536.
56. V. D. Bartenev et al., Trans. of the International Conference on High-Energy Physics Apparatus [in Russian], Dubna (1970).
57. V. D. Bartenev et al., Trans. of the International Conference on High-Energy Physics [in Russian], Kiev (1970).

LARGE-AREA SCINTILLATION COUNTERS

1. PROPORTIONAL SCINTILLATION COUNTERS

B. B. Govorkov and V. S. Chukin

The properties of large-area proportional scintillation counters (LASC) are reviewed. Questions of light collection in scintillators, different types of light guides, and substances for optical contacts are discussed in detail. A nonuniformity of ±5% and an energy resolution of ± 13% can be obtained by using twisted strip light guides.

1. INTRODUCTION

The devices now arbitrarily classed as large-area scintillation counters (LASC) include counters in which the large face of the scintillator has $S \geq 0.1$ m^2 and light is collected at a small face [1, 2]. The restriction $S \geq 0.1$ m^2 is due to some extent to the particle energies attained and utilized on accelerators. In the case of synchrotrons with energies of hundreds of MeV detectors with $S \sim 0.01$ m^2 have been regarded as LASC [3-6]. In the energy range 1-10 GeV the transverse dimensions of scintillators increase to 0.05-0.1 m^2 [1, 2, 7-11]. For energies of tens or hundreds of GeV the transverse dimensions of the detector increase to 1-10 m^2.

The strong dependence of the cross sections of processes involving high-energy particles on the transverse momentum transfer and the increase in ionization mean-free paths and particle decay lengths with energy increase lead to a characteristic elongation of experimental devices in the direction of the primary particle beams of a modern accelerator. On the other hand, the reduction of interaction cross sections with energy (e.g., the cross section of the charge-exchange process $\pi^- p \rightarrow \pi^0 n$ is proportional to 1/p [12], where p is the π meson momentum) and the multiplicity of particles necessitate the construction of devices with high geometric efficiency (with 2π geometry). The experimental requirements of axial length and high geometric efficiency can be satisfied simultaneously only by detectors with sufficiently large transverse dimensions. For instance, on the 76-GeV Serpukhov proton accelerator of the Institute of High-Energy Physics the experimental apparatuses, which use nuclear electronic techniques, are 50-100 m long and have detectors with areas of 1-5 m^2.

In such apparatuses LASC are used: 1) to establish the fact of passage of a charged particle through the scintillator (nonproportional scintillation counters); 2) to measure the specific energy loss dE/dx of the particles in passing through the counter (proportional counters); 3) to measure the time of flight t of particles between the counters and determine the coordinates of particles in the counter (time scintillation counters); 4) to determine the kind of particle, its energy, and other characteristics by the use of LASC in some "sandwich" arrangement.

In this review we discuss some physical properties of large-area proportional scintillation counters. The review does not deal with questions of the electronics of scintillation counters since, firstly, this subject has already been dealt with in the literature (see [13, 14], for instance) and, secondly, a large proportion of the electronic equipment required for LASC must be obtained in the form of commercially produced units.

P. N. Lebedev Physics Institute, Academy of Sciences of the USSR. Translated from Problemy Fiziki Élementarnykh Chastits i Atomnogo Yadra, Vol. 2, No. 3, pp. 763-800, 1972.

II. PROPORTIONAL SCINTILLATION COUNTERS

1. General Remarks

In proportional scintillation counters the pulse amplitude A is proportional to the number of photons N_0 emitted in the scintillation:

$$A = kN_0, \tag{1}$$

and is independent of the position of the luminescent flash in the scintillator. In this treatment we do not specify the kind of particle nor the nature of the scintillator. It is known [15-17] that the proportionality factor k is given by the product of the light collection factor (k_1), the quantum yield of the photocathode (k_2) averaged over its surface, and the photomultiplier gain (k_3): $k = k_1k_2k_3$. The variation of the light collection factor k_1 with the position of the scintillation flash is mainly responsible for the nonuniformity of a scintillation counter. The uniformity of an LASC can be quantitatively characterized by the amplitude nonuniformity factor corresponding to the maximum of the amplitude spectrum:

$$\Delta x_m = \pm \frac{(x_m)_{max} - (x_m)_{min}}{(x_m)_{max} + (x_m)_{min}}, \tag{2}$$

where $(x_m)_{max}$ and $(x_m)_{min}$ are the greatest and least amplitudes x_m when different points in the scintillator are irradiated by monochromatic particles. The nonuniformity factor $\Delta\bar{x}$ for the most probable amplitude ($\bar{x}$) of the spectrum is defined in a similar way. In addition, the dispersions $D(x_m)$ and $D(\bar{x})$ are determined in the usual way from measurements of x_m and $\bar{x}$ for different points in the scintillator. The parameters Δx_m and $\Delta\bar{x}$ are indices of the greatest possible nonuniformity. The values of Δx and $D(x)$ together give a complete and accurate picture of the uniformity of light collection in a counter.

Proportional counters are characterized by their energy resolution, as well as by coefficient k. The presently used index of the energy resolution of LASC is the ratio of the total width at half-height Δ (TWHH) of the peak of the amplitude spectrum to the maximum amplitude:

$$\delta = \Delta / x_m. \tag{3}$$

It is understood here that the amplitude spectrum is measured for the case of relativistic particles traversing the counter centrally perpendicular to the largest face and that x_m depends linearly on the particle energy. The energy resolution is often characterized by the ratio of the standard deviation $\sigma(\bar{x})$ to the mean value $\bar{x}$.

The energy resolution of LASC depends mainly on the fluctuations of ionization energy loss of the particle in the scintillator, which have a Landau distribution [18], and fluctuations in the number of photoelectrons from the PM photocathode. Figure 1, taken from [17], shows the ratio of the half-width of the ionization loss distribution $\tilde{\Delta}$ to the most probable loss ΔE_p as a function of the thickness of an organic scintillator for singly charged particles with different velocities. The design of a proportional counter must ensure that the uncertainty of the amplitude due to nonuniformity of light collection is less than the smearing due to ionization loss fluctuations. The energy resolution of a device, which depends on the fluctuations of dE/dx, can be improved by a factor of $n^{1/2}$ by convolution of the amplitude distributions from n proportional counters of the same kind.

An evaluation of the fluctuations of the number of photoelectrons requires a knowledge of N_0, i.e., the number of photons in the luminescent flash. The number of photons N_0 is calculated from the most probable energy loss ΔE_p of the particle in the scintillator [we note that for relativistic electrons $\Delta E_p \approx$ 2 MeV/(g · cm^2)], the luminescence energy yield $\eta = \Delta E_s/E_p$, where ΔE_s is the energy of the flash, and the spectral distribution of the luminescence. For commercial plastic scintillators (NE, Pilot, Phillips, etc.) the mean wavelength of the scintillation peak is $\bar{\lambda}_s$ = 420-440 nm, and the half-width $\Delta\lambda_s$ = 50-80 nm. The value of N_0 can be obtained by introducing a mean energy $\bar{E}_{ph}$ of the flash photons, and then $N_0 = \Delta E/\bar{E}_{ph}$. We note that the energy loss of a relativistic particle in the formation of one photon with wavelength 420 nm in a polystyrene scintillator activated with p-terphenyl and α-NPO is 170-180 eV.

2. Scintillator Optics

Escape Cones. In the consideration of the passage of charged particles through a scintillation counter it is usually assumed that: a) at every point within the scintillator the luminescence is produced isotropi-

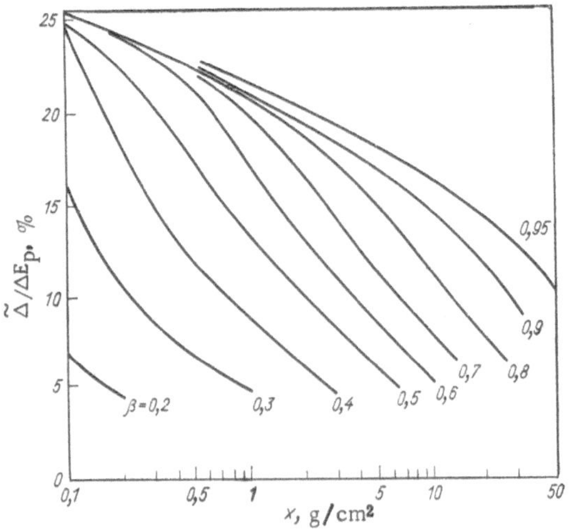

Fig. 1

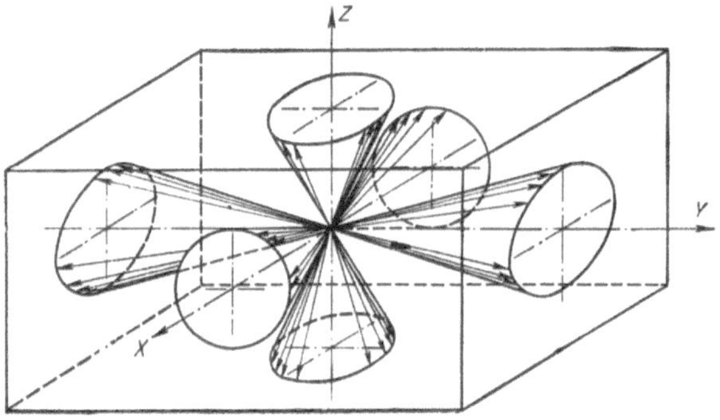

Fig. 2

Fig. 1. Half-width of spectral curve of ionization loss $\tilde{\Delta}/\Delta E_p$ as function of thickness of organic scintillator x for different particle velocities β.

Fig. 2. "Escape cones" for light in a rectangular parallelepiped with $n > \sqrt{2}$.

cally; b) the scintillator is homogeneous and isotropic; c) the ratio of the smallest dimension of the scintillator to the longest wavelength of the luminescence is very large. The last assumption justifies the use of the concepts of geometric optics in the treatment of the passage of light through a scintillator. Since the light rays diverge widely in the scintillator of an LASC, lens optics are not suitable for the evaluation of light collection. In this case we have to use the concept of "escape cones" [19-21]. For uniform collection of light throughout the scintillator volume it is essential that the fraction of light entrapped by total internal reflection is independent of the position of the luminescent flash. As was shown in [3, 19], the only geometrical shape in which this is possible is a rectangular parallelepiped. In fact, from any point in a perfectly polished and absolutely transparent scintillator in the shape of a rectangular parallelepiped the luminescence can emerge only through six cones with a common apex at this point and with an angle of $2\varphi_c$ at the apex, where φ_c is the angle of total internal reflection at the scintillator-air boundary.

The rest of the light is entrapped by total internal reflection (TIR). Figure 2 shows the six escape cones. We should stipulate that not all the light enclosed in the escape cones leaves the scintillator in the first encounter with a face. The probability of reflection of rays with $\varphi < \varphi_c$ is calculated from the Fresnel formulas (for instance, for a scintillator with n = 1.58 and a ray with $\varphi = \varphi_c - 0.5°$ the probability of reflection is 0.455). However, even if such a ray is reflected from the face of the scintillator, it strikes the opposite face at the same angle, and so on. It is highly probable that as a result of several successive encounters with the faces a ray with $\varphi < \varphi_c$ emerges from the scintillator. (In the example considered, the probability of emergence of a ray within six encounters is 99%). This justifies the use of the concept of "escape cones" for the description of light propagation in a scintillator. Henceforth for simplicity we assume that the probability of a photon with $\varphi < \varphi_c$ leaving the scintillator is unity.

We consider the relationship between the fraction of light entrapped by TIR, $F = \Omega_{TIR}/4\pi$, and the refractive index n in several specific cases.

a. For a rectangular parallelepiped the value of F is divided into three intervals, depending on n of the scintillator [19]:

$$\text{for} \quad n \leqslant 1.225 \qquad F \equiv 0;$$
$$\text{for} \quad n \geqslant 2 = 1.414 \qquad F = 3\cos\varphi_c - 2;$$
$$\text{for} \quad 1.225 \leqslant n \leqslant 1.414 \qquad F = 3\cos\varphi_c - 2 + \frac{3}{\pi}\Omega_{oc},$$

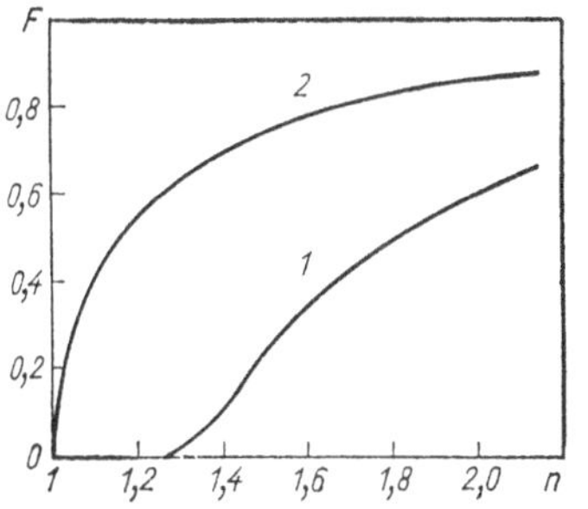

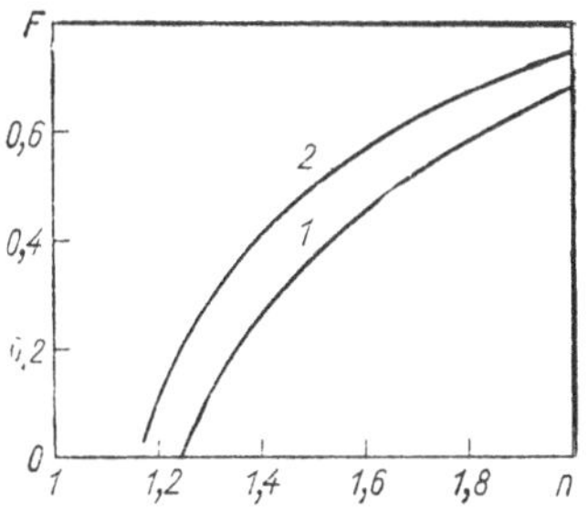

Fig. 3 Fig. 4

Fig. 3. Fraction of luminescence entrapped by TIR as a function of refractive index n for a rectangular parallelepiped (curve 1) and for infinite plane-parallel plates (curve 2).

Fig. 4. Fraction of luminescence collected through one (curve 1) or two (curve 2) faces of rectangular parallelepiped as function of n.

where Ω_{oc} is the solid angle common to two overlapping cones:

$$\Omega_{oc} = 4\left[\operatorname{arctg}(\sin^2\varphi_c - \cos^2\varphi_c)^{1/2} - \cos\varphi_c \operatorname{arctg}\frac{(\sin^2\varphi_c - \cos^2\varphi_c)^{1/2}}{\cos\varphi_c}\right]. \tag{4}$$

b. For infinite plane-parallel plates with $n > 1$, $F = \cos\varphi_c$. The values of F(n) for a parallelepiped (curve 1) and for infinite plane-parallel plates (curve 2) are shown in Fig. 3.

c. For a parallelepiped in which the light is collected through one face

$$F = (3\cos\varphi_c - 2) + \frac{(1-\cos\varphi_c)}{2} = \frac{1}{2}(5\cos\varphi_c - 3). \tag{5}$$

It is understood in this case that all the light entrapped by TIR eventually passes through the collecting face to the photosensitive layer. In addition, it is assumed that the scintillation occurs at a distance $r \geq D\tan\varphi_c$ from the photosensitive layer, where D is the diagonal of the emergence face of the parallelepiped [3]. This condition is necessary to ensure that the contribution of the escape cone directed towards the collecting surface is independent of the position of the scintillation flash. This condition determines the length of the light guide from the scintillator to the PM photocathode.

d. For a parallelepiped with two oppositely placed collecting faces (or one collecting face, and the opposite face specularly reflecting)

$$F = 2\cos\varphi_c - 1. \tag{6}$$

The relationship F(n) for (5) and (6) is shown in Fig. 4. For a plastic (liquid) scintillator n = 1.58 and, hence, the light collection for case d exceeds the light collection for case c by a factor of 1.3. If the face of the parallelepiped opposite the collecting face is made perfectly black, the light collection in case c will then be half of that in case d [22, 23].

Calculation of Light Collection from Small Face of a Large-Area Scintillator. Detailed calculations of the optical efficiency of an LASC were carried out in [3, 1]. Below we calculate the light collection in a scintillator of large area $L \times L$ and thickness d with a light guide in the form of M strips of width d in optical contact with the collecting face. As will be shown in the last part of the review, LASC with such light guides have the best characteristics.

Thus, let a flash be produced at point P(x, y) in the scintillator. This results in the isotropic emission of $N_0(\lambda)$ photons with wavelength in the interval $[\lambda, \lambda + d\lambda]$. We calculate the fraction of light entering the m-th light guide. This requires a knowledge of the spectral distribution of the luminescence, and the absorption spectra of the scintillator, light guide, and substance providing optical contact between the scintillator and light guide.

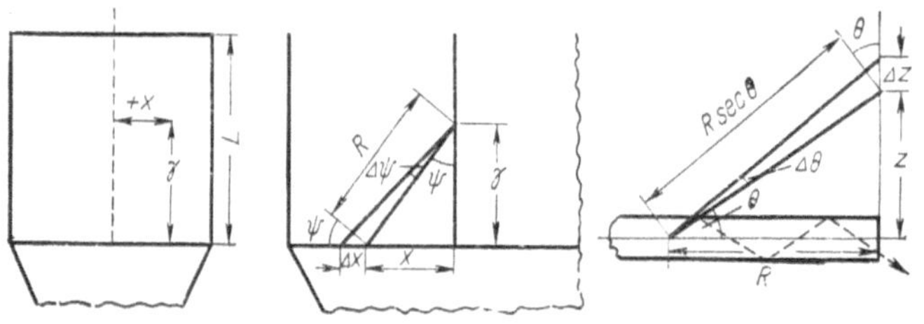

Fig. 5. Coordinate system for calculation of light collection.

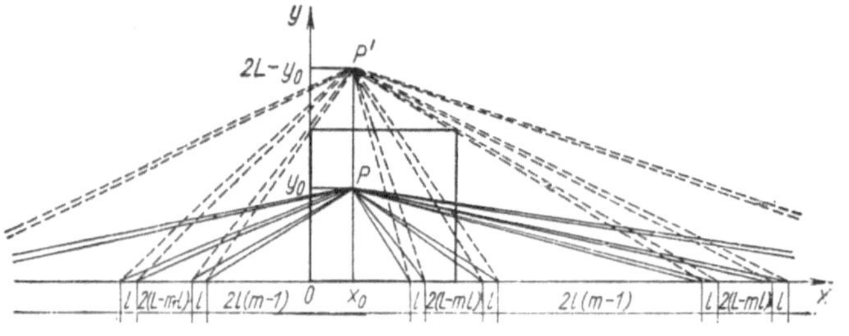

Fig. 6. Contributions of light rays from real (P) and virtual (P') sources in m-th light guide of width l.

We denote the coordinates by x, y and the angles of the light ray by θ, φ (Fig. 5). We assume that the width l of the light guide is small and $l = \Delta x$. The number of photons which reach an element $\Delta x \Delta z$ of the collecting face of the scintillator from point (x, y) without reflection from the upper and side faces is

$$d^2 N_0(\lambda)\, d\lambda = \frac{N_0(\lambda)\, d\lambda d\Omega}{4\pi} = \frac{N_0(\lambda)\, d\lambda}{4\pi} \frac{\Delta x \cos\varphi \Delta z \cos\theta}{(R/\cos\theta)^2}, \tag{7}$$

since $R = y/\cos\varphi$, we obtain

$$\frac{d^2 N_0(\lambda)\, d\lambda}{\Delta x} = \frac{N_0(\lambda)\, d\lambda}{4\pi} \cdot \frac{\cos^3\varphi \cos^3\theta \Delta z}{y^2}. \tag{8}$$

For a given x we integrate with respect to z:

$$\frac{dN_0(\lambda)\, d\lambda}{\Delta x} = \frac{N_0(\lambda)\, d\lambda}{2\pi} \int_0^z \frac{\cos^3\varphi \cos^3\theta dz}{y^2} = \frac{N_0(\lambda)\, d\lambda}{2\pi} \int_0^{\theta_c} \frac{\cos^2\varphi \cos\theta d\theta}{y} = \frac{N_0(\lambda)\, d\lambda}{2\pi} \frac{\cos^2\varphi \sin\theta_c}{y}, \tag{9}$$

where $\theta_c = \varphi_c$ is the angle of total internal reflection. In addition to contribution (9) we need to take into account the contribution of rays which go from the point P into the upper halfplane, are reflected from the upper face, and reach an element $\Delta x \Delta z$ of the emergence face. We also take into account the contributions of rays directed towards the side walls. After reflection from the walls these rays can also reach the collecting surface. The rays directed into the upper halfplane are taken into account by the introduction of a point source P' with coordinates $(x, y' = 2L - y)$ (Fig. 6), which is the mirror image of source P relative to the upper face of the scintillator. We determine the limits of integration with respect to φ for rays directed to the left and right faces of the scintillator from the real (P) and virtual (P') sources. The rays directed towards the left face will intersect the x axis at the points

$-(x + KL - lm)$ and $-[x + KL - (m-1)\, l]$ for K even and positive;

$-[x + (K-1)L + l\,(m-1)$ and $-[x + (K-1)\, L + lm]$ for K odd and positive.

Here K is the number of times the ray meets the side walls. Rays directed towards the right face intersect the x axis at points

$$KL + l(m-1) - x \text{ and } KL + ml - x \qquad \text{for K even and positive;}$$

$$(K+1)L - ml - x \text{ and } (K+1)L - l(m-1) - x \text{ for K odd and positive.}$$

The contributions of the rays directed towards the left and right faces can be combined. To do this we give the number of collisions for rays directed towards the left face a negative sign, i.e., we regard K as a negative quantity. The coefficient (K − 1) for odd K can then be replaced by (K + 1). The limits of integration with respect to φ then become

$$\left[\operatorname{arctg}\frac{KL - x + l(m-1)}{y},\ \operatorname{arctg}\frac{KL - x + lm}{y}\right] \text{ for even K;} \tag{10}$$

$$\left[\operatorname{arctg}\frac{(K+1)L - x - lm}{y},\ \operatorname{arctg}\frac{(K+1)L - x - l(m-1)}{y}\right] \text{ for odd K.} \tag{11}$$

For the virtual source the limits with respect to φ are obtained from (10) and (11) by replacement of y by 2L − y. It is easy to show that the limits of integration with respect to θ for the contributions of rays directed towards the side faces will be as follows:

$$\left[\arccos\frac{\cos\theta_C}{\sin\varphi},\ \theta_C\right].$$

We now consider the absorption of light in the scintillator and the loss due to reflections at the surfaces. Let $\alpha_{sc}(\lambda)$ be the absorption coefficient, and $\gamma(\lambda)$ the reflection coefficient, of the scintillator surfaces for photons with wavelength $[\lambda, \lambda + d\lambda]$. [We assume that $\gamma(\lambda) \approx 1$.] The path of a ray with angles θ, φ in the scintillator before entry into the light guide is

$$S = \begin{cases} y/\cos\varphi\cos\theta & \text{for the real source;} \\ (2L-2)/\cos\varphi_M\cos\theta & \text{for the virtual source.} \end{cases} \tag{12}$$

The absorption in the scintillator is allowed for by multiplying the probability of entry of the ray into the light guide by $\exp[-\alpha_{sc}(\lambda)S]$.

For a ray with coordinates θ, φ the number of encounters with the scintillator faces is

$$\beta = \begin{cases} \dfrac{y}{d}\,\dfrac{\operatorname{tg}\theta}{\cos\varphi} + \dfrac{y\operatorname{tg}\varphi}{L} & \text{for the real source;} \\ \dfrac{2L-y}{d}\,\dfrac{\operatorname{tg}\theta}{\cos\varphi_M} + \dfrac{2L-y}{L}\operatorname{tg}\varphi_M + 1 & \text{for the virtual source.} \end{cases}$$

The loss due to reflections is allowed for by multiplying the probability of entry of the ray into the m-th light guide by $[\gamma(\lambda)]^{\beta}$.

We now write a general expression for the fraction of the scintillation light which enters all the light guides from point (x, y):

$$N(\lambda)\,d\lambda = \frac{N_0(\lambda)\,d\lambda}{2\pi}\sum_{m=1}^{L/l}\left\{\int_0^{\theta_C}\int_{\varphi_1}^{\varphi_2} e^{-\frac{\alpha_{sc}y}{\cos\theta\cos\varphi}}\cdot\gamma^{\frac{y}{d}\frac{\operatorname{tg}\theta}{\cos\varphi}}\,d\varphi\cos\theta\,d\theta\right.$$

$$+\sum_{\substack{K=-\infty \\ K-\text{odd}}}^{+\infty}\ \int_{\arccos\frac{\cos\theta_C}{\sin\varphi}}^{\theta_C}\ \int_{\varphi_3}^{\varphi_4} e^{-\frac{\alpha_{sc}y}{\cos\theta\cos\varphi}}\,\gamma^{\left(\frac{y}{d}\frac{\operatorname{tg}\theta}{\cos\varphi}+\frac{y}{L}\operatorname{tg}\varphi\right)}\,d\varphi\cos\theta\,d\theta$$

$$+\sum_{\substack{K=-\infty \\ K\text{-even}}}^{+\infty}\ \int_{\arccos\frac{\cos\theta_C}{\sin\varphi}}^{\theta_C}\ \int_{\varphi_5}^{\varphi_6} e^{-\frac{\alpha_{sc}y}{\cos\theta\cos\varphi}}\,\gamma^{\left(\frac{y}{d}\frac{\operatorname{tg}\theta}{\cos\varphi}+\frac{y}{L}\operatorname{tg}\varphi\right)}\cdot d\varphi\cos\theta\,d\theta$$

$$+\sum_{\substack{K=-\infty \\ K-\text{odd}}}^{+\infty} \int_{\arccos \frac{\cos\theta_c}{\sin\varphi_M}}^{\theta_c} \int_{\varphi_7}^{\varphi_8} e^{-\frac{\alpha_{sc}(2L-y)}{\cos\theta\cos\varphi_M}} \gamma^{\left(\frac{2L-y}{d}\frac{\operatorname{tg}\theta}{\cos\varphi_M}+\frac{2L-y}{d}\operatorname{tg}\varphi_M+1\right)} d\varphi_M \cos\theta d\theta$$

$$+\sum_{\substack{K=-\infty \\ K-\text{even}}}^{+\infty} \int_{\arccos \frac{\cos\theta_c}{\sin\varphi_M}}^{\theta_c} \int_{\varphi_9}^{\varphi_{10}} e^{-\frac{\alpha_{sc}(2L-y)}{\cos\theta\cos\varphi_M}} \gamma^{\left(\frac{2L-y}{d}\frac{\operatorname{tg}\theta}{\cos\varphi_M}+\frac{2L-y}{d}\operatorname{tg}\varphi_M+1\right)} d\varphi_M \cos\theta d\theta, \quad (13)$$

where

$$\left.\begin{aligned}
\varphi_1 &= \operatorname{arctg}\frac{(m-1)l-x}{y}; \quad \varphi_2 = \operatorname{arc\,tg}\frac{ml-x}{y};\\
\varphi_3 &= \operatorname{arctg}\frac{(K+1)L-x-lm}{y};\\
\varphi_4 &= \operatorname{arctg}\frac{(K+1)L-x-l(m-1)}{y};\\
\varphi_5 &= \operatorname{arctg}\frac{KL-x+l(m-1)}{y};\\
\varphi_6 &= \operatorname{arctg}\frac{KL-x+lm}{y};\\
\varphi_7 &= \operatorname{arctg}\frac{(K+1)L-x-lm}{2L-y};\\
\varphi_8 &= \operatorname{arctg}\frac{(K+1)L-x-l(m-1)}{2L-y};\\
\varphi_9 &= \operatorname{arctg}\frac{KL-x+l(m-1)}{2L-y};\\
\varphi_{10} &= \operatorname{arctg}\frac{KL-x+lm}{2L-y}
\end{aligned}\right\} \quad (14)$$

for the mirror image $\varphi_M = 2\pi - \varphi$.

In expression (13) the first term gives the direct contribution from the real source, the second and third terms give the contributions of rays directed towards the side faces of the scintillator (for $K > 0$ the contribution of rays directed towards the right in Fig. 6, for $K < 0$ the contribution of rays directed to the left). The fourth and fifth terms give the contributions due to the mirror image of the source, i.e., the contributions of rays directed into the upper halfplane from point P.

Expression (13) allows an accurate calculation of light collection in the scintillator. The only thing to remember is that $N_0(\lambda)d\lambda$ is a function of the photon wavelength. In the deduction of (13) it was assumed that in the interval $[\lambda, \lambda + d\lambda]$ all the quantities which depend on λ can be regarded as constant.

In the paper by D. Brini et al. [3] a graphical method of integrating expression (13) with respect to x and z was used. The relative fraction of light $\Delta R = N(\lambda)/N_0(\lambda)$ entering the PM was regarded as the sum of the contributions from the real (ΔR_t) and virtual (ΔR_b) sources:

$$\Delta R(\lambda) = \Delta R_t(\lambda) + \Delta R_b(\lambda). \quad (15)$$

The quantities ΔR_t and ΔR_b were calculated in relation to the absorption parameter

$$A(y, \lambda) = \alpha_{sc}(\lambda)y + \alpha_{lg}(\lambda)l, \quad (16)$$

where α_{sc} and α_{lg} are the absorption coefficients of the scintillator and light guide. We note that the loss of light in the light guide was allowed for in [3] by multiplication of (13) by the exponent $\exp[-\alpha_{sc}(\lambda)l]$, where l is the length of the light guide.

$\Delta R_t(A, \lambda)$ was evaluated by calculating the distribution of illumination of a plane $\pi(x, z)$ situated at a distance y from the site of the scintillation flash (Fig. 7). The collecting face of the scintillator was divided into three equal elements (squares with $\Delta x = \Delta z = d$). By corresponding specular reflections the whole plane π was divided into a set of squares equal to the elements of the collecting surface. The mean number of photons arriving at an elementary area dxdz on plane π from point P is

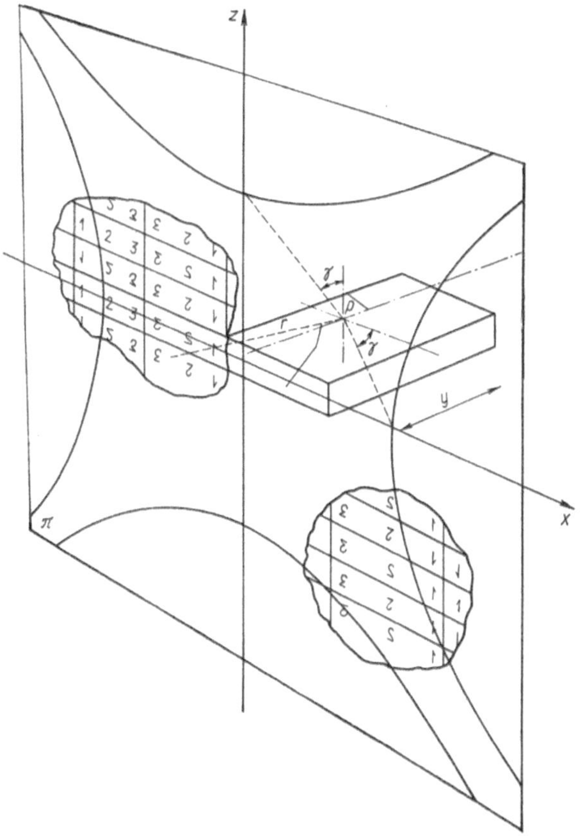

Fig. 7. Auxiliary plane π for graphic integration of light collection in scintillator.

$$dN_i = \frac{N_0}{4\pi} \frac{y\,dx\,dz}{(x^2+y^2+z^2)^{3/2}}. \tag{17}$$

The total amount of light collected by one of the three elements of the emergence face is

$$N_i = \frac{N_0}{4\pi} \sum_{j=1}^{n} \int_{z_{1j}}^{z_{2j}} \int_{x_{1j}}^{x_{2j}} \frac{y\,dx\,dz}{(x^2+y^2+z^2)^{3/2}}, \quad i=1, 2, 3, \tag{18}$$

where x_{1j}, x_{2j}, z_{1j}, and z_{2j} are the coordinates of the vertices of the elementary areas on plane π.

The summation limits are given by the four branches of hyperbolas which are formed by the intersection with plane π of the "escape cones" from point P with axes parallel to the plane xz. The equations of the hyperbolas are: $x^2 \tan^2 \varphi_c = y^2 + x^2$ and $z^2 \tan^2 \varphi_c = y^2 + x^2$.

The values of $\Delta R_t(A, \lambda)$ and $\Delta R_b(A, \lambda)$ given in Fig. 8 were calculated by graphic integration of expression (18). Curves $\Delta R_t(A, \lambda)$ and $\Delta R_b(A, \lambda)$ can be used to evaluate the nonuniformity and light collection efficiency of any scintillator. As an example we consider a scintillator with a large face of area $L \times L = 50 \times 50$ cm. We assume that the luminescence is monochromatic and that the absorption coefficient of the scintillator for this radiation is $\alpha_{sc} = 0.003\ \text{cm}^{-1}$ (plastic scintillator of Pilot Y type, etc.). From Fig. 8 we find for y = 50 cm, $\Delta R_t = 20\%$ and $\Delta R_b = 9\%$, so that ΔR (50 cm) = 29%; for y = 0 cm, $\Delta R_t = 24.5\%$, but $\Delta R_b = 6.7\%$, since the contribution of ΔR_b must be evaluated for $y' = 2L - y$. As a result, ΔR_b (0 cm) = 31.2%.

From ΔR(50 cm) and ΔR (0 cm) we obtain a nonuniformity coefficient $\Delta x_m = \pm 3.6\%$. As an estimate of the light collection efficiency we can use the mean value k = 30.1%.

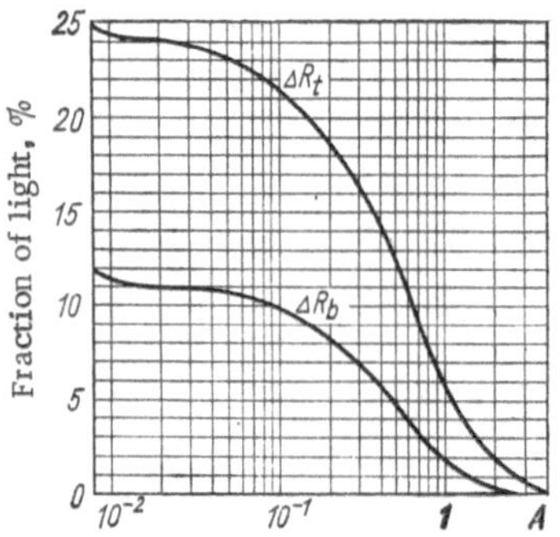

Fig. 8

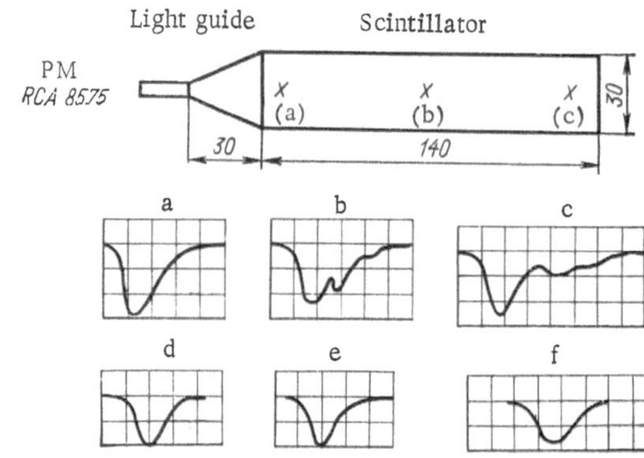

Fig. 9

Fig. 8. Fractions of light reaching PM from real (ΔR_t) and virtual (ΔR_b) sources as function of absorption parameter $A(\lambda, y, \alpha_{sc}, \alpha_{lg})$.

Fig. 9. Oscillograms of counter pulses due to passage of 2-GeV electrons through different points in scintillator. One division is 10 nsec [22].

Greater uniformity can be obtained if the light is collected from two opposite faces of the scintillator. In this case

$$\Delta R = \Delta R_{t1} + \Delta R_{t2}, \tag{19}$$

ΔR_{t1} being calculated for $A(y, \lambda)$ and ΔR_{t2} for $A(L-y, \lambda)$. In the case of a specular surface opposite the collecting face of the scintillator

$$\Delta R = \Delta R_{t1} + \gamma \Delta R_{t3}. \tag{20}$$

Here γ is the reflection coefficient of the specular surface; ΔR_{t1} is evaluated for $A(y, \lambda)$, and ΔR_{t3} for $A(2L-y, \lambda)$.

Approximate calculations of the light collection in a large-area scintillator were also made in [1]. Expression [13] was integrated with respect to variables θ and φ in the limits

$$0 < \theta_c, \quad \varphi < \varphi_c.$$

The last restriction is by no means exact, since rays with $\varphi > \varphi_c$ reflected from the side faces can reach the collecting surface.

The absorption of light in the scintillator was taken into account by the factors $\exp(-\alpha_{sc} s)$ and replacement of $\sec\theta$ and $\sec\varphi$ in them by their average values in the considered intervals of angles θ and φ:

$$\left.\overline{\sec\theta} = \frac{1+\sec\theta_c}{2}; \quad \overline{\sec\varphi} = \frac{1+\sec\varphi_c}{2}.\right\} \tag{21}$$

For a plastic scintillator $\varphi_c = \theta_c \approx 39°$ (NE 102, etc.), so that the angle intervals are fairly large. Hence, the validity of such a procedure is not at all obvious.

The total contribution of the scintillation to an element dx of the collecting surface is given by the following expression:

$$\frac{dN_0}{dx} = \frac{N_0(\lambda)\,d\lambda}{2\pi}\sin\theta_c\left\{\frac{y}{x^2+y^2}\exp\left[\left(-\frac{\alpha_{sc} y}{4}\right)(1+\sec\theta_c)(1+\sec\varphi_c)\right] + \gamma\,\frac{2L-y}{x^2+(2L-y)^2}\right.$$
$$\left.\times\exp\left[-\frac{1}{4}\alpha_{sc}(2L-y)(1+\sec\theta_c)(1+\sec\varphi_c)\right]\right\}, \tag{22}$$

where the first term is the contribution of the cone directed towards the collecting surface, and the second is the contribution of the cone directed towards the opposite face and specularly reflected from that face onto the collecting surface.

TABLE 1. Some Properties of Plastic Scintillators Produced by Pilot Chemicals and Nuclear Enterprises [22]

Type of plastic scintillator	$\lambda_{1/2}$, cm	Decay constant τ, nsec	Yield relative to anthracene, %	Cost per mm² (2.5 cm thick), dollars
«Pilot M»	175	2,0	68	930
«Pilot Y»	170	3,0	64	680
«Pilot B»	153	1,7	68	930
«Pilot F»	132	2,0	65	520
«Pilot A»	120	3,0	61	520
«NE-110»	132	3,3	60	618
«NE-102A»	87,5	2,4	65	541

Thus, the optical characteristics of an LASC (light collection efficiency, nonuniformity, etc.) can be calculated exactly from formula (13). In addition, these parameters can be evaluated sufficiently accurately by using the graphic solution of Brini et al. [3] (curves of ΔR_t and ΔR_b in Fig. 8) or the approximate analytical solution (formula (22) of Crabb et al. [1]).

Absorption Constants of Plastic Scintillators. To complete our discussion of scintillator optics we make some comments on absorption constants. The absorption of the luminescence is characterized not only by the coefficient $\alpha_{sc}(\lambda)$, but also by the absorption length λ_a, in which the number of photons is reduced by a factor e, and the half-absorption length $\lambda_{a/2}$, in which the number of photons is halved.

In the case of an exponential absorption law these quantities are connected by the simple relationships

$$\alpha_{sc}\lambda_a = 1; \quad \lambda_a \ln 2 = \lambda_{a/2}.$$

It should be noted that the parameters α_{sc}, λ_a, $\lambda_{a/2}$ measured in different investigations have a fairly considerable scatter. This is due to the difference in measurement conditions. The parameters characterizing the absorption in a scintillator are usually measured on specimens in the form of long strips with a blackened edge [22, 23]. The need for such blackening is illustrated in Fig. 9, which shows oscillograms of the pulses of a counter with a (140 × 30) × 2.5 cm scintillator irradiated with a beam of monochromatic 2-GeV electrons [22]. As the figure shows, blackening of the scintillator edge away from the PM prevents the superposition of the contributions of the rays from the real and virtual luminescence sources. The measured values of $\lambda_{a/2}$, together with certain other parameters of scintillators of different types, are given in Table 1. The values of $\lambda_{a/2}$ for Pilot Y and Pilot M scintillators allow the construction of scintillation counters with area 1.5 × 1.5 m and a nonuniformity $\Delta x_m \leq \pm 10\%$.

Another recent investigation of plastic scintillators is [23], in which the absorption parameters were measured with electrons from radioactive sources. Isolation of electrons with energy in the range 0.5-3.5 MeV was effected with an error of 3% by means of a small permanent magnet.

Information relating to the parameters of plastic scintillators produced by various organizations in the USSR can be found in the collections [24]. We note, however, that at present we have no plastics corresponding to Pilot Y, Pilot M, and NE 110. In addition, exact measurements of the absorption constants of plastic scintillators produced in the USSR by the blackened edge method have not been carried out. This makes it difficult to select plastic scintillators for the construction of LASC with prescribed properties.

A new approach to the construction of large-area plastic scintillators, developed in [25-27], merits attention. It consists essentially in separation of the functions of the plastic scintillator components: The primary polymer is chosen to secure high transparency and good mechanical properties, and not for efficient conversion of the energy lost by the particles on ionization to luminescence energy. The energy conversion function is transferred to the "secondary solvent" and luminescent additives. This approach means that scintillators based on vinyl-aromatic polymers (polystyrene, etc.) can be replaced by scintillators based on acrylic polymers (methyl methacrylate, etc.). The All-Union Scientific-Research Institute of Single Crystals, Scintillating Materials, and Very Pure Substances (Kharkov) has now mastered the experimental fabrication of plastic scintillators, based on Plexiglas, with dimensions 1.5 × 1.5 × 2 cm. The secondary solvent is naphthalene (15% by weight); POPOP is used as a luminescent additive. The light yield of

Fig. 10. Twisted strip light guide.

such a scintillator is 60-80% of the light yield of a standard polystyrene-based scintillator. The luminescence decay time is 8-10 nsec; $\lambda_{a/2}$ is different for different specimens and varies from 100 to 200 cm, depending on the polymerization conditions and the finish of the scintillator faces [28]. The feasibility of fabricating elastic scintillators based on styrene copolymers and methyl methacrylates has also been demonstrated [27]. The fabrication of such large-area scintillators would facilitate the construction of various target "anticoincidence jackets," etc.

3. Light Guides

One of the problems encountered in the design of LASC is the efficient transfer of light from the scintillator to the PM photocathode without loss of uniformity of light collection throughout the scintillator. This problem can be solved by the use of light guides operating on the total internal reflection principle.

For uniformity of light collection in the scintillator the length of the light guide must satisfy the condition [3, 10]

$$l \gg D \operatorname{tg} \varphi_c, \tag{23}$$

where D is the diagonal of the collecting face; φ_c is the TIR angle. It is assumed here that the light guide is a direct extension of the scintillator. Moreover, if the refractive index n_{lg} of the light guide material differs from the refractive index n_{sc} of the scintillator, the escape cone is altered. The angle φ_c in this case is replaced by φ_c', given by

$$\varphi_c' = \arcsin \frac{(n_{lg}^2 - 1)}{n_{sc}}. \tag{24}$$

For instance, for a polystyrene-based scintillator with $n_{sc} = 1.58$ and a Plexiglas light guide with $n_{lg} = 1.49$ the angle $\varphi_c = 50.6°$ is replaced by $\varphi_c' = 44°$, so that the solid angle of the escape cone is reduced by a factor of 1.3. There will be no loss if a methyl methacrylate-based scintillator and a Plexiglas light guide are used, since in this case $n_{lg} = n_{sc}$.

The transfer of light from a scintillator to a light guide without loss in the case of $n_{sc} > n_{lg}$ necessitates the use of a light guide with increasing cross-sectional area. In this case the angle of inclination of the light guide surface to the normal to the collecting face of the scintillator will be

$$\alpha = 1 - \frac{(1 + n_{lg}^2 - n_{sc}^2)^{1/2}}{(n_{sc}^2 - 1)^{1/2}}. \tag{25}$$

If conditions (23) and (25) are fulfilled, the light guide will not affect the uniformity of the scintillation counter.

In LASC the light guide has to ensure efficient transfer of light from the long narrow edge of the scintillator to the circular cross section of the PM photocathode. The choice of shape of light guide to secure this transition is based on the Garwin condition [29].

A light guide with a cross section of one shape will transmit without loss all the light which undergoes TIR to a light guide with a cross section of different shape if the transition is completely adiabatic, i.e., the area of the cross section remains constant and its outline changes slowly.

From a consideration of the passage of light through an adiabatic light guide Garwin obtained the following relationship for individual rays:

$$A(z) \sin^2 \theta(z) = \text{const}, \tag{26}$$

where A(z) is the cross-sectional area of the light guide at a distance z along the axis, and θ is the angle of inclination of the ray to the light guide axis. The constant in relationship (26) is given by the initial

TABLE 2. Twisted Strip Light Guides

A_1, cm^2	A_2, cm^2	Size of strip, cm^3	No. of strips	J	J/J_Δ	Ref.
40,6×0,6=24,4	5×5=25	0,6×5×45,7	8	0,25	≥2	[9]
15×0,31=4,64	πr^2=19,6	0,31×2,5×25	6	—	2,8	[33]
45,7×0,6=27,4	πr^2=15,2	0,6×5×50	9	0,60	≥3	[1]

conditions, i.e., A(0) and $\theta(0)$. If at z = 0 the light rays are distributed uniformly in a cone with $\theta = \theta_c$, then when A(z) = const all the light will be transmitted through the light guide without loss. If the cross-sectional area decreases [A(z) < A(0)] the angle of the cone is increased, since a fraction of the rays, equal to A(z)/A(0), will escape from the light guide.

The operation of light guides and light collectors of various shapes is based on the consequences of relationship (26).

Gorenstein and Luckey [30] suggested the use of gross fiber optics to achieve a transition from a long narrow strip to a square or circular cross section. The light guide consists of separate Plexiglas strips in the form of a fan. The wide end of the light guide is coupled to the face of the plastic scintillator. A photograph of a strip light guide constructed by V. G. Dolgalev and N. G. Kotel'nikov in the P. N. Lebedev Physics Institute, Academy of Sciences of the USSR is shown in Fig. 10 [31]. Such light guides are made by twisting and bending strips of Plexiglas heated to 140-150°C. The method of fabricating twisted strip light guides is described in detail in [32].

It was pointed out in [33] that a bent and twisted strip is an adiabatic light guide. We do not think that this is correct. The bending and twisting must destroy the adiabaticity of the strip, since they reduce the solid angle of the rays which reach the exit section. In [1] the factor by which the solid angle of a strip light guide was reduced was taken as 1 − G(x) = 0.6.

The performance of twisted strip light guides was investigated in [30, 33]. The characteristics of such light guides were compared with the characteristics of common triangular (wedge) or hyperbolic (fishtail) light guides. Table 2 gives the parameters of twisted strip guides: A_1 and A_2 are the cross-sectional areas at the beginning and end of the light guide; J is the efficiency of the light guide; J/J_Δ is the ratio of the efficiencies of strip and triangular (hyperbolic) light guides. The table shows that the efficiency of twisted strip light guides with a length of approximately 50 cm is about 0.5 and is several times greater than the efficiency of other forms of light guides. It was shown in [58] [for small (10 × 10) cm scintillators] that the use of strip light guides twisted through 90° leads to much more uniform light collection than in a counter in which fine fiber optics is used for light collection.

As was shown in [29, 34], the angular distribution of light crossing any plane normal to the axis of a light guide operating on the TIR principle is contained within a cone of half angle

$$\theta_{max} = \frac{\pi}{2} - \theta_c,$$

where θ_c is the critical TIR angle. Therefore, if a conventional light guide is extended by the addition of a short section of the same material with specularly reflecting walls which enlarges the angle of the cone from θ_{max} to $\pi/2$ the light can be concentrated on a smaller area than the cross-sectional area of an adiabatic light guide. The permissible reduction of the exit area is given by the ratio

$$\frac{\sin^2\theta_2}{\sin^2\theta_1} = \frac{1}{\sin^2\theta_{max}} = \frac{n_{lg}^2}{n_{lg}^2 - 1}. \quad (27)$$

For a Plexiglas light collector it is 1.82.

Light collectors of this type can be made in the form of right cones [35] or modified paraboloids [36]. The optics of a cone channel with specularly reflecting walls were thoroughly discussed in [37, 38]. Williamson indicated a simple method of determining the rays entering the cone which attain the exit aperture. Figure 11 shows the figure formed by reflection of a cone in its walls. The exit aperture and its mirror images form a polygon.

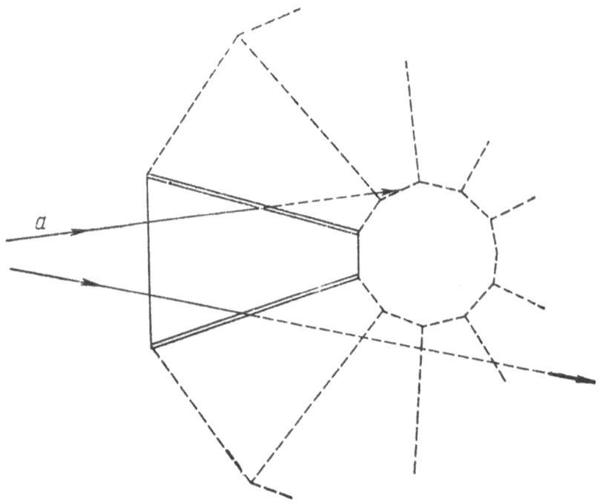

Fig. 11. Williamson's method of determining the passage of light rays through a cone channel [37].

All the rays which enter the main cone and intersect the polygon reach the exit aperture (ray a). In this case the number of intersections of the ray with the walls of the cone and with their mirror images gives the number of reflections. By replacing the polygon with the circumscribed circle Williamson obtained a relationship between the ratio of the entrance (T) and exit (t) apertures and the limiting angle θ of the ray in the center of the entrance aperture of the cone (sine theorem condition in standard optics):

$$\sin\theta = t/T.$$

A three-dimensional treatment of a cone channel was given in [38]. A conical light collector was used in [32].

Hinterberg and Winston proposed and thoroughly investigated a light collector in the form of a modified paraboloid [36], which effects maximum conversion of the aperture for minimum length L:

$$L = \left[\frac{1 + \frac{1}{\sin\theta_{max}}}{2\operatorname{tg}\theta_{max}}\right] t = \left[\frac{n + (n^2 - 1)^{1/2}}{n^2 - 1}\right] \frac{t}{2}, \tag{28}$$

where t is the exit aperture of the light collector (diameter of PM photocathode).

In the case of right circular cones the limit is attained at infinite length. Figure 12 shows the reduced L/t of the light collector as a function of the ratio of the entrance and exit apertures, from which it is easy to determine the dimensions of the light collector for each specific case.

The operation of a parabolic light guide with aluminized walls and an entrance/exit ratio of 1.8 was experimentally tested in [36]. It was found that such a light collector transmits 95% of the light incident on its entrance face from a twisted strip light guide. The 5% loss of light is due to absorption by the reflecting coating.

Thus, the use of twisted strip guides in conjunction with parabolic or conical light collectors secures the transmission of practically all the light from the exit face of a scintillator to the smaller [by a factor of $n^2/(n^2 - 1)$] face of a PM photocathode without contravening Garwin's relation. It is assumed in this case, of course, that there is no loss of light due to the nonadiabaticity of the light guide or to absorption in the light guide and collector material.

Such losses of light do occur in reality and must be taken into consideration in the calculation of light collection. These effects can be included by simple multiplication of expressions (13) and (22) by the following factors [1]: a) the factor A_1/A_2, the ratio of the entrance and exit areas of the light guide; b) the factor $\exp[-\alpha_{lg}(\lambda)l]$, which allows for absorption of luminescent light in its path l in the light guide; c) the factor $1 - G(x)$, which allows for loss of light due to the bending and twisting of the strip light guide or the "taper" of a triangular or hyperbolic light guide.

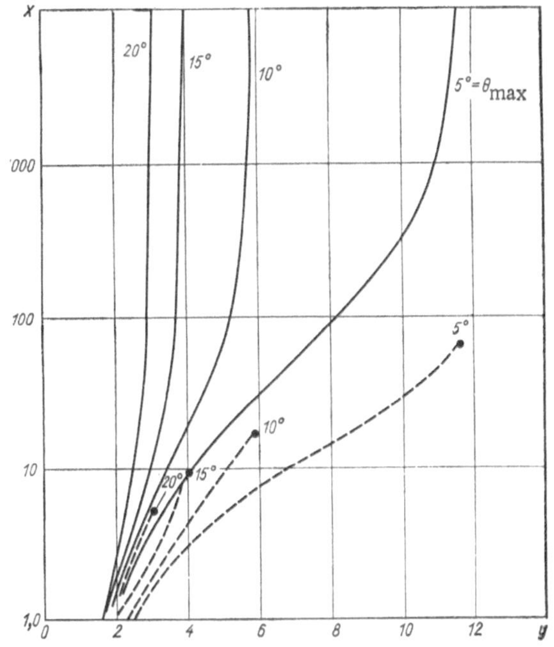

Fig. 12. Relative length L/t of light collector as function of aperture ratio y = T/t [35]; – – –) modified parabolic cone; ——) right circular cone; •) theoretical aperture limit, $1/\sin\theta$.

We conclude with a brief discussion of the materials used for the fabrication of LASC light guides. The common materials are polymethyl methacrylates (Plexiglas, Perspex, Lucite, etc.).

The results of measurements of the wavelengths of light transmitted by organic glasses of various kinds are shown in Fig. 13 [39]. As the figure shows, PMMA organic glass, obtained by polymerization of very pure methyl methacrylate with azobisisobutyronitrile as an initiator, is the most transparent. The effective absorption coefficient of PMMA organic glass for the luminescence of a plastic scintillator (p-terphenyl in polystyrene + POPOP) is 0.0025-0.0030 cm^{-1} [40]. PMMA-A organic glass is at present produced experimentally in the form of flat sheets (1.5 × 1.5) m^2 × (1-2) cm.

4. Substances and Adhesive Compounds for Optical Coupling of Components of Counter Optical System

An important factor in the development of scintillation counter design is the choice of substance for optical coupling of the components of the counter optical system. The use of immersion fluids or optical cements greatly reduces the loss of light due to reflections at the contacting faces of the optical components (detector–light guide and light guide–PM window). The light loss at these junctions is minimized if the refractive index of the immersion medium has a value between the refractive indices of the coupled materials.

Important characteristics determining the suitability of a particular optical coupler in a scintillation counter are the optical transmission in a thin layer (40 ± 10 μ)* in the range of luminescence wavelengths, the absence of chemical interaction with the coupled materials, which can lead to reduced transmission at the optical union, and the stability of the properties of the union in time. In addition, in most LASC designs the immersion material must have adhesive properties to ensure reliable coupling of the optical components.

We will discuss briefly the results of investigations of the properties of various substances and adhesive compounds which are useful in the construction of LASC.

*A layer of this thickness is usually formed between the components of a counter optical system when the specific pressure applied in cementing does not exceed 1 kg/cm^2.

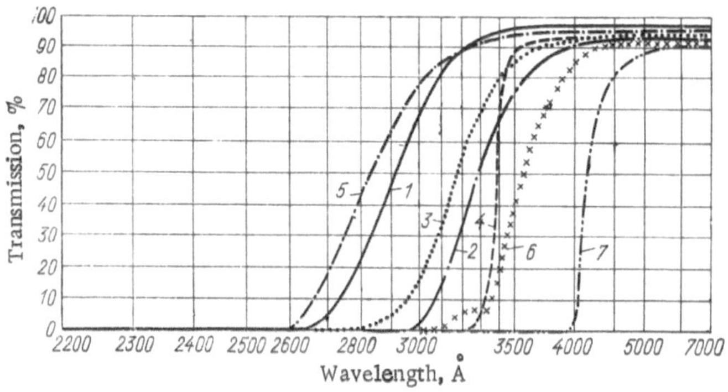

Fig. 13. Transmission spectra of different organic glasses: 1) BS-3 filter; 2) ICI Perspex (plasticized); 3) ICI Perspex (nonplasticized); 4) Dupont Lucite; 5) PMMA organic glass; 6) SAD organic glass; 7) Sh-35 organic glass.

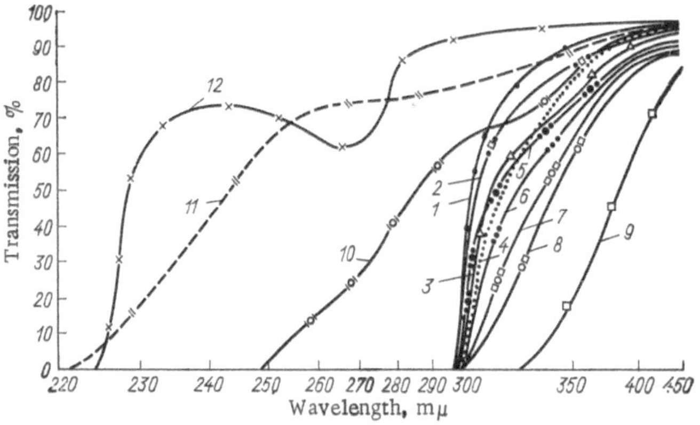

Fig. 14. Transmission spectra of epoxy resins and substances used in adhesive compounds: 1) É-33; 2) ÉD-5; 3) ÉD-6; 4) Epoxy 1200; 5) É-44; 6) É-40; 7) É-49; 8) É-41; 9) DÉG-1; 10) polyethylene polyamine; 11) hexamethylenediamine; 12) R5 solvent.

Epoxy Resins. The transmission spectra of the commonest epoxy resins and solid compositions derived from them are shown in Fig. 14 [41]. These substances have high transmission in the visible region of the spectrum. In the ultraviolet region the light transmission of epoxy resins and cements is cut off at $\lambda = 300$ mμ.

It should be noted that the water-soluble cement based on DÉG-1 resin (curve 9), which was used in [42, 43] and which allows the separation of cemented components, has low light transmission. A disadvantage of optical cements derived from epoxy resins (including OK-50, OK-72, etc. [47, 48]) is the high physiological activity (toxicity) of their constituents. In addition, it should be noted that the properties of cements can change with time. This can result in loss of optical contact between components with different thermal expansion coefficients (e.g., between an organic-glass light guide and a silicate-glass PM window).

Adhesive Compounds Based on Monomeric Cyanoacrylates. Adhesive compounds based on monomeric cyanoacrylates ("Tsiakrin" cements) can be recommended for the cementing of acrylic plastics (organic glass, high-transparency scintillating plastic, etc.). Data on the light transmission of some monomeric cyanoacrylates, two cements, and various additives are given in [44].

Figure 15 shows the transmission spectra of recommended cements based on ethyl cyanoacrylate. The stability of the properties of the optical union in time can be gauged from the fact that no changes

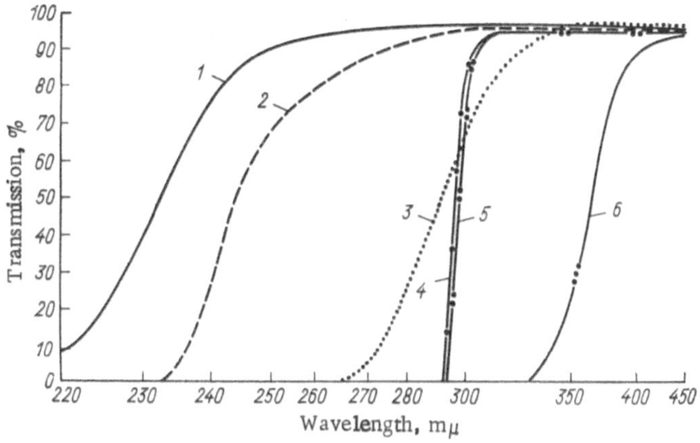

Fig. 15. Transmission spectra of cyanoacrylates and plasticizers (the transmission spectrum of a BS-3 filter and Canada balsam are shown for comparison): 1) First ethyl cyanoacrylate-based compound; 2) second ethyl cyanoacrylate-based compound; 3) BS-3 filter; 4) dibutyl phthalate; 5) diallyl phthalate; 6) Canada balsam.

have been observed in the light transmission curve of specimens (organic glass–high-transparency scintillating plastic) cemented with "Tsiakrin" in 1967.

Organosilicon Liquids and Vaselines. The optical and technical properties of some organosilicon compounds are discussed in detail in [45]. Distinctive features of these products are: chemical inertness, nontoxicity, wide temperature range of application (−50 to +200°C), and high transmission in a wide range of wavelengths extending to the far ultraviolet.

Figure 16 shows the transmission spectra of organosilicon liquids and organosilicon vaselines. The vaselines include KV-3 vaseline, which is made from PMS-400 polymethylsiloxane liquid with 10 wt.% Aerosil. This product can be used at temperatures from −65 to +200°C with hardly any change in its highly viscous consistency.

Figure 17 shows the transmission spectra of adhesive compounds based on polyorganosiloxysilazanes [products L-24-7/3, KO-961 p (previously M-10), KT-30 and polyorganosiloxane rubbers (Élastosil 11-02 and Élastosil 43-04)].

The Élastosil products [41, 46] are very attractive technologically, since they are single-component cements which are vulcanized at room temperature in the presence of atmospheric moisture and are subsequently converted to a rubberlike film with a tensile strength of 3.5-5 kg/cm^2, and a strength of 250-300%. The high elasticity of the adhesive layer allows reliable union of materials with different thermal expansion coefficients and operation in the temperature range −55 to +250°C. In conclusion Table 3 gives the mechanical and technological properties of some optical cements produced by our industry [41, 46].

Reflecting Coatings. Published data on the reflectivity of materials and different kinds of coatings used in the optical systems of scintillation counters and Cerenkov counters are disparate and are sometimes inconsistent, especially for the ultraviolet region. The reason for this is probably the difference in measurement techniques and conditions, and in the methods of preparing the reflecting surface of the investigated materials. A summary of published data on the reflectivity of coatings and materials used in spectrometric equipment is given in [49]. In this investigation the conditions of preparation and measurement of the test specimens were the same. The measurements were made with an EPS-3T Hitachi spectrophotometer in conjunction with an integrating sphere in the wavelength range 210-2600 mμ in the P. N. Lebedev Physical Institute. Figure 18 shows the measured reflection spectra of pigment and pigment filler powders used as binders and as components of reflecting compounds. The reflection spectra of multilayer coatings developed in the State Scientific Research and Development Institute of the Paint and Varnish Industry are shown in Fig. 19. The high reflectivity of VL-548 (previously VL-115-1) enamel in the visible region of the spectrum is in good agreement with the data of [50, 51].

TABLE 3. Mechanical and Technological Properties of Some Optical Cements

Optical cement	Refractive index n_D	Transmission region, Å	Density ρ, g/cm³	Temperature range of application, °C	Specific pressure during cementing, kg/cm²	Time under pressure, min	Strength of cement union, kg/cm²
ÉD-5 epoxy resin compound	1,5	3000–26 000	1,18–1,2	+5–70*	0,5–5,0	1440 (1 day)	1,1†
Tsiakrin-ÉO	1,44	2200–26 000	1,1	–60–60†	0,1–1,0	0,1–2	25†, ‡
Élastosil 43-04	1,42–1,45	2000–26 000	1,0–1,15	–80–+200	0,1–1,0	120–360	4,5†

* Cemented materials – organic glass and silicate glass.
† Organic glass and scintillating plastic.
‡ Fracture occurs in material.

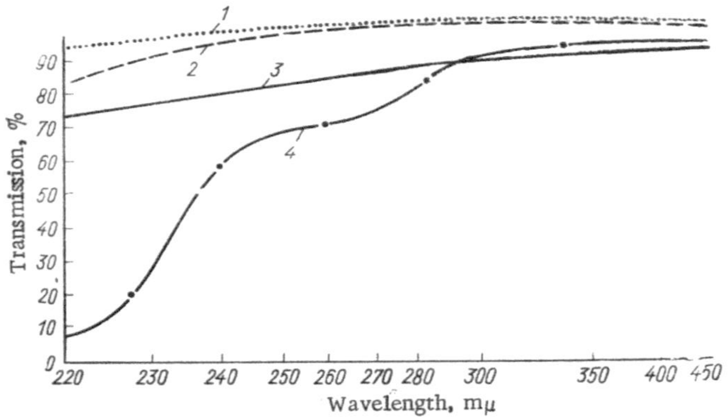

Fig. 16

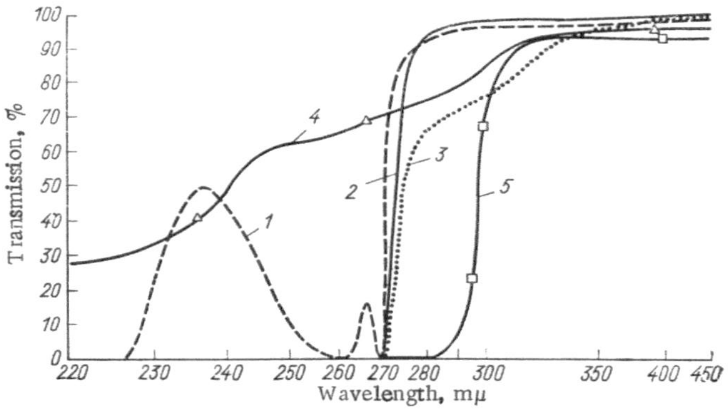

Fig. 17

Fig. 16. Transmission spectra of organosilicon liquids and vaselines: 1) FS-15 liquid; 2) PMS-15 liquid; 3) KV-3 vaseline; 4) KV-É vaseline.

Fig. 17. Transmission spectra of polyorganosilazane products with adhesive properties: 1) L24-7-3; 2) KO-961-11 (previously M-10); 3) KT-30; 4) Élastosil 43-04; 5) Élastosil 11-02.

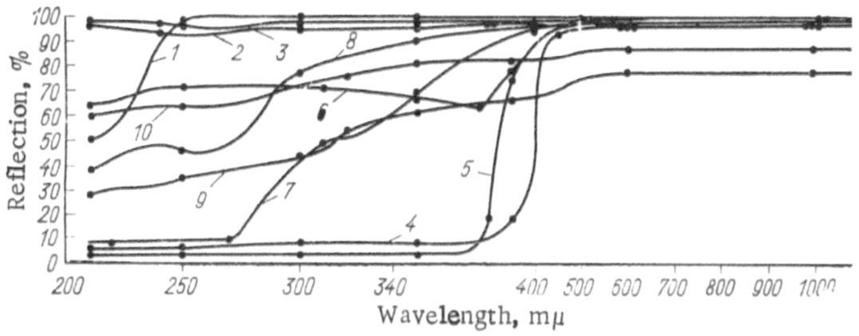

Fig. 18. Reflection spectra of white pigment and pigment filler powders: 1) MgO; 2) $MgCO_3$; 3) $BaSO_4$; 4) TiO_2; 5) ZnO; 6) $ZnAl_2O_4$; 7) $MgTiO_3$; 8) Al_2O_3; 9) SiO_2; 10) BN.

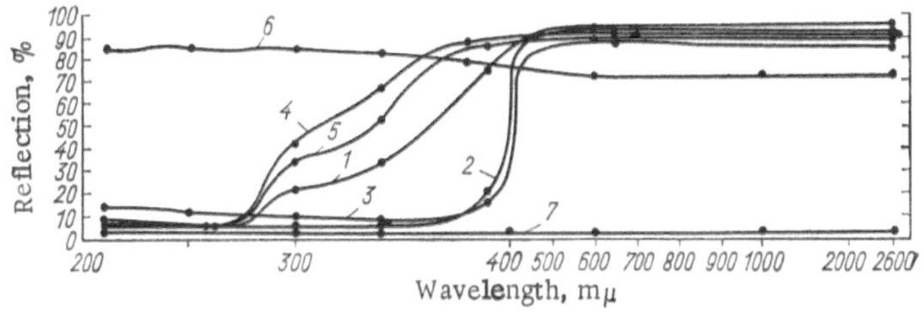

Fig. 19. Reflection spectra of some coatings: 1) VL-548 2) VL-55; 3) AS-81; 4) F_p-580; 5) 1M70; 6) Ko-88; 7) AK-243.

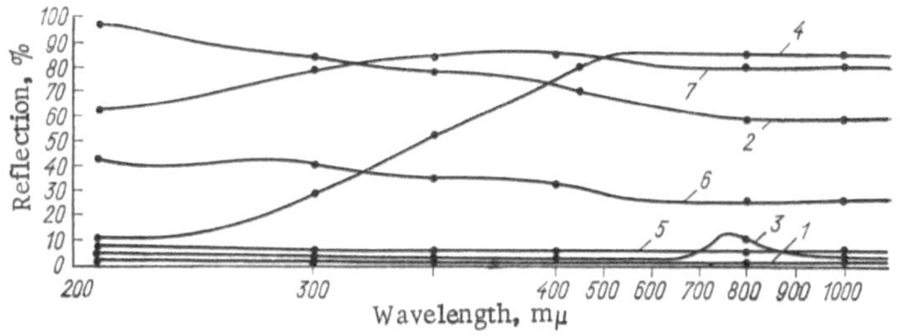

Fig. 20. Reflection spectra of some structural materials: 1) CuO (galvanic blacking); 2) Al (etched); 3) Al (galvanic blacking); 4) paper (Whatman); 5) black paper (photographic); 6) ground silicon single crystal; 7) Teflon.

Some LASC require coatings with an absorption coefficient close to unity. The enamel AK-243 (see Fig. 19) can be recommended as such a coating. Reflectivity data for some materials used in the construction of scintillation counters are given in Fig. 20. Compositions of reflecting coatings, which can be reproduced in the laboratory, can be found in [52-54].

5. Characteristics of Some Large-Area Scintillation Counters

The results of recent investigations of LASC are given in Table 4.

LASC have been investigated most thoroughly in [1, 11]. The authors of [1] investigated counters with $(45.7)^2 \times 0.6$ cm scintillators and three different light guides: strip, triangular, and fishtail. They found that the uniformity and energy resolution with the twisted strip light guide were three times better than with the other types of light guides. In addition, there was no difference in the performance of a simple triangular light guide and the more complicated fishtail guide.

TABLE 4. Characteristics of Some Large-Area Scintillation Counters

Counter characteristic	References				
	[1] (1966)	[9] (1963)	[57] (1965)	[11] (1967)	
Scintillator	*NE* 102A	*NE* 102A	*NE* 102A	*NE* 102A	*NE* 102 A
Scintillation dimensions, cm^3	$(45,7)^2 \times 0,6$	$(45,7 \times 40,6) \times 0,6$	$(138 \times 55) \times 3,5$	$(50)^2 \times 2$	$(100 \times 50) \times 2$
Light guide	Strip Plexiglas	Strip Plexiglas	Triangular Plexiglas	Triangular (acute-angle) Plexiglas	Triangular (right-angle) Plexiglas
Length of light guide, cm	45,7	45,7	27	39	22,5
No. of light guides	1	1	2	1	2
Type of PM	*EMI* 6097 *S*	—	*EMI* 9583 *B*	56 *AVP*	56 *AVP*
No. of PMs	1	1	2	1	2
Photocathode diameter, cm	4,4	4,4	3,7	5	5
Radiation source	Cosmic rays (μ mesons)	Electrons (100 MeV)	Cosmic rays (μ mesons)	Cosmic rays	Cosmic rays
Nonuniformity Δx_{max}, %	±12	±6	±32	±17	±15,6
Energy resolution, TWHH/x_{max}	0,65	0,50	0,50	Poor	0,70
Light collection coefficient, %	1,9	—	—	—	—

Table 4. Continued

Counter characteristic	References					
	[8] (1960)	[2] (1967)		[55] (1968)	[7] (1958)	[10] (1963)
Scintillator	*NE* 102A	*SPD*		Pilot J	Liquid scintillator	Liquid scintillator
Scintillation dimensions, cm^3	$(55 \times 17,5) \times 3,8$	$(50)^2 \times 5$		$(182 \times 91) \times 2,54$	$(76 \times 76) \times 5$	$(160 \times 100) \times 14$
Light guide	Triangular, flat	Rectangular parallelepiped, Plexiglas		No light guide	No light guide	No light guide
Length of light guide, cm	17	—	—	—	—	—
No. of light guides	1	1	2	—	—	—
Type of PM	*EMI* 9514 *B*	61*PK*412	61*PK*412	*RCA* 8055	*RCA* 5819	54 *AVP*
No. of PMs	1	1	2	2	8	4
Photocathode diameter, cm	—	4,2	4,2	12,7	—	12,7
Radiation source	Cosmic rays	^{137}Cs		Cosmic rays	Cosmic rays	π mesons
Nonuniformity Δx_{max}, %	±16	±4 (±2,7)*	(±2,1)*	~±9	±6	(a)
Energy resolution, TWHH/x_{max}	0,32	—	—	1,7	0,41	0,2–0,3
Light collection efficiency, %	—	—	—	—	—	~1

*Except for a region close to the PM.

(a) It was shown in [10] that if a light guide 30-40 cm long is used the nonuniformity will not exceed ±(5-10)%.

In [11] LASC with triangular organic-glass (Perspex) light guides were investigated. Counters with scintillators of different area and different thickness were examined. The results for two types of counters with the highest uniformity are given in Table 4. A comparison of the data of [1, 9] and the data of [11] again indicates that the best results as regards counter uniformity are obtained with twisted strip light guides.

Original methods of light collection in large-area plastic scintillators were used in [2, 55]. In [2] a $50 \times (5)^2$ cm Plexiglas bar was attached to a $(50)^2 \times 5$ cm scintillator (Fig. 21). The bar and the scintillator were in optical contact. A 61 PK 412 photomultiplier was fitted to the end of the bar with a small adapter. "Topograms" of the light collection in the scintillator for cases in which one and two PM's were used are shown in Fig. 22 (left and right, respectively). In Gillespie's investigation [55] light collection from (182 ×

Fig. 21

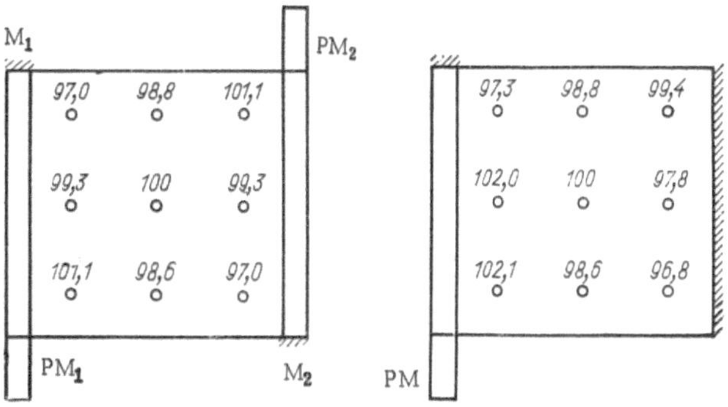

Fig. 22

Fig. 21. Connection of scintillator, Plexiglas bar, and PM in [2].

Fig. 22. "Topograms" of counter uniformity: PM) photomultiplier; M) mirror.

$X=29,5$ $X_p=17,6$	28,5 16,0	28,5 16,6	27,3 14,5	30,0 17,8	26,5 16,9	26,6 14,1	27,6 16,4
30,2 17,7	27,4 15,0	27,8 14,9	27,4 15,4	28,9 16,5	27,5 15,8	26,3 14,2	29,2 16,5
29,9 17,8	27,1 14,5	27,9 15,3	28,4 16,1	27,9 15,6	27,6 14,7	26,6 14,3	28,9 16,8
28,6 16,6	27,2 14,6	28,2 15,8	29,6 17,0	29,0 16,5	28,8 16,0	28,4 15,8	28,6 16,2

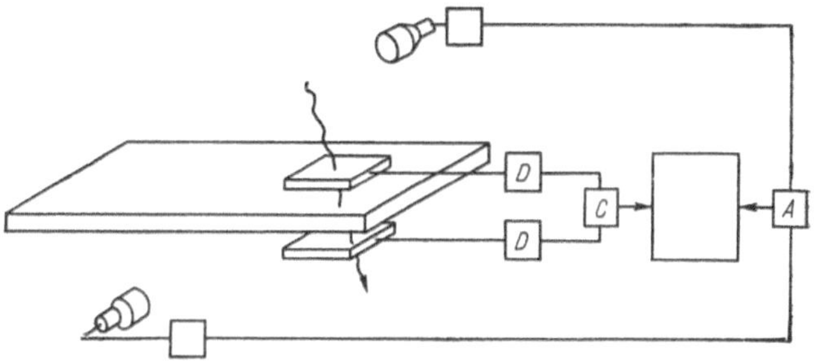

Fig. 23. Arrangement of scintillator and photomultipliers [55] and results of measurements for two scintillators and two PM's.

91) × 2.54 cm plastic scintillators was effected without light guides. A schematic drawing of the arrangement of the scintillator and the photomultipliers for measurement of the uniformity of light collection, and the results of measurements for the case of two scintillator sheets and two photomultipliers are shown in Fig. 23.

The nonuniformity for two sheets was $\Delta x_m = \pm 11.5\%$, and for a single sheet $\Delta x_m = \pm 9\%$, which is less than the nonuniformity values for most counters with light guides.

Thus, as Table 4 shows, counters with scintillators of area $(50)^2$-$(100)^2$ cm^2 (with absorption length $\lambda_a \approx$ 120-200 cm) can be obtained with a nonuniformity $\Delta x_m = \pm(5\text{-}10)\%$ by using twisted strip light guides or the light collection arrangements used in [2, 55].

It is much more difficult to design a proportional scintillation counter with good energy resolution. As Table 4 shows, a typical resolution for LASC at present is $\pm(20\text{-}25)\%$.

The energy resolution of scintillation counters has been thoroughly discussed in the monographs [16, 17, 56]. As already mentioned, the energy resolution of LASC depends mainly on two factors: statistical fluctuations of the energy loss dE/dx of the particles in the scintillator and statistical fluctuations of the number of photoelectrons from the PM photocathode. The statistical fluctuations of the energy loss are not greatly affected by the scintillator thickness. For instance, the ratio $\tilde{\Delta}/\Delta E_p$ changes from 0.23 to 0.18 when d changes from 0.5 to 5 g/cm^2. Hence, the connection of several proportional scintillation counters in series is a more effective means of improving the energy resolution of a counter due to ionization loss fluctuations than an increase in its thickness. The situation is different in the case of fluctuations of the number of PM photoelectrons. The number of photoelectrons and its fluctuations depends on the number of photons N_0 produced in the luminescent flash, the light collection efficiency k_1, and the photocathode quantum yield k_2. For a rough idea of the number of photoelectrons produced by the passage of a relativistic particle through a thin scintillator we consider the following example. Let a relativistic electron pass through a polystyrene-based scintillator 0.6 cm thick. Then 1.2 MeV/170 eV = 7000 photons with wavelength $\lambda \approx 410$ mμ will be formed in the scintillation flash. We assume that the optical efficiency of the counter is $k_1 \simeq 2\%$ [1]. The PM photocathode will then receive approximately 140 photons, which will give rise to 21 photoelectrons if the photocathode quantum yield is 0.15.

Fluctuations of this number of photoelectrons lead to an uncertainty $\Delta_{pe}/\Delta E_p \approx 0.58$ in energy measurement. If we assume that the fluctuations of energy loss in the scintillator and of the number of photoelectrons from the PM photocathode are independent, the energy resolution of LASC can be obtained by summing the dispersions of the distributions. For $\Delta_L/\Delta E_p = 0.23$ (half-width due to the Landau distribution) we obtain for the overall resolution $\delta = 0.63$.

As Table 4 shows, for most of the investigated LASC the energy resolution δ is 0.5-0.6, i.e., depends mainly on the fluctuations of the number of photoelectrons from the PM photocathode. In [1, 9] the thickness (0.6 cm) of the scintillator was not sufficient for the best resolution. In addition, in [1] the areas at the entrance and exit of the twisted strip light guide differed by a factor of 2. For the LASC used in [8, 11, 57] the energy resolution was poor even with scintillators with thickness d = 2-3.5 cm. This is due to the fact that the use of solid triangular or hyperbolic light guides leads to a low light collection efficiency.

The same applies to the LASC of Gillespie [55], in which no light guides were used.

Thus, we can draw the following conclusions from Table 4: a) the decisive factor governing the energy resolution of an LASC is the optical efficiency (light collection efficiency); b) the only way of improving the energy resolution of LASC at present is by an increase in the thickness of the scintillator to 2-3 cm in counters with twisted strip light guides.

CONCLUSIONS

A consideration of light collection in a scintillator, light guides, and other components of LASC and an examination of the characteristics of tested LASC indicate that if

1) the selected material is a scintillating plastic 2-3 cm thick with absorption length $\lambda_a = 200$ cm and high conversion efficiency;

2) the light is collected by a twisted strip light guide of highly transparent organic glass (PMMA-A) with $\lambda_a = 300$ cm;

3) a PM with a high quantum yield (up to 0.2-0.25) is selected;

4) Élastosil rubber cement, Tsiakrin, etc., is used to secure effective optical contact between the counter components, it is possible to construct a large-area (0.5 × 0.5 m or 1 × 1 m) proportional scintillation counter with better parameters than those in Table 4, viz., a counter with nonuniformity $\Delta x_m = \pm 5\%$ and energy resolution $\delta = \text{TWHH}/x_m = 30$, i.e., with $\sigma = \pm 13\%$.

LITERATURE CITED

1. D. G. Crabb et al., Nucl. Instr. Methods, 45, 301 (1966).
2. M. Chudy and M. Seman, Fys. Cas., 17, 82 (1967).
3. D. Brini et al., Nuovo Cimento, Suppl., 11, 1048 (1955).
4. V. V. Krivitskii, Pribory i Tekh. Éksperim., 1, 35 (1956).

5. V. K. Kosmachevskii and M. S. Ainutdinov, Pribory i Tekh. Éksperim., 4, 49 (1956).
6. V. F. Grushin and A. N. Zinevich, Pribory i Tekh. Éksperim., No. 2, 29 (1958).
7. C. H. Miller, E. P. Hinks, and G. C. Hanna, Canad. J. Phys., 36, 54 (1958).
8. C. F. Barnaby and J. C. Barton, Proc. Phys. Soc., A, 76, 745 (1960).
9. P. Gorenstein and D. Luckey, Rev. Sci. Instr., 34, 196 (1963).
10. H. Faissner et al., Nucl. Instr. Methods, 20, 289 (1963).
11. P. K. Grieder, Nucl. Instr. Methods, 55, 295 (1967).
12. R. J. Eden, High-Energy Collisions of Elementary Particles, Univ. Press, Cambridge (1967).
13. Yu. K. Akimov et al., Fast Electronics for Nuclear Particle Detection [in Russian], Atomizdat, Moscow (1970).
14. A. A. Sanin, Electronic Devices for Nuclear Physics [in Russian], Fizmatgiz, Moscow (1961).
15. J. B. Birks, Scintillation Counters, McGraw-Hill, New York (1953).
16. V. O. Vyazemskii et al., The Scintillation Method in Spectrometry [in Russian], Gosatomizdat, Moscow (1961).
17. Yu. K. Akimov, Scintillation Methods of Detecting High-Energy Particles [in Russian], MGU (1963).
18. D. J. Landau, J. Phys. USSR, 8, 201 (1944).
19. W. A. Shurcliff and R. J. Clark-Jones, J. Opt. Soc. Am., 39, 912 (1949).
20. R. H. Gillette, Rev. Sci. Instr., 21, 294 (1950).
21. D. Brini et al., Nuovo Cimento, 11, 655 (1954).
22. J. K. Walker, Nucl. Instr. Methods, 68, 131 (1969).
23. B. Jean-Marie, Nucl. Instr. Methods, 75, 287 (1969).
24. Single Crystals, Scintillators, and Organic Phosphors [in Russian], Nos. 1-5, Izd-vo VNIIM, Kharkov (1968).
25. V. D. Bezuglyi and S. A. Mukhina, Pribory i Tekh. Éksperim., No. 2, 82 (1967).
26. O. A. Gunder et al., Pribory i Tekh. Éksperim., No. 3, 66 (1969).
27. O. A. Gunder and S. A. Malinovskaya, in: Single Crystals, Scintillators, and Organic Phosphors [in Russian], No. 4, Izd-vo VNIIM, Kharkov (1968).
28. S. A. Malinovskaya, Dissertation, Moscow (1970).
29. R. L. Garwin, Rev. Sci. Instr., 23, 755 (1952).
30. P. Gorenstein and D. Luckey, Rev. Sci. Instr., 34, 196 (1963).
31. A. S. Belousov et al., Preprint FIAN, No. 71 (1970).
32. P. Dougan et al., Nucl. Instr. Methods, 78, 317 (1970).
33. P. A. Piroué, IEEE Trans. Nucl. Sci., 13, No. 3, 1 (1966).
34. H. Hinterberger and R. Winston, Rev. Sci. Instr., 39, 419 (1968).
35. S. L. Linder and J. E. Jackson, Rev. Sci. Instr., 37, 1094 (1966).
36. H. Hinterberger and R. Winston, Rev. Sci. Instr., 37, 110 (1966).
37. D. E. Williamson, J. Opt. Soc. Am., 42, 712 (1952).
38. W. Witte, Infrar. Phys., 5, 179 (1965).
39. M. A. Rubtsov, M. I. Frolova, and V. S. Chukin, Pribory i Tekh. Éksperim., 4, 59 (1969).
40. M. I. Frolova, I. V. Chekmodeeva, and A. G. Khabakhpashev, Pribory i Tekh. Éksperim., No. 4, 239 (1967).
41. V. S. Chukin, Preprint FIAN SSSR, No. 96 (1967); V. S. Chukin, Proceedings of All-Union Seminar on "Glues and Glue-Based Compounds" [in Russian], Vol. 2, Izd-vo MD MTP, Moscow (1970), p. 75.
42. Yu. A. Aleksandrov, A. V. Kutsenko, M. N. Maikov, and V. V. Pavlovskaya, Pribory i Tekh. Éksperim., No. 3, 221 (1966).
43. L. A. Permyakova, Pribory i Tekh. Éksperim., No. 6, 207 (1969).
44. A. M. Polyakova et al., Dokl. Akad. Nauk SSSR, 178, No. 2, 370 (1968).
45. R. P. Antonov et al., Pribory i Tekh. Éksperim., No. 6, 180 (1969).
46. E. I. Minsker, V. V. Severnyi, and V. S. Chukin, "Optical elastic cements derived from low-molecular polyorganosiloxane rubbers," Pribory i Tekh. Éksperim. (in press).
47. S. G. Gudkevich, O. V. Lebedev, and N. S. Selyaninova, Pribory i Tekh. Éksperim., No. 1, 198 (1961).
48. G. S. Dragun, Pribory i Tekh. Éksperim., No. 4, 241 (1967).
49. L. M. Vinogradova, L. M. Khlopina, and V. S. Chukin, Pribory Tekh. Éksperim., No. 3, 191 (1971).
50. A. G. Khabakhpashev and V. A. Tseluikin, Pribory i Tekh. Éksperim., No. 1, 202 (1964).
51. V. I. Logachev, V. S. Sinitsyna, and V. S. Chukin, Preprint FIAN SSSR (1966); in: Cosmic Rays [in Russian], No. 11, Nauka, Moscow (1969), p. 185.

52. I. Harris and K. Ogilvie, Rev. Sci. Instr., 27, 113 (1956).
53. A. R. Abrosimov, Pribory i Tekh. Éksperim., No. 6, 48 (1960).
54. G. L. Schnuzmacher, Rev. Sci. Instr., 32, 1380 (1961).
55. C. R. Gillespie, Rev. Sci. Instr., 39, 1724 (1968).
56. N. O. Chechik, S. M. Fainshtein, and T. M. Lifshits, Electron Multipliers [in Russian], Gostekhizdat, Moscow (1957).
57. F. Ashton et al., Nucl. Instr. Methods, 37, 181 (1965).
58. I. Lehraus and R. Matthewson, Nucl. Instr. Methods, 81, 85 (1970).